Advanced Metamaterials for Engineers

Online at: https://doi.org/10.1088/978-0-7503-5754-8

Advanced Metamaterials for Engineers

Edited by
Lulu Wang
Biomedical Device Innovation Center, Shenzhen Technology University, Shenzhen 518118, People's Republic of China

Muharrem Karaaslan
Department of Electrical and Electronics Engineering, Natural Sciences and Engineering Faculty, Iskenderun Technical University, Iskenderun, Hatay 31200, Türkiye

IOP Publishing, Bristol, UK

ISBN 978-0-7503-5754-8 (ebook)
ISBN 978-0-7503-5752-4 (print)
ISBN 978-0-7503-5755-5 (myPrint)
ISBN 978-0-7503-5753-1 (mobi)

DOI 10.1088/978-0-7503-5754-8

Version: 20231201

IOP ebooks

British Library Cataloguing-in-Publication Data: A catalogue record for this book is available from the British Library.

Published by IOP Publishing, wholly owned by The Institute of Physics, London

IOP Publishing, No.2 The Distillery, Glassfields, Avon Street, Bristol, BS2 0GR, UK

US Office: IOP Publishing, Inc., 190 North Independence Mall West, Suite 601, Philadelphia, PA 19106, USA

Contents

Editor biographies

Lulu Wang

Lulu Wang, Shenzhen Technology University, China,
Email: wanglulu@sztu.edu.cn, or lwang381@hotmail.com;

Lulu Wang is a Distinguished Biomedical Engineering Professor at Shenzhen Technology University. She received ME (First class Hons.) and PhD degrees from the Auckland University of Technology, New Zealand, in 2009 and 2013, respectively. Dr Wang is currently honored as a Fellow Member of the American Society of Mechanical Engineers (ASME Fellow), a Professional Member of SigmaXi Science Honors Society and Royal Society of New Zealand (PRSNZ), a foreign high-level talent of Guangdong Province (Chinese green card holder), and Shenzhen overseas high-level talent Program (Peacock Plan Class-B). Dr Wang is a member of the Institute of Electrical and Electronics Engineers (IEEE), the American Association for the Advancement of Science (AAAS), and a friend member of the Organization of Women Scientists in Developing Countries (OWSD).

Dr Wang's research interests include medical devices, electromagnetic sensing and imaging, electromagnetic materials, and computational mechanics. Over the past five years, Dr Wang has authored over 80 peer-reviewed publications, three books, four editorial books, 11 book chapters, and 25 issued patents. She is an active reviewer of 26 international journals. She has given more than 40 invited talks at international conferences and workshops. She has received multiple National and International Awards from various professional societies and organizations. She has been selected as one of the World's Top 2% Scientists in 2021 (by Stanford University).

Muharrem Karaaslan

Muharrem Karaaslan, İskenderun Technical University, Hatay, Türkiye
Email: muharrem.karaaslan@iste.edu.tr, or dr.muharremkaraaslan@gmail.com;

Muharrem Karaaslan is an Electrical and Electronics Engineering Professor at Iskenderun Technical University. He received an ME from Mustafa Kemal University and a PhD from Cukurova University, Turkey, in 2003 and 2009, respectively. Dr Karaaslan is currently head of Electrical and Electronics Engineering at Iskenderun Technical University.

Dr Karaaslan's research interests include antennas, microwaves, electromagnetic sensing, electromagnetic materials, and metamaterials. Over the past five years, Dr Karaaslan has authored over 180 peer-reviewed publications. He is an active reviewer of over 60 international journals. He has been selected as one of the World's Top 2% Scientists in 2021 (by Stanford University).

List of contributors

Yadgar I Abdulkarim
Physics Department, College of Science, Charmo University, Chamchamal, Sulaimania, Iraq

Fatih Özkan Alkurt
Department of Electrical and Electronics Engineering, Faculty of Engineering and Natural Sciences, Iskenderun Technical University, Iskenderun, Hatay, Turkey

Olcay Altintaş
Department of Avionics Iskenderun Technical University, Iskenderun, Hatay, Turkey

Mehmet Bakir
Department of Computer Engineering, Yozgat Bozok University, Yozgat, Turkey

Muhamad Abuzar Baqir
Department of Electrical and computer Engineering, COMSATS University Islamabad, Pakistan

Aysu Belen
Hybrid and Electric Vehicle Technology, Iskenderun Vocational School of Higher Education, Iskenderun Technical University, Hatay, Turkey

Mehmet A Belen
Department of Electrical Electronic Engineering, İskenderun Technical University, Hatay, Turkey

Luis A Espitia
Department of Mechanical Engineering University of Córdoba, Montería, Colombia

Arlet P Franco
Researcher at Fundación SAHED, Montería, Colombia

Musa N Hamza
Department of Physics, College of Science, University of Raparin, Sulaymaniyah, Iraq

Yunus Kaya
Department of Electricity and Energy, Bayburt University, Turkey

Peyman Mahouti
Department of Avionics Electronic, Yıldız Technical University, İstanbul, Turkey

Rashid Mirzavand
Electrical and Computer Engineering Department, University of Alberta, Edmonton, Canada

Zafer Özer
Mersin University Mersin, Vocational School, Mersin, Turkey

Muhammad Saqlain
Department of Electrical and computer Engineering, COMSATS University Islamabad, Pakistan

Mohamed Taouzari
Hassan First University of Settat, Morocco

Özlem Tari
Department of Electrical Electronic Engineering, İstanbul Arel University, İstanbul, Turkey

Erkan Tetik
Department of Mathematics and Science Education, Usak University, Usak, Turkey

Emin Ünal
Department of Electrical and Electronics Engineering, Faculty of Engineering, Necmettin Erbakan University, Meram, Konya, Turkey

Fatima-Ezzahra Zerrad
Hassan First University of Settat, Morocco

IOP Publishing

Advanced Metamaterials for Engineers

Lulu Wang and Muharrem Karaaslan

Chapter 1

Characterization of metamaterials

Arlet Patricia Franco and Luis Armando Espitia

Symbols

v	Velocity of the EM wave into material (m s^{-1})
μ_{abs}	Magnetic permeability (H m^{-1})
ε_{abs}	Electric permittivity (F m^{-1})
c	Speed of light in the free space (3×10^8m s^{-1})
μ_0	Magnetic permeability of the free space ($4\pi \times 10^{-7}$ H m^{-1})
ε_0	Electric permittivity of the free space (8.854×10^{-12} F m^{-1})
ε'	Dielectric constant (–)
ε''	Loss factor (–)
n	Refractive index (–)
ω	Angular frequency (rad s^{-1})
σ	Electrical conductivity (S m^{-1})
r	Radius (m)
a	Distance between wires (m)
ε_{eff}	Effective permittivity (–)
z	Impedance (Ω)
R	Resistance (Ω)
X	Reactance (Ω)
G	Geometrical factor (Ω)
T	Transmission coefficient (–)
Γ	Reflection coefficient (–)
d	Metamaterial slab thickness (m)
λ	Wavelength (m)
k_0	Wave number in the free space (1/λ)
η	Filling factor (%)
$\tan\delta$	Loss tangent (–)
f_0	Resonant frequency (Hz)
Q	Quality factor (–)
γ	Propagation constant (rad m^{-1})

Metamaterials (MTMs) are novel artificial composite materials developed for specific electric and magnetic responses, which are not available in traditional materials [1, 2]. As artificial media, these materials must be composed of elements for which the unit cell structure is smaller than the electromagnetic wavelength [3], which confers remarkable attributes. Among the reported characteristics for metamaterials are the negative refraction index, electric permittivity and magnetic permeability [4, 5]. These properties enable the use of MTMs on specific electronic devices and therefore their characterization is mandatory to predict the electromagnetic response depending on the operational parameters. The characterization of metamaterials has been widely performed by several measurement techniques and mathematical approaches for different material arrangements at frequencies from low microwaves to optical range [1, 6] due to diverse application fields, including aerospace and car industries, energy harvesting, biomedical and communications, among others [7–9]. Material behavior for applications at frequencies from low microwave to optical ranges are characterized by electromagnetic (EM) parameters known as dielectric and magnetic properties, which determine the response to an incident EM field. The EM energy propagation is governed by Maxwell´s equations, which establish the amount and distribution of energy inside the material [10]. The velocity of the EM waves into the material depends on its characteristics and is defined as follows [11, 12]:

$$v = \frac{1}{\sqrt{\mu_{\text{abs}} \times \varepsilon_{\text{abs}}}} \tag{1.1}$$

$$c = \frac{1}{\sqrt{\mu_0 \times \varepsilon_0}} \tag{1.2}$$

where v is the velocity of the EM wave into the material (m s^{-1}), μ_{abs} is the magnetic permeability (H m^{-1}) and ε_{abs} is the electric permittivity of the material (F m^{-1}). μ_0 and ε_0 are the magnetic permeability and electric permittivity of the free space, respectively, $4\pi \times 10^{-7}$ H m^{-1} and 8854×10^{-12} F m^{-1} [10]. The complex permittivity (ε) is composed of the relative electric permittivity or dielectric constant ε' and the dielectric loss factor ε'', whereas the complex permeability μ is composed of the relative permeability of the material μ' and the magnetic loss factor μ'' and they are defined as [13, 14]:

$$\varepsilon = \varepsilon' - j\varepsilon'' \tag{1.3}$$

$$\mu = \mu' - j\mu'' \tag{1.4}$$

where $j = \sqrt{-1}$. For the free space and several materials $\mu' = 1$ [15]. The refractive index (n) or index of refraction of a material characterizes its optical properties. It is related to the ability to measure the light waves bending when passing from one medium to another. For a specific medium, n depends on chemical composition, geometry, magnetic and electric properties and is defined as [16–19]:

$$n^2 = \varepsilon\mu. \tag{1.5}$$

1.1 Classification of metamaterials

Metamaterials can be sorted according to dielectric and magnetic properties into four main categories: double positive, epsilon negative, mu negative and double negative materials [20, 21]. The schematic classification of metamaterials is presented in figure 1.1.

1.1.1 Double positive (DPS) materials

DPS materials are natural, characterized by positive permittivity and permeability values simultaneously [20–22]. Most dielectric materials exhibit these features and usually they are transparent to the incident electromagnetic waves and electric vector (*E*), magnetic vector (*B*) and the wave vector (*k*) [23]. They are also known as right-handed materials (RHMs) [24].

1.1.2 Epsilon negative (ENG) materials

ENG materials are characterized by outstanding properties such as a negative refraction index, negative phase velocity, reverse Doppler effect and Cerenkov effect, among others. This category is composed of ceramic and polymer matrix composites, metals, doped semiconductors and plasmas [20, 22, 23]. For microwave applications, the most common ENG materials are metal wires arranged as an infinite long parallel square matrix embedded into a dielectric medium. The electromagnetic wave propagation in this case is similar to the propagation in plasma. The permittivity of these materials is negative at frequency $\omega < \omega_p$, where ω_p

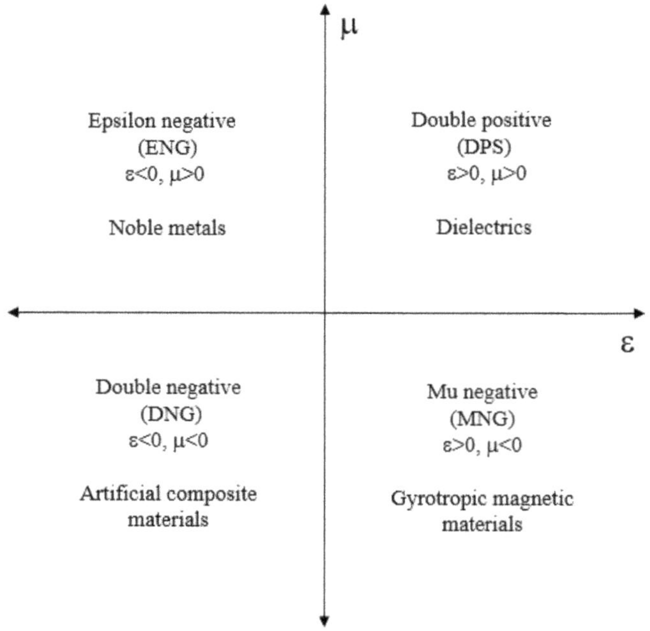

Figure 1.1. Classification of materials according to permittivity and permeability values.

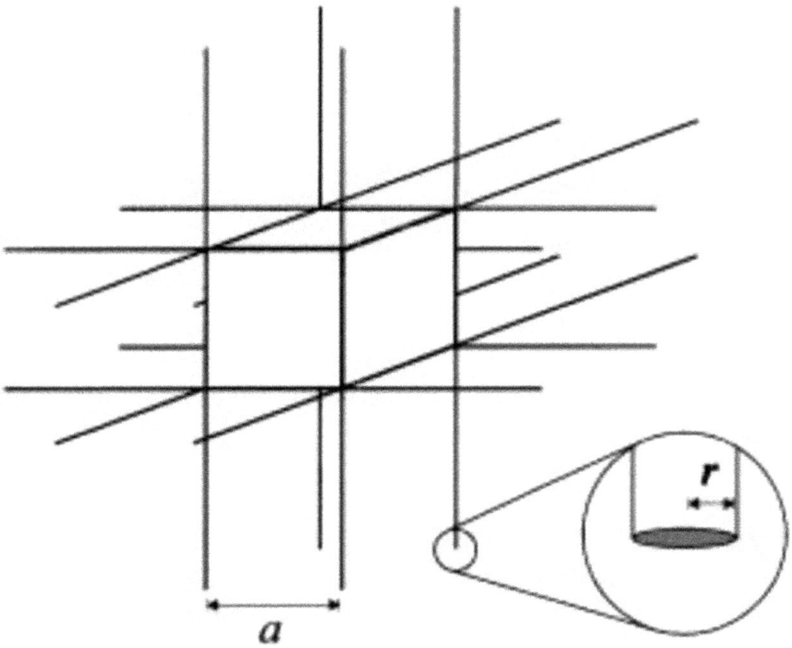

Figure 1.2. Schematic diagram of periodic metal wires proposed by Pendry. Reprinted figure with permission from [25], Copyright (1996) by the American Physical Society.

is the plasma frequency of the structure, which can be controlled because it depends on the radius r (m). The materials' conductivity (σ) denotes the electrical conductivity (S m^{-1}) [9, 23, 25]. According to this, the effective permittivity can be defined as [9]:

$$\varepsilon_{\text{eff}} = 1 - \frac{\omega_p^2}{\omega\left[\omega - \frac{i\left(\omega_p^2 a^2 \varepsilon_0\right)}{\sigma\pi r^2}\right]} \tag{1.6}$$

where a is the distance between wires (m), being $r \ll a$. The schematic illustration of this structure is presented in figure 1.2.

1.1.3 Mu negative (MNG) materials

This group comprises composite materials with negative permeability. As materials with one negative parameter only, they are also known as single negative (SNG). Some natural materials such as ferrites present these characteristics [20, 26]. Since MNG materials show remarkable behavior, they have been developed for artificial structures as resonators (split-ring resonators (SRRs), spiral resonators (SRs), S- and Ω-shaped resonators) [27]. SRRs can present circular or square geometry and they are characterized as a high conductive resonant structure [9]. The unit cells for a SRR with circular and square geometry is presented in figure 1.3.

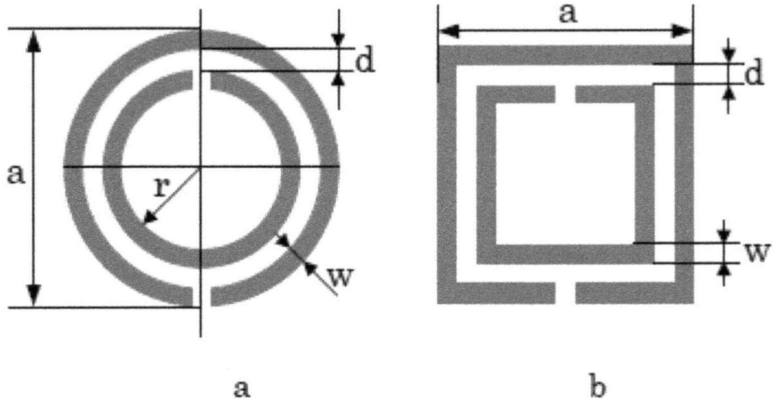

Figure 1.3. Unit cell for MNG material. Circular geometry (a) and square geometry (b). Reproduced from [9].

1.1.4 Double negative (DNG) materials

DNG materials do not have a natural formation, they are artificially developed to display negative permittivity and permeability simultaneously. They were initially reported theoretically by Veselago in 1968 [19] and named left-handed media. According to their properties, DNG materials are also known as negative-index materials (NIMs) or backward (BW) media [28]. Materials with these characteristics can be developed by using small scatterers, composed of metallic or magnetodielectric materials [1, 6, 9, 28]. This is the most interesting metamaterials category, since by attending to these characteristics, different media for several applications can be developed. DNG materials can be mainly composed of three structures: thin wires and SRR, transmission lines or mushroom structures [9]. The combination of thin metal wires and SRR is the most commonly employed structure and is shown in figure 1.4.

1.2 Types of MTM

There are several heterogeneous materials artificially developed with different applications, evidencing superior behavior when compared with natural materials. These materials can be also categorized according to their properties and uses as artificial dielectrics, artificial magnetics, chiral, plasmonic, omega shape and tunable materials, among others.

1.2.1 Artificial dielectrics

Artificial dielectrics (ADs) were developed to meet the requirement of lighter materials on microwave lens fabrication. These kinds of materials are widely studied for radar technology application [30, 31]. ADs are composed of miniature conducting particles or surfaces immersed in an insulating dielectric media according to a regular pattern [31, 33]. Among the outstanding properties of these materials are colossal permittivity values and a high effective refractive index, which can be

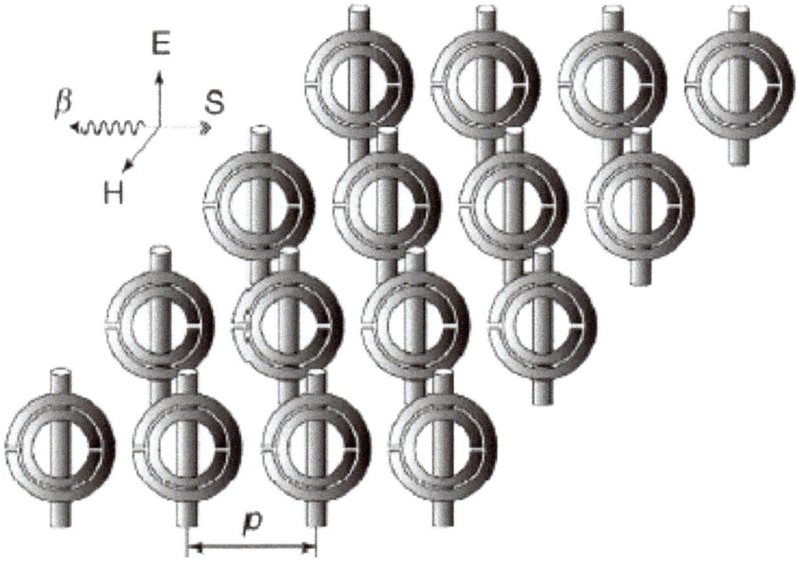

Figure 1.4. DNG material structure composed of thin wires and a split-ring resonator (SRR). Reprinted from [29], Copyright (2009), with permission from Elsevier.

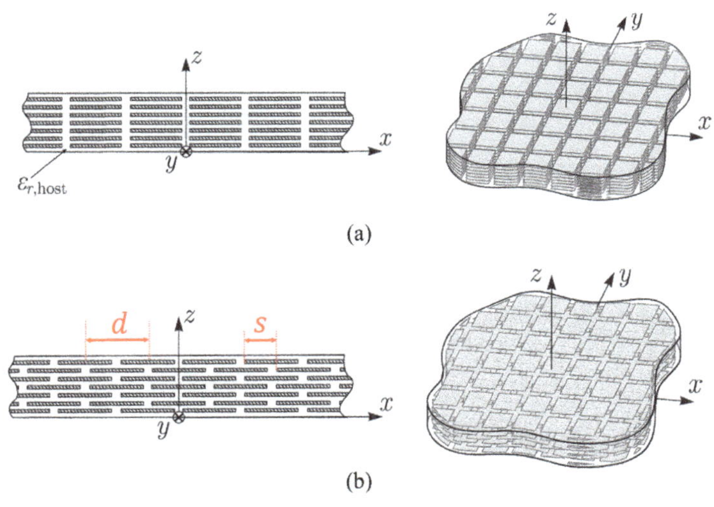

Figure 1.5. 2D side view and 3D prospective view of AD slabs with (a) aligned layers and (b) layers alternatively shifted in both x and y. © [2017] IEEE. Reprinted, with permission, from [31].

regulated with the material array and geometry [16]. Several studies and numerical approaches have been performed in order to describe the AD behavior as a function of conducting and host material, geometry and layer distribution [16, 31]. It has been reported that an increase in permittivity values has been achieved by shifting the metallic patterns in adjacent layers as presented in figure 1.5. According to this, the method of layers being alternatively shifted introduces an additional capacitance when compared to an aligned configuration [31, 33].

1.2.2 Artificial magnetics

Artificial magnetic media (AMM) are materials formed by an ensemble of small metallic inclusions, arranged with a periodic or not periodic distribution. They show magnetic behavior when exposed to an applied electromagnetic field. In order to improve the magnetic properties, several geometrical configurations and designs for numerous applications have been evaluated [34, 35]. Among the reported configurations for artificial magnetic media are split-ring resonators (SRRs) [21, 34, 36], modified split-ring resonators (MSRRs) [36], Swiss roll resonators (SR-Rs) [35, 37], *n*-turn spiral resonators (n-SRs) [38] and metasolenoids [35]. The SRR is formed by two concentric metallic split rings placed on a dielectric material [34] whereas the MSRR replaces one of the rings by another ring located behind the other ring, at the opposite side of the dielectric media [36]. In the case of the SR-R, it is composed of rolled metal sheets. Although this configuration presents strong magnetic properties, the operational frequency is limited [35, 37]. For the *n*-turn spiral configuration, the main advantage is related to simple manufacturing and control of electromagnetic properties [38]. Metasolenoids appear as an alternative configuration to improve the magnetic properties of artificial media, employing rectangular shapes. It is possible to achieve a successful performance at a wider frequency range [35]. Common configurations for artificial magnetic media are presented in figure 1.6.

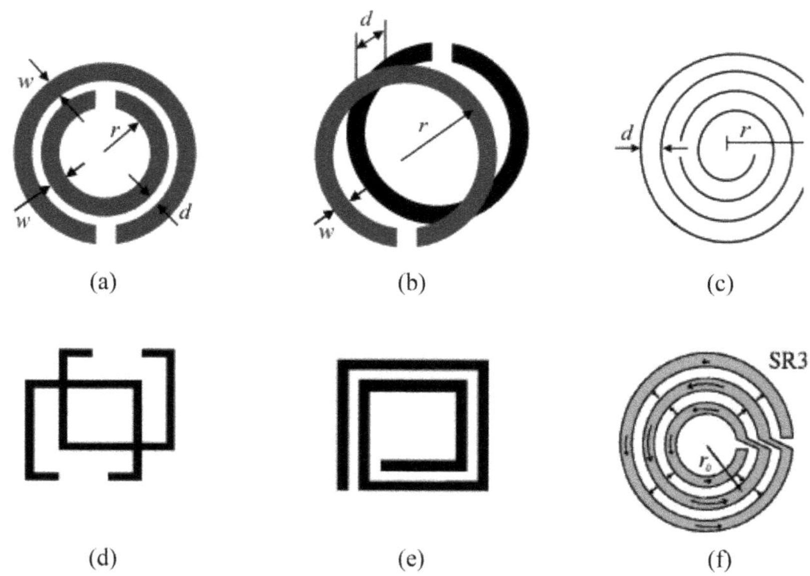

Figure 1.6. Split-ring resonator (SRR) (a), modified split-ring resonator (MSRR) (b), cross section for Swiss roll resonator (c). (Reproduced from [35] courtesy of The Electromagnetics Academy.) Meta solenoid (d), split spiral (e). (Reprinted figure with permission from [38], Copyright (2004) by the American Physical Society.) Three turns spiral resonator SR3 (f). (© [2007] IEEE. Reprinted, with permission, from [39].)

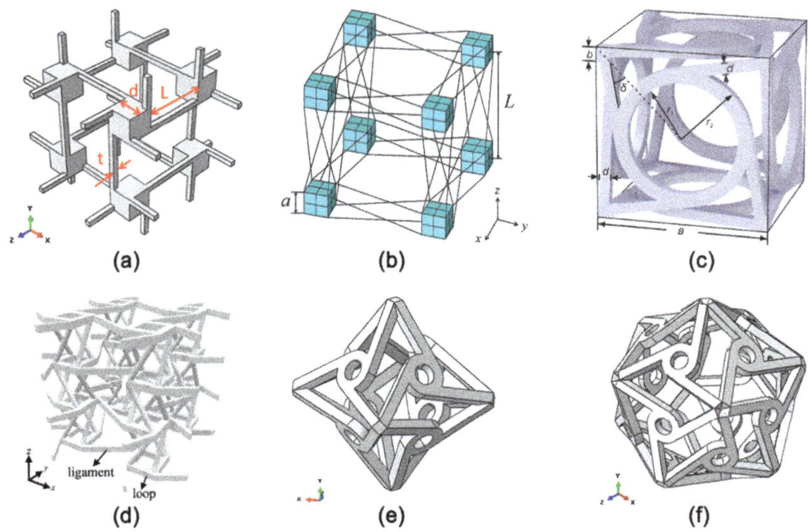

Figure 1.7. (a) Classification of 3D chiral mechanical metamaterials. 3D anti-tetrachiral mechanical metamaterials, (b) 3D chiral metastructures, (c) compression-twist chiral mechanical metamaterials, (d) 3D cellular metamaterials with planar tetrachiral, (e) 3D chiral dodecahedron lattice and (f) 3D chiral regular icosahedrons lattice. (Reprinted from [40], Copyright (2019), with permission from Elsevier.)

1.2.3 Chiral materials

Chiral MTMs are based on natural structures, widely presented by vegetal and animal organisms. The term chiral means that an object cannot be superposed to its mirror image [40]; for instance, the human hands are an excellent example of chiral structures, therefore, the right hand cannot be superposed on the left hand [20]. They are composed of periodical chiral microstructures that can display prominent mechanical properties such as negative values for Poisson's ratio and stiffness [40–42], resulting in vibration attenuation, impact energy absorption and negative coefficient of thermal expansion [40, 41]. According to this, several structures of unit cells' development for chiral metamaterials have been assessed including concave honeycomb, rotating polygon, lattice, origami, among others [41]. The main advantage of this kind of structure is that negative values of permittivity and permeability are not required to attain a negative refractive index, which can be reached with a strong chirality [43, 44]. Furthermore, several materials have been studied to improve the chiral metamaterials' performance; among these materials are graphene [42], carbon nanotubes, [45], metals [3, 41] and polymers [46]. Different configurations for chiral structures have been also proposed for large applications [47–50]. Some chiral structures are presented in figure 1.7.

1.2.4 Plasmonic materials

Plasmonic materials have been developed for application in the microelectronic components of optics. The predominant metals for these kinds of metamaterials'

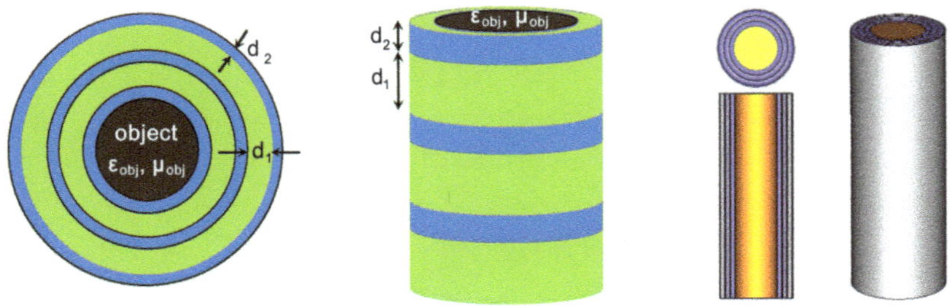

(a) (b) (c)

Figure 1.8. (a) Cylindrical cloak made of concentric shells of different materials, (b) cylindrical cloak made of stacked shells of different materials and (c) cylinder surrounded by a cloak made of concentric shells of different materials. (© [2010] IEEE. Reprinted, with permission, from [112].)

manufacturing are silver and gold due to their high conductivity at radio frequency range and high reflectivity at optical range [51]. These kinds of materials are widely studied for applications in photonic devices because the element sizes can be reduced to the nanometer scale [52]. Other interesting applications for plasmonic materials are related to data storage, solar cells, sensors and cloaking device development [52]. Plasmonic structures with tunable and controllable dimensions with nanometer precision, based on chiral principles, can be successfully attained by the arrangement of several metallic nanoparticles with different chemical compositions and geometry [53]. Structures with different applications can be obtained including plasmonic and non-plasmonic materials to attain diverse responses according to the required operational parameters. In figure 1.8 different arrangements of plasmonic and non-plasmonic materials are presented for cloak applications [21].

1.2.5 Omega shape materials

Omega type metamaterials are composed of omega shape elements, which are formed by a C-shaped ring with two wires connected to both culminations [54]. This kind of wire configuration can achieve structures with a reduced bianisotropy [55], negative permeability and refractive index [56, 57]. Uses for omega shape structures have been reported in several applications such as sensors for fuel [58], liquid chemical detection [8], microwave absorbers [56], antennas [57], among others, employing topologies such as omega shape resonators [56–58] and omega shape resonator integrated transmission lines [8]. Applications in light-controlled switchers and optical diodes in on-chip optical systems and communication systems can also be explored [59]. The performance of these structures can be improved by a combination of reverse omega shapes as presented in figure 1.9.

1.2.6 Tunable materials

A tunable metamaterial has the capability of adjusting its refractive index according to the electromagnetic wave frequency [22]. This remarkable characteristic has increased the demand for these kinds of multifunctional electronic components within the electromagnetic spectrum [60]. In order to meet these requirements,

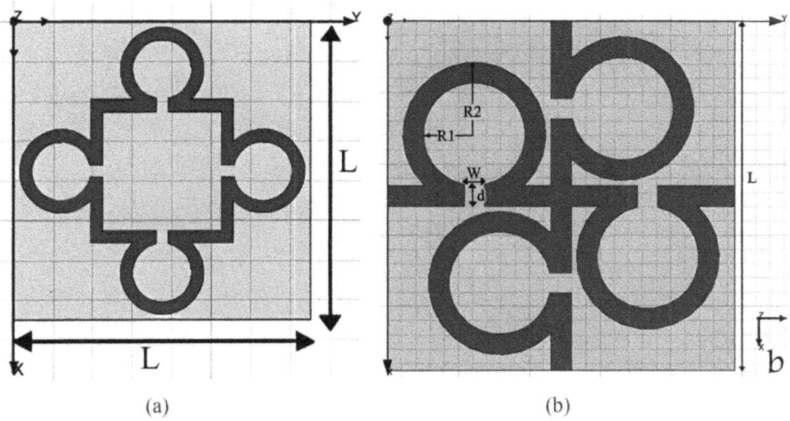

Figure 1.9. (a) Top view of a new metamaterial design using four omega type elements and (b) second symmetrical scheme using four omega type elements [55] John Wiley & Sons. [2011].

several methods have been assessed to enlarge their operational bandwidth through active inclusions of semiconductors [61], ferroelectrics [62, 63], graphene [64, 65] liquid crystal [66, 67], varactor diodes [68], and so on into the metamaterial matrix. Different metamaterial topologies (e.g. split-ring resonators (SRRs), electric-field coupled-LC (ELC) resonators, omega-ring materials, etc) and tuning mechanisms (e.g. phase-transition materials, microelectromechanical systems (MEMs), 2D material, thermal annealing, etc) have been also studied [61, 68, 69]. Depending on the phenomenon that promotes the change, tunable metamaterials can be classified in four categories: thermal, optical, mechanical and electrical effects [64]; therefore, the tuning behavior of metamaterials must be modulated according to the application requirements. A tunable spatial light modulator with crystal liquid inclusion is presented in figure 1.10.

1.3 Metamaterials' properties dependence

As metamaterials are usually composed of matrices adding two or more diverse substances, they can attain remarkable properties that are different from elementary constituents [23]. Therefore, their properties can present colossal variations depending on the proportion of different substances in the matrix, in that permittivity and permeability values can be manipulated from the microwave to optical region by adjusting their shapes, sizes and arrangement, among others [59]. These topics were briefly introduced previously and will be discussed below.

1.3.1 Frequency

Usually, the material properties show a considerable variation as a response to an applied electromagnetic field (EF). Depending on a material's characteristics, the response might vary substantially. For dielectric materials without an EF, the dipoles are randomly oriented and the polarization is zero [71]; nevertheless, under

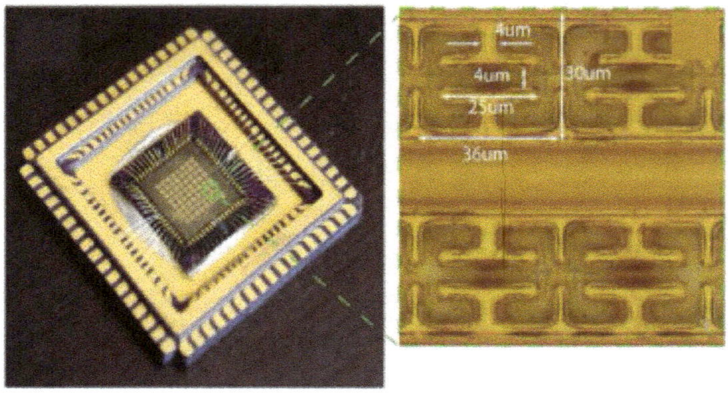

Figure 1.10. Camera image of an MMA spatial light modulator device coated with liquid crystal (left); close-up of MMA unit cells and corresponding dimensions (right), [70] John Wiley & Sons. [2014].

an EF influence, they turn trying to accompany the EF until the critical frequency. The electrical permittivity or dielectric constant usually decreases with increasing frequency at microwave range, which can be explained by the reduced ability of the dipolar molecules to follow the alternating electric field at higher frequencies [72], whereas the loss factor behavior can increase or decrease at a specific frequency range. When the frequency exceeds the relaxation point (which occurs at critical frequency), the dielectric molecule gradually loses the ability to follow the rapidly alternating field, reducing both the relative permittivity and the dielectric loss factor. The relaxation frequency of the dielectric materials increases with temperature, consequently the peak in the loss factor curve moves to higher frequencies. Permeability also presents a frequency dependent behavior for both the real and imaginary part [73–75]. As reported by Osipov [74], Rozanov [73] and Chowdary [75], the real and imaginary part of magnetic permeability presented similar behavior to the electrical permittivity and dielectric loss factor, respectively, as can be seen in figure 1.11.

1.3.2 Geometry and size

Geometrical dimension and shape of a metamaterial structure is relevant for ensuring a homogenous EF distribution inside the material and the required EF magnitude and potential voltage [76]. Different metamaterials for several applications have been developed based on these parameters and considering other factors such as density of arrangement and alignment, which also determine the effective properties of the assembled medium [77]. According to this, Illias [78] reported for a cable joint, the reduction in the electric field magnitude with the increase of both the central axis of the conductor and the insulation of material. Shape of inclusion is also relevant for the location of the dispersive region; therefore, their spread into a material sample affects permittivity and permeability [74]. On the other hand, size is also an important parameter in the design of metamaterials, where the unit cell

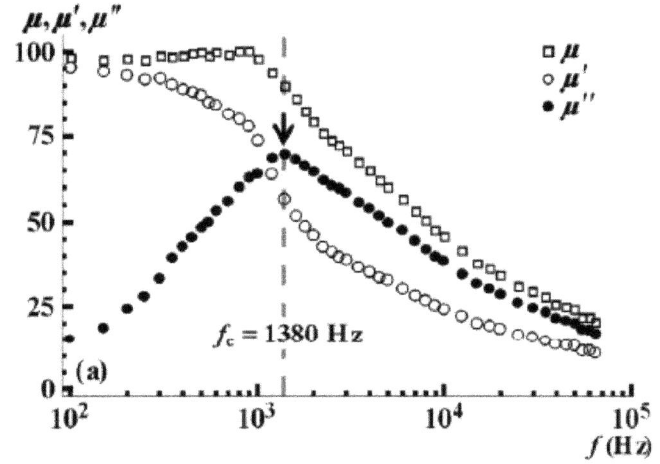

(b)

Figure 1.11. Complex permeability spectrum for $Fe_{10}Co_{90}$ at 273 K. (Reproduced from [75]. © IOP Publishing Ltd. All rights reserved.)

dimension must be inferior to wavelength to guarantee a suitable behavior at operational conditions [3, 79]. Furthermore, the requirement for smaller devices is a challenge for materials science, hence researchers are focused on the development of nano elements.

1.3.3 Temperature

Temperature has a strong effect on the behavior of materials. It affects several properties of metamaterials' structures such as electrical conductivity due to an increase in ion mobility, permittivity due to thermal agitations of molecules inside the material [72] and other physical and mechanical properties related to thermal expansion of materials due to temperature changes. The temperature dependence of metamaterials' properties have been reported by different authors; for instance, Varadan [80] explained theoretically the effect of temperature on the resonance frequency of split-ring resonators and thin wires between 25°C and 400°C, employing an FR4 and low temperature co-fired ceramic substrates. Permittivity and permeability are also temperature dependent. Chowdary [75] studied the temperature dependence of the permeability of $Fe_{10}Co_{90}$ at frequencies between 10^2 and 10^5 Hz, reporting a decreasing behavior at temperatures between 77 and 873 K, whereas Bobicki [81] presented an assortment of permittivity changes depending on substances and temperature ranges. On the other hand, the temperature dependence of metamaterials also allows one to develop new composites with tuning behavior, as previously demonstrated. The temperature effect on the complex permittivity of bauxite is presented in figure 1.12.

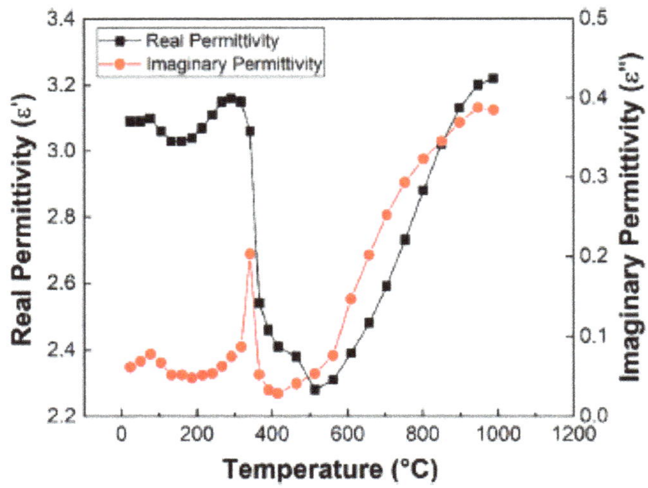

Figure 1.12. Complex permittivity of bauxite as a function of temperature at 912 MHz K. (Reprinted from [81], Copyright (2020), with permission from Elsevier.)

1.3.4 Homogenity

A homogeneous medium is fundamental to ensure a uniform electromagnetic field distribution inside the material. Particle concentration or inclusions along the metamaterial, instead of being regular distribution, can result in the generation of non-uniform fields inside the medium. According to this, the effective parameters of composites and metamaterials are highly dependent on inclusions' distribution and volume fraction [74, 77]. In order to ensure the homogeneity of the metamaterial, mixing rules and homogenization models have been developed [74, 77, 82]. Although the homogeneity of the metamaterial is a desirable characteristic, it is not always required; therefore, other researchers have been working on non-homogeneous materials. For instance, inhomogeneous thermal laminate metamaterials have been developed to concentrate and twist the heat flow [83]. This kind of material can also have applications for cloaking and optics devices [4].

1.4 Techniques of characterization of MTMs

The characterization of MTMs is fundamental to determine the response to an applied electromagnetic field. As previously presented, parameters such as electrical and magnetic properties of metamaterials must be measured to design verification and adjustment. Since MTMs are usually non-uniform and anisotropic, their characterization can be considered a challenge, and several experimental techniques are available including the resonator, free space, the waveguide, S-parameters and Nicolson–Ross–Weir. These methods can successfully retrieve the constitutive parameters of MTMs [2, 84]. However, advantages and disadvantages have been reported in the literature; therefore, the selection of the most suitable technique depends on material, operation frequency range, temperature, accuracy, sample size and thickness, among others [84, 85]

1.4.1 Resonator methods

Resonant methods usually present higher accuracy and sensitivity than non-resonant methods. Nevertheless, they provide reliable results for dielectric materials with low losses and require careful sample preparation [86, 87]. Resonant methods include the dielectric resonator method and cavity perturbation method [88], employing the transverse magnetic (TM) or transverse electric (TE) mode, presented in figure 1.13 [10]. In the dielectric resonator method, the sample under test (SUT) resonates in the measurement system and the dielectric properties are deduced from the resonant properties [88, 89].

On the other hand, the cavity perturbation method is based on the shift in resonant frequency and the change in absorption characteristics of a tuned resonant cavity due to insertion of the SUT [10, 90]. The electromagnetic characteristics of the SUT can be estimated from the changes of the quality factor Q_f (ratio of energy stored to energy dissipated) and resonant frequency of the cavity [10, 91]. The cavity size must be designed for the frequency required and it needs an initial calibration, but after development of calibration curves, the calculation process is simple. Electromagnetic properties employing this method have been reported at temperatures between −35°C and 140°C [10]. These methods are recommended for low loss dielectric materials with a permeability equivalent to μ_0. They are not suitable for broadband measurements of electromagnetic properties of metamaterials, in that, their accuracy in materials characterization is performed over a discrete number of frequencies [86, 87]. For retrieval of parameters, validated approaches such as those presented by Hakki in 1960 [92] are employed. According to this, Alimenti [93–95] employed a dielectric loaded resonator composed of a metallic cavity and loaded with a dielectric crystal. In this configuration, the electromagnetic field is focused inside the crystal itself, reducing the conduction losses in the lateral walls of the

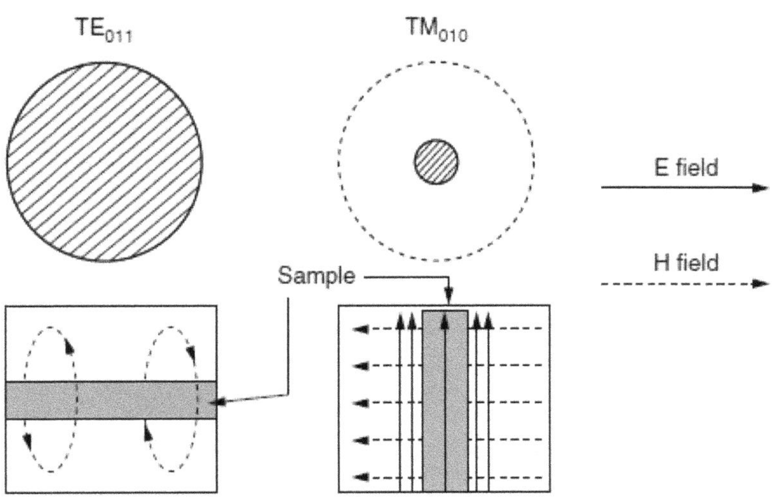

Figure 1.13. Schematic of a circular perturbation cavity in simple TE and TM modes [10] [2005], reprinted by permission of the publisher (Taylor & Francis Ltd, http://www.tandfonline.com.).

cavity and therefore increasing the sensitivity. This effect can be enhanced with higher permittivity crystals [94]. For metallic samples, they are inserted into the cavity, and the surface impedance of the sample Z_s (equation (1.7)) is related to the measured resonant frequency f_0 and quality factor Q of the dielectric resonator, according to equations (1.8) and (1.9) [93].

$$Z_s = R_s + iX_s \tag{1.7}$$

$$\frac{1}{Q} = \sum_j \frac{R_j}{G_j} + \eta \tan\delta = \frac{R_s}{G_s} + \left[\frac{R_m}{G_m} + \eta \tan\delta\right] \tag{1.8}$$

$$\frac{\Delta f_0}{f_{0,\,\mathrm{ref}}} = -2\sum_j \frac{\Delta X_j}{G_j} - 2\eta\,\frac{\Delta\varepsilon'}{\varepsilon'} = -2\frac{\Delta X_s}{G_s} + \left[-2\frac{\Delta X_m}{G_m} - 2\eta\frac{\Delta\varepsilon'}{\varepsilon'}\right] \tag{1.9}$$

where R_j, X_j and G_j are the surface resistance, reactance and geometrical factor of the jth conductive surface in the cavity, respectively, and η, $\tan\delta$, ε' are the filling factor ($\eta \sim 1$), the loss tangent and the real part of the complex permittivity of the crystal, respectively. Δx indicates the variation of the quantity x with respect to a reference value x_{ref}. The subscripts m and s are related to the metal cavity walls and to the superconductive sample, respectively. The terms in brackets represent the instrumental background [93]. According to Pompeo [96] and Krupka [97], the cavity accuracy can be enhanced at the cryogenic environment employing a resonator performed in the TE011 mode, with circular, planar currents induced on the conducting bases, based on Hakki–Coleman geometry [92, 96].

1.4.2 S-parameter

Permittivity and permeability values of a metamaterial can be retrieved from the reflection (S_{11}) and transmission (S_{21}) data, measured for a plane electromagnetic wave incident normally on a metamaterial slab [98, 99]. The S-parameters are used to extract the effective constitutive parameters by solving the inverse problem, derived from known transmission line equation [99]. Although this method presents some limitations related to the accuracy in the effective slab thickness or the reflection and transmission coefficients, which might be small in magnitude value [98], it is employed by several researchers for metamaterials' characterization, considering both homogeneous and inhomogeneous slabs [98, 100, 101]. According to this, parameters such as refractive index (n) and wave impedance (z) can be extracted by this method. The relationship between these parameters is presented in equations (1.10) and (1.11) [55, 100].

$$\varepsilon = \frac{n}{z} \tag{1.10}$$

$$\mu = nz \tag{1.11}$$

For passive materials, Re(z) and Im(n) $\geqslant$ 0 [100]. The S-parameters, related to refractive index n and impedance z, are defined as [98]:

$$S_{11} = \frac{R_{01}(1 - e^{i2nk_0d})}{1 - R_{01}^2 e^{i2k_0d}} \tag{1.12}$$

$$S_{21} = \frac{(1 - R_{01}^2)e^{ink_0d}}{1 - R_{01}^2 e^{ink_0d}} \tag{1.13}$$

where $R_{01} = z - 1/z + 1$, S_{ij} are the scattering matrix elements, d is the metamaterial slab thickness and k_0 is the wave number in the free space [98].

$$z = \pm\sqrt{\frac{(1 + S_{11})^2 - S_{21}^2}{(1 - S_{11})^2 - S_{21}^2}} \tag{1.14}$$

$$e^{i2nk_0d} = X \pm i\sqrt{1 - X^2} \tag{1.15}$$

where $X = 1/2S_{21}(1 - S_{11}^2 + S_{21}^2)$. The value of the refractive index can be determined from equation (1.15) as [98]:

$$n = \frac{1}{k_0d}\left\{[[\ln\,(e^{ink_0d}]'' + 2m\pi] - i[e^{ink_0d}]'\right\} \tag{1.16}$$

where m is an integer related to the branch index of n'. Equations (1.14) and (1.15) can be used for homogeneous slabs. Nevertheless, for inhomogeneous slabs, these correlations must be carefully addressed [98].

1.4.3 Waveguide method

The waveguide method is suitable for metamaterials' characterization because it requires an easier sample preparation and allows one to perform broadband measurements. Although rectangular waveguide methods were limited to characterization of isotropic samples, they might be successfully employed for electromagnetic characterization of bianisotropic biaxial metamaterials [86]. In this method, the parameters can be retrieved from the TE_{10} excitation performing four independent measurements employing a rectangular waveguide as reported by Chen [84]. For the proposed experimental set up, a slab sample with thickness d must fill the waveguide cross section completely. At these controlled conditions, the incident, reflected and transmitted waves are standing waves. With the origin matching with a face of the slab, S_{11} is equal to the reflection coefficient and S_{21} is related to the transmission coefficient T according to [84]:

$$S_{21} = Te^{ik_{0z}d} \tag{1.17}$$

where k_{0z} is the longitudinal wave number of the incident wave. After the S-parameter measurements, it is possible to retrieve the permeability and permittivity tensors. Additionally, equations (1.14) and (1.15) can be employed to determine the impedance z and the refractive index n. According to the reported results, the proposed set up successfully characterizes different kind of materials, suggesting applications such as homogenization of metamaterials [84].

1.4.4 Nicolson–Ross–Weir method

The Nicolson–Ross–Weir (NRW) method is based on the measurement or numerical simulation of the reflection and transmission coefficients of a planar material sample [102, 103]. This method is a widely employed technique for measuring the permittivity and permeability of homogeneous, isotropic materials [103] due to its theoretical simplicity and ease of algorithmic implementation. Despite that, this method presents some limitations associated to the material impedance and wave velocity into the material [87] that must be considered during parameters' retrieval [102, 104, 105]; therefore, several researchers have worked on alternative ways to extract parameters in order to improve the method's reliability, reducing its ambiguous results [102, 104, 105]. For instance, a simple parameters' extraction from an omega shaped resonator, employing the NWR method, was performed by Labidi (2017), through the following equations:

$$n_{\text{eff}} = \sqrt{\varepsilon_{\text{eff}} \times \mu_{\text{eff}}} \tag{1.18}$$

The NRW algorithm was based on transmission (S_{21}) and reflection (S_{11}) coefficients to extract the material properties as follows [57]:

$$S_{11} = \frac{\Gamma(1 - T^2)}{1 - \Gamma^2 T^2} \tag{1.19}$$

$$S_{21} = \frac{T(1 - \Gamma^2)}{1 - \Gamma^2 T^2} \tag{1.20}$$

where Γ and T are the reflection of the air–sample interface and the transmission of the wave propagation, respectively, which are given by:

$$\Gamma = \frac{z - 1}{z + 1} \tag{1.21}$$

$$T = e^{\gamma d} \tag{1.22}$$

γ and z are the propagation constant of the sample and the normalized characteristic impedance, respectively. They are related to μ and ε according to the following relationships:

$$\gamma = \gamma_0 \sqrt{\mu\varepsilon} \tag{1.23}$$

$$z = \frac{\mu}{\varepsilon} \tag{1.24}$$

$$\gamma_0 = \frac{j2\pi}{\lambda_0} \tag{1.25}$$

where γ_0 is the constant of the free-space propagation and λ_0 is the free-space wavelength (m). From equations (1.18) and (1.19), Γ and T can be described as [57]:

$$\Gamma = k \pm \sqrt{k^2 - 1} \tag{1.26}$$

$$T = \frac{S_{11}S_{21} - \Gamma}{1 - (S_{11} + S_{21})\Gamma'} \tag{1.27}$$

$$k = \frac{S_{11}^2 - S_{21}^2 + 1}{2S_{11}} \tag{1.28}$$

where the auxiliary parameter k is a function of the scattering parameters [88]. From equations (1.20) and (1.23), μ and ε can be established.

$$\sqrt{\frac{\mu}{\varepsilon}} = \frac{1 - \Gamma}{1 + \Gamma} \tag{1.29}$$

$$\varepsilon = \frac{\gamma}{\gamma_0}\left(\frac{1 - \Gamma}{1 + \Gamma}\right) \tag{1.30}$$

$$\mu = \frac{\gamma}{\gamma_0}\left(\frac{1 + \Gamma}{1 - \Gamma}\right) \tag{1.31}$$

As can be seen, experiments on the original formulation of the NRW method have introduced some variations oriented to improve the accuracy and to reduce the noise measurements [94, 106], reduce the use of additional algorithms [102] or to avoid ambiguity in the results [104]. According to this, the characterization of each developed material requires a rigorous parameter retrieval in order to adapt the method to the specific metamaterial conditions.

1.4.5 Free-space method

This technique offers several advantages when compared to waveguide and resonant methods because it is suitable for a wide frequency range and it is more flexible concerning sample parameters such as size, shape and consistency. Although the preparation of the sample is also easier [107], it requires plane wave approximation, and therefore non-uniformity in the sample's traverse dimensions might result in undesired scattering and diffraction effects [86]. In order to enhance the accuracy of measurement, some alternatives such as an advanced calibration and a high-quality anechoic chamber can be implemented [107]. The free-space method set up consists of two antennas connected to the vector network analyzer (VNA) ports. This configuration allows one to characterize different materials, based on transmission and reflection concepts. For sample characterization, it must be placed between the horns, as represented in figure 1.14 [108].

For the experimental set up presented in figure 1.14, the antennas and the space between them can be considered as a two port network, whose behavior is

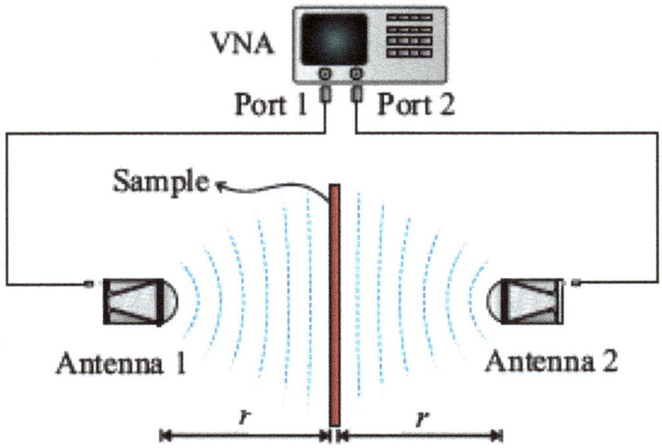

Figure 1.14. Representation of the free-space method. The sample is placed between two antennas, connected to the ports of the VNA K. Reproduced from [109]. CC BY 4.0.

Table 1.1. The main advantages and disadvantages for the most employed techniques for electromagnetic properties' measurement [84, 85, 108].

Method	Advantages	Disadvantages
Resonator	High accuracy High sensitivity Reliable results for materials with low losses	Sample requires careful preparation Measurements must be performed at specific frequencies
Waveguide	High frequency measurements Suitable for anisotropic materials	Is not suitable for low frequency measurements Sample must fill the cross section waveguide
Free space	Easier sample preparation High frequency range Non-destructive measurements Suitable for high loss materials	Requires plane wave approximations Additional signal processing is often required Special environments such as an anechoic chamber are suggested

characterized by its four S-parameters: S_{11}, S_{12}, S_{21} and S_{22}, that will be employed to extract the electromagnetic properties of the sample [108]. Hence, the electric permittivity and magnetic permeability must be retrieved through validated mathematical algorithms. As can be seen, the non-resonant methods are based on the measurement of the S-parameters or mathematical calculation. Therefore, a high accuracy technique for the measurement of dielectric properties must be chosen. The main advantages and disadvantages of the most commonly employed techniques for electromagnetic properties' measurement are summarized in table 1.1.

1.5 Results and discussion

As previously presented, the characterization of MTMs is a relevant issue to design and manufacture artificial materials with superior performance for several applications. Even though this science has accomplished structures with remarkable behavior as waveguides [9], sensing [8, 109, 110], super lenses [21, 111], cloaking [4, 112], absorbers [113], energy harvesting devices [20, 114], antenna and antenna elements [9, 57], it still presents some challenges which need to be met to guarantee MTMs' scalability and use at the industrial scale. Table 1.2 presents some of the challenges of MTMs to be addressed for specific applications.

Table 1.2. Some of the challenges of MTMs to be addressed for specific applications.

Year	Authors	Application	Challenges
2011	Soukoulis and Wegener [115]	Photonic metamaterials	To develop techniques for fabricating large-scale 3D isotropic metamaterials
2016	Yoon *et al* [116]	Fabrication of practical metamaterials	Mass production of metamaterials for use in practical applications
2018	De Marco *et al* [117]	Silicon-based dielectric metamaterials	A fabrication method to produce uniform and crystalline Si blocks ranging from 75 to 200 nm in diameter, containing few impurities and little to no porosity
2018	Sun *et al* [118]	Wireless power transfer metamaterials	To enhance transfer efficiency and transfer distance To promote research on commercial application such as electric vehicles, implantable medical devices and enlarge the transfer power
2018	Zhang *et al* [119]	Wireless power transfer metamaterials for electric vehicles (EVs)	To achieve the negative permeability of metamaterials at KHz range To evaluate the dissipation and minimize the loss of metamaterials by optimizing structures and resonant unit cells To decrease volume and weight of the current metamaterial- based wireless power transfer technique to be used on EV contactless charging systems
2018	Palma *et al* [120]	Acoustic metamaterials in aeronautics	The inclusion of moving fluids in models and simulations towards the utilization of metamaterial concepts in aeronautical applications

			To develop numerical methods to simulate the acoustic material behavior under aeronautical operating conditions
2019	Wu *et al* [121]	Additive manufacturing for metamaterials' development	To increase resolution on manufacturing techniques to achieve typical unit cell size ranging from few millimeters to few nanometers. To overcome the limitations associated with production efficiency, costs, limited accuracy, anisotropy, heterogeneity on microstructure and material properties, residual stresses, among others
2020	Fischer *et al* [122]	Mechanical metamaterials	To scale metamaterials from laboratory to industrial applications
2022	Lvov *et al* [123]	Auxetic metamaterials for biomedical devices	To evaluate the long-term performance of auxetic-based medical devices and the potential benefits of auxetic structures for medical instruments and implantable devices
2023	Sinha *et al* [124]	Programmable multi-physical mechanical metamaterials	Real-time reconfigurability and functionality programming. 4D printing and scalability
2023	Zhou *et al* [125]	3D/4D printing of mechanical metamaterials	To increase the research on 3D/4D printing of mechanical metamaterials because the potential of this technology on metamaterials' manufacturing has not been completely explored The development of high-resolution 3D printing technology for several materials To study the disadvantages associated with 3D printing technology such as residual stress and bond strength between printed layers
2023	Aijaz and Ruchi [126]	Metamaterial-inspired antennas	More research about the use of metamaterial on antenna applications

As previously summarized in table 1.2, the main challenge for MTMs science is to scale the manufactured structures from laboratory to real applications, which is related to the lack of information about raw material and processing cost of the developed materials. Forthcoming technologies such as additive manufacture and

3D/4D printing emerge as an alternative to produce MTMs at the industrial scale. Nevertheless, the technologies and materials employed for these processes must be carefully studied in order to solve some of the limitations related to residual stress and manufacturing resolution. As can be seen, there is a long way to go in MTMs science and therefore researchers must aim to develop scalable materials in terms of raw materials' prices and process replication.

1.6 Conclusions

As MTMs have been developed to meet the specific requirements of particular applications, their characterization is mandatory to evaluate their dielectric and magnetic properties and, consequently, be capable of predicting the material response to an applied electromagnetic field at a wide frequency range. There are different characterization techniques and parameter retrieval methods, the most suitable among them will depend on both material composition and operation frequency range. According to their properties, they are classified in double positive, epsilon negative, mu negative and double negative metamaterials. They can also be categorized depending on their use as artificial dielectrics, artificial magnetics, chiral, plasmonic, omega shape and tunable materials. In particular, chiral materials have shown outstanding mechanical behavior suggesting an additional study field on mechanical and construction science, due to the possibility of modulating the response to crash and seismic events.

According to this, the manufacturing of MTMs must be considered as a high impact science, owing to the theorical basis on biology, physics and chemistry. Therefore, MTMs must be assessed with a multidisciplinary approach to understand not only their electromagnetic response but also their mechanical properties, as well as the relationship among the chemical and physical properties of the materials used in the manufacturing process. Materials cost is another matter that must be evaluated to determine the feasibility of MTM development at the industrial scale. In this way, additive manufacturing and 3D/4D printing could be alternative processes to be explored in this field.

Bibliography

[1] Holloway C L, Dienstfrey A, Kuester E F, O'Hara J F, Azadd A and Taylor A J 2009 A discussion on the interpretation and characterization of metafilms/metasurfaces: the two-dimensional equivalent of metamaterials *Metamaterials* **3** 100–12

[2] Yousefi L, Boybay M S and Ramahi O M 2011 Characterization of metamaterials using a strip line fixture *IEEE Trans. Antennas Propag.* **59** 1245–53

[3] Yang W, Lu J, Zhuang W, Qi J, Wang C, Wang H and Su G *et al* 2023 PtS_2 metamaterials: fabrication and physical properties *Appl. Surf. Sci.* **614** 156277

[4] Kadic M, Bückmann T, Schittny R and Wegener M November 2013 Metamaterials beyond electromagnetism *Rep. Prog. Phys.* **76**

[5] Kshetrimayum R S 2005 A brief intro to metamaterials *IEEE Potentials* **23** 44–6

[6] Eleftheriades G V and Engheta N 2011 Metamaterials: fundamentals and applications in the microwave and optical regimes *Proc. IEEE* **99** 1618–21

[7] Li J, Zhang Z Y, Liu H T and Wang Y B 2022 Design and characterization of novel bi-directional auxetic cubic and cylindrical metamaterials *Compos. Struct.* **299** 116015
[8] Abdulkarima Y *et al* 2020 Design and study of a metamaterial based sensor for the application of liquid chemicals detection *J. Mater. Res.* **9** 10291–304
[9] Buriak I A, Zhurbal V O, Vorobjov G S, Kulizhko V R, Kononov O K and Rybalko O 2016 Metamaterials: theory, classification and application strategies (Review) *J. Nano-Electron. Phys.* **8** 0408801 0240811
[10] Datta A K, Sumnu G and Raghavan G 2005 Dielectric properties of food *Engineering Properties of Foods* (Boca Raton, FL: CRC Press)
[11] Halliday D and Resnick R 1973 *Física II* (Brazil: Livros Técnicos e Científicos, Rio de Janeiro)
[12] Nelson S O 2010 Fundamentals of dielectric properties measurements and agricultural applications *J. Microwave Power Electromagn. Energy* **44** 98–113
[13] Mudgett R E 1985 Dielectric properties of food *Microwaves in the Food Processing Industry* (Orlando: Academic)
[14] Debye P 1929 *Polar Molecules* (Nova York: The Chemical Catalog Company)
[15] Mehdizadeh M 2010 *Microwave/RF Applicators and Probes for Material Heating. Sensing and Plasma Generation* (Oxford: Elsevier)
[16] Pimenov A and Loidl A 2006 Experimental demonstration of artificial dielectrics with a high index of refraction *Phys. Rev.* B **74** 193102
[17] Singh S 2002 Refractive index measurement and its applications *Phys. Scr.* **65** 167–80
[18] Ramakrishna S A 2005 Physics of negative refractive index materials *Rep. Prog. Phys.* **68** 449–521
[19] Veselago V G 1968 The electrodynamics of substances with simultaneously negative values of ε and μ *Sov. Phys. Usp.* **10** 509–14
[20] Liu Y and Zhang X 2011 Metamaterials: a new frontier of science and technology *Chem. Soc. Rev.* **40** 2494–507
[21] Bilotti F and Sevgi L 2012 Metamaterials: definitions, properties, applications, and FDTD-based modeling and simulation *Int. J. RF Microwave Comput. Aided Eng.* **22** 422–38
[22] Mendhe S E and Kosta Y P 2011 Metamaterial properties and applications *Int. J. Inform. Technol. Knowl. Manag.* **4** 85–9
[23] Fan G, Sun K, Hou Q, Wang Z, Liu Y and Fan R 2021 Epsilon-negative media from the viewpoint of materials science *EPJ Appl. Metamater.* **8** 1–19
[24] Jokanovic B, Geschke R, Beukman T S and Milosevic V 2010 Metamaterials: characteristics, design and microwave applications *Afr. Inst. Electr. Eng.* **101** 82–92
[25] Pendry J B, Holden A J, Stewart W J and Youngs I 1996 Extremely low frequency plasmons in metallic mesostructures *Phys. Rev. Lett.* **76** 4773–6
[26] Grigorenko A N, Geim A K, Gleeson H F, Zhang Y, Firsov A A, Khrushchev I Y and Petrovic J 2005 Nanofabricated media with negative permeability at visible frequencies *Nature* **438** 335–8
[27] Tamandani A R, Ahmadi-Shokouh J and Tavakoli S 2013 Wideband planar split ring resonator-based metamaterials *Prog. Electromagn. Res.* M **28** 115–28
[28] Holloway C L, Kuester E F, Baker-Jarvis J and Kabos P 2003 A double negative (DNG) composite medium composed of magnetodielectric spherical particles embedded in a matrix *IEEE Trans. Antennas Propag.* **51** 2596–603
[29] Caloz C 2009 Perspectives on EM metamaterials *Mater. Today* **12** 13–20

[30] Shamonina E and Solymar L 2007 Metamaterials: how the subject started *Metamaterials* **1** 12–8

[31] Cavallo D and Felita C 2017 Analytical formulas for artificial dielectrics with nonaligned layers *IEEE Trans. Antennas Propag.* **65** 5303–11

[32] Sihvola A 2007 Metamaterials in electromagnetics *Metamaterials* **1** 2–11

[33] Barzegar-Parizi S and Rejaei B 2015 Calculation of effective parameters of high permittivity integrated artificial dielectrics *IET Microwave Antennas Propag.* **9** 1287–96

[34] Kabiri A, Yousefi L and Ramahi O M 2010 On the fundamental limitations of artificial magnetic materials *IEEE Trans. Antennas Propag.* **58** 2345–53

[35] Maslovski S, Ikonen P, Kolmakov I and Tretyakov S 2005 Artificial magnetic materials based on the new magnetic particle: metasolenoid *Prog. Electromagn. Res., PIER* **54** 61–81

[36] Marques R, Medina F and Rafii-El-Idrissi R 2002 Role of bianisotropy in negative permeability and left-handed metamaterials *Phys. Rev. B* **65** 144441–6

[37] Pendry J B, Holden A J, Robbins D J and Stewart W J 1999 Magnetism from conductors and enhanced nonlinear phenomena *IEEE Trans. Microwave Theory Tech.* **47** 2075–84

[38] Baena J D, Marques R and Medina F 2004 Artificial magnetic metamaterial design by using spiral resonators *Phys. Rev. B* **69** 144021–144 025

[39] Yousefi L and Ramahi O M 2007 New artificial magnetic materials based on fractal Hilbert curves *Proc. IWAT07* pp 237–40

[40] Wu W, Hu W, Qian G, Liao H, Xu X and Berto F 2019 Mechanical design and multifunctional applications of chiral mechanical metamaterials: a review *Mater. Des.* **180** 107950

[41] Li J, Yang Q, Huang N and Tao R 2021 A novel mechanical metamaterial with tailorable Poisson's ratio and thermal expansion based on a chiral torsion unit *Smart Mater. Struct.* **30** 115004

[42] Wang J, Honga L and Jiaoa P 2023 Graphene-reinforced shape memorable chiral metamaterials: theoretical analysis with experimental and numerical validations *Mater. Des.* **226** 111648

[43] Dalgac S, Bakır M, Karadag F, Karaaslan M, Akgoland O and Unal E 2020 Microfluidic sensor applications by using chiral metamaterial *Mod. Phys. Lett.* B **34** 2050031

[44] Wang B, Zhou J, Koschny T, Kafesaki M and Soukoulis C M 2009 Chiral metamaterials: simulations and experiments *J. Opt.* A **11** 114003

[45] Niu X, Yang X, Li H, Liu J, Liu Z and Wang K 2020 Application of chiral materials in electrochemical sensors *Microchim. Acta* **187** 676

[46] Yuan C, Mu X M, Dunn C K, Haidar J, Wang T J and Qi H J 2018 Thermomechanically triggered two-stage pattern switching of 2D lattices for adaptive structures *Adv. Funct. Mater.* **28** 1705727

[47] Xia R, Song X K, Sun L J, Wu W W, Li C L, Cheng T B and Qian G 2018 Mechanical properties of 3D isotropic anti-tetrachiral metastructure *Phys. Status Solidi* B **255** 1700343

[48] Farrugia P S, Gatt R and Grima J N 2019 A novel three-dimensional anti-tetrachiral honeycomb *Phys. Status Solidi* B **256** 1800473

[49] Frenzel T, Kadic M and Wegener M 2017 Three-dimensional mechanical metamaterials with a twist *Science* **358** 1072–4

[50] Fu M H, Liu F M and Hu L L 2018 A novel category of 3D chiral material with negative Poisson's ratio *Compos. Sci. Technol.* **160** 111–8

[51] Kurgin J 2015 How to deal with the loss in plasmonics and metamaterials *Nat. Nanotechnol.* **10** 2–6
[52] Zhang J, Zhang L and Xu W 2012 Surface plasmon polaritons: physics and applications *J. Phys.* D **45** 113001
[53] Wang Z, Cheng F, Winsor T and Liu Y 2016 Optical chiral metamaterials: a review of the fundamentals, fabrication methods and applications *Nanotechnology* **27** 412001
[54] Aydin K, Li Z, Hudlicka M, Tretyakov S A and Ozbay E 2007 Transmission characteristics of bianisotropic metamaterials based on omega shaped metallic inclusions *New J. Phys.* **9** 326
[55] Basiry R, Abiri H and Yahaghi A 2011 Electromagnetic performance analysis of Omega-type metamaterial radomes *Int. J. RF Microw. Comput. Aided Eng.* **21** 665–73
[56] Bilal R M H, Baqir M A, Iftikhar A, Ali M M, Rahim A A, Niaz M, Mughal A M J and Naqvi. S A 2021 A novel omega shaped microwave absorber with wideband negative refractive index for C-band applications *Optik* **242** 167278
[57] Labidi M, Salhi R and Choubani F 2017 A design of metamaterial multi-band bowtie antenna based on omega-shaped resonator *Appl. Phys.* A **123** 313
[58] Altıntas O, Aksoy M and Ünal. E 2019 Design of a metamaterial inspired omega shaped resonator-based sensor for industrial implementations *Phys.* E **116** 113734
[59] Wang Y, Kim I, Jin R, Jeong H, Li J, Dong Z and Rho J 2018 Experimental verification of asymmetric transmission in continuous omega-shaped metamaterials *RSC Adv.* **8** 38556
[60] Turpin J P, Bossard J A, Morgan K L, Werner D H and Werner P L 2014 Reconfigurable and tunable metamaterials: a review of the theory and applications *Int. J. Antennas Propag.* **2014** 429837
[61] Degiron A, Mock J J and Smith D R 2007 Modulating and tuning the response of metamaterials at the unit cell level *Opt. Express* **15** 1115–27
[62] Hand T H and Cummera S A 2008 Frequency tunable electromagnetic metamaterial using ferroelectric loaded split rings *J. Appl. Phys.* **103** 066105
[63] Wang Z, Guo Z, Li Z and Zeng K 2023 Design, manufacture, and characterisation of hierarchical metamaterials for simultaneous ultra-broadband sound-absorbing and superior mechanical performance *Virtual Phys. Prototyp.* **18** e2111585
[64] Bang S, Kim J, Yoon G, Tanaka T and Rho J 2018 Recent advances in tunable and reconfigurable metamaterials *Micromachines* **9** 560
[65] Wang B, Wang L, Wang G, Wang L, Zhai X, Li X and Huang W 2013 A simple nested metamaterial structure with enhanced bandwidth performance *Opt. Commun.* **303** 13–4
[66] Shrekenhamer D, Chen C and Padilla W 2013 Liquid crystal tunable metamaterial absorber *Phys. Rev. Lett.* **110** 177403
[67] Xu J, Yang R, Fan Y, Fu Q and Zhang F 2021 A review of tunable electromagnetic metamaterials with anisotropic liquid crystals *Front. Phys.* **9** 633104
[68] Zhao J, Cheng Q, Chen J, Qi M Q, Jiang W X and Cui T J 2013 A tunable metamaterial absorber using varactor diodes *New J. Phys.* **15** 043049
[69] Xu R J and Lin Y S 2022 Actively MEMS-based tunable metamaterials for advanced and emerging applications *Electronics* **11** 243
[70] Savo S, Shrekenhamer D and Padilla W 2014 Liquid crystal metamaterial absorber spatial light modulator for THz applications *Adv. Opt. Mater.* **2** 275–9
[71] National Research Council 1994 *Microwave Processing of Materials* (Washington, DC: The National Academies Press)

[72] Franco A P, Yamamoto L Y, Tadini C C and Gut J A W 2015 Dielectric properties of green coconut water relevant to microwave processing: effect of temperature and field frequency *J. Food Eng.* **155** 69–78
[73] Rozanov K N, Sipov A V O, Petrov D A, Starostenko S N and Yelsukov E P 2009 The effect of shape distribution of inclusions on the frequency dependence of permeability in composites *J. Magn. Magn. Mater.* **321** 738–41
[74] Osipov A V, Rozanov K N, Simonov N A and Starostenko S N 2002 Reconstruction of intrinsic parameters of a composite from the measured frequency dependence of permeability *J. Phys.: Condens. Matter* **14** 9507–23
[75] Chowdary K M and Majetich S A 2014 Frequency-dependent magnetic permeability of $Fe_{10}Co_{90}$ nanocomposites *J. Appl. Phys.* **47** 175001
[76] Salem A A, Abd-Rahman R, Kamarudin M S, Ahmed H, Jamail N A M, Othman N A, Yousof M F M, Ishak M T and Al-Ameri 2019 The effect of insulator geometrical profile on electric field distributions *Indones. J. Electr. Eng. Comput. Sci.* **14** 618–27
[77] Stebe K J, Lewandowski E and Ghosh M 2009 Oriented assembly of metamaterials *Science* **325** 159–61
[78] Illias H A, Lee Z H, Bakar A H A, Mokhlis H and Ariffin A M 2012 Distribution of electric field in medium voltage cable joint geometry *IEEE Int. Conf. on Condition Monitoring and Diagnosis* pp 1052–4
[79] Hess O, Pendry J B, Maier S A, Oulton R F, Hamm J M and Tsakmakidis K L 2012 Active nanoplasmonic metamaterials *Nat. Mater.* **11** 573–84
[80] Varadan V and Ji L 2010 Temperature dependence of resonances in metamaterials *IEEE Trans. Microwave Theory Tech.* **58** 2673–81
[81] Bobicki E R, Pickles C A, Forster J, Marzoughi O and Hutcheon R 2020 High temperature permittivity measurements of selected industrially relevant ores: review and analysis *Miner. Eng.* **145** 106055
[82] Simovski C R 2011 On electromagnetic characterization and homogenization of nanostructured metamaterials *J. Opt.* **13** 013001
[83] Cipolla J, Gokhale N, Norris A N and Nagy A 2011 Design of inhomogeneous pentamode metamaterials for minimization of scattering *J. Acoust. Soc. Am.* **130** 2332
[84] Chen H, Zhang J, Bai Y, Luo Y, Ran L, Jiang Q and Kong J A 2006 Experimental retrieval of the effective parameters of metamaterials based on a waveguide method *Opt. Express* **14** 12945–9
[85] Jilani M, Rehman M, Khan A M, Khan M T and Ali S M 2012 A brief review of measuring techniques for characterization of dielectric materials *Int. J. Inform. Technol. Electr. Eng.* **1** 1–5
[86] Hasar U C, Muratoglu A, Bute M, Barroso J J and Ertugrul M 2017 effective constitutive parameters retrieval method for bianisotropic metamaterials using waveguide measurements *IEEE Trans. Microwave Theory Tech.* **65** 1488–97
[87] Costa F, Borgese M, Degiorgi M and Monorchio A 2017 Electromagnetic characterisation of materials by using transmission/reflection (T/R) devices *Electronics* **6** 95
[88] Chen L, Ong C K and Tan B T G 1999 Cavity perturbation technique for the measurement of permittivity tensor of uniaxially anisotropic dielectrics *IEEE Trans. Instrum. Meas.* **48** 1023–30
[89] Kobayashi Y and Tanaka S 1980 Resonant modes of a dielectric rod resonator short-circuited at both ends by parallel conducting plates *IEEE Trans. Microwave Theory Tech.* **28** 1077–85

[90] Buffler C 1995 Advances in dielectric measurement of foods *Characterization of Food: Emerging Methods* (Marlborough: Elsevier Science)

[91] Liu W, Sun H and Xu L 2018 A microwave method for dielectric characterization measurement of small liquids using a metamaterial-based sensor *Sensors* **18** 1438

[92] Hakki B W and Coleman P D 1960 A dielectric resonator method of measuring inductive capacities in the millimeter range *IRE Trans. Microwave Theory Tech.* **8** 402–10

[93] Alimenti A, Torokhtii K, Silva E and Pompeo N 2019 Challenging microwave resonant measurement techniques for conducting material characterization *Meas. Sci. Technol.* **30** 065601

[94] Alimenti A, Torokhtii K, Pompe N, Piuzzi E and Silva E 2020 Characterisation of dielectric 3D-printing materials at microwave frequencies *Acta Imeko* **9** 26–32

[95] Alimenti A, Pitella E, Torokhtii K, Pompe N, Piuzzi E and Silva E 2023 A dielectric loaded resonator for the measurement of the complex permittivity of dielectric substrates *IEEE Trans. Instrum. Meas.* **72** 6001009

[96] Pompeo N, Torokhtii K and Silva E 2014 Dielectric resonators for the measurements of the surface impedance of superconducting films *Meas. Sci. Rev.* **14** 164–70

[97] Krupka J and Mazierska J 2000 Single-crystal dielectric resonators for low-temperature electronics applications *IEEE Trans. Microwave Theory Tech.* **48** 1270–4

[98] Chen X, Grzegorczyk T M, Wu B, Pacheco J and Kong J A 2004 Robust method to retrieve the constitutive effective parameters of metamaterials *Phys. Rev.* E **70** 016608

[99] Cohen D and Shavit R 2015 Bi-anisotropic metamaterials effective constitutive parameters extraction using oblique incidence S-parameters method *IEEE Trans. Antennas Propag.* **63** 2071–8

[100] Smith D R, Schultz S, Markos P and Soukoulis C M 2002 Determination of effective permittivity and permeability of metamaterials from reflection and transmission coefficients *Phys. Rev.* B **65** 195104

[101] Hasar U, Ozturk G, Bute M and Ertugrul M 2020 Method for electromagnetic property extraction of sublayers in metal-backed inhomogeneous metamaterials *IEEE Access* **8** 151705–18

[102] Luukkonen O, Maslovski S I and Tretyakov S A 2011 A stepwise Nicolson–Ross–Weir-based material parameter extraction method *IEEE Antennas Wirel. Propag. Lett.* **10** 1295–9

[103] Rothwell E J, Frasch J L, Ellison S M, Chahal P and Ouedraogo R O 2016 Analysis of the Nicolson–Ross–Weir method for characterizing the electromagnetic properties of engineered materials *Prog. Electromagn. Res.* **157** 31–47

[104] Angiulli G and Versaci M 2021 Retrieving the effective parameters of an electromagnetic metamaterial using the Nicolson–Ross–Weir method: an analytic continuation problem along the path determined by scattering parameters *IEEE Access* **9** 77511–25

[105] Angiulli G and Versaci M 2022 Extraction of the electromagnetic parameters of a metamaterial using the Nicolson–Ross–Weir method: an analysis based on global analytic functions and Riemann *Surf. Appl. Sci.* **12** 11121

[106] Vicente A N, Dip G M and Junqueira. C 2011 The step by step development of NRW method *SBMO/IEEE MTT-S International Microwave and Optoelectronics Conference (IMOC 2011)* (Natal, Brazil) pp 738–42

[107] Kemptner E and Thurner S 2012 Free space material characterization for microwave frequencies *6th European Conf. on Antennas and Propagation (EUCAP) (Prague)* (Czech Republic) pp 3513–5

[108] Gonçalves F J F, Pinto A G M, Mesquita R C, Silva E J and Brancaccio A 2018 Free-space materials characterization by reflection and transmission measurements using frequency-by-frequency and multi-frequency algorithms *Electronics* **7** 260

[109] Du P, Tan K and Xing X 2012 A novel binary tree support vector machine for hyperspectral remote sensing image classification *Opt. Commun.* **285** 3054–60

[110] Altintas O, Aksoy M, Akgol O, Unal E, Karaaslan M and Sabah C 2017 Fluid, strain and rotation sensing applications by using metamaterial based sensor *J. Electrochem. Soc.* **164** B 567

[111] Fan N, Lee H, Sun C and Zhang X 2005 Sub-diffraction-limited optical imaging with a silver superlens *Science* **308** 534–7

[112] Bilotti F, Tricarico S and Vegni L 2010 Plasmonic metamaterial cloaking at optical frequencies *IEEE Trans. Nanotechnol.* **9** 55–61

[113] Bilal R M H, Baqir M A, Iftikhar A, Ali M M, Rahim A A, Akhtar M, Mughal M J and Naqvi. S A 2021 A novel omega shaped microwave absorber with wideband negative refractive index for C-band applications *Optik—Int. J. Light Electron Optics* **242** 167278

[114] Du H, Lin X, Zheng H, Qu B, Huang Y and Chu D 2016 Colossal permittivity in percolative ceramic/metal dielectric composites *J. Alloys Compd.* **663** 848

[115] Soukoulis C and Wegener M 2011 Past achievements and future challenges in the development of three-dimensional photonic metamaterials *Nat. Photonics* **5** 523–30

[116] Yoon G, Kim I and Rho J 2016 Challenges in fabrication towards realization of practical metamaterials *Microelectron. Eng.* **163** 20

[117] De Marco M L, Semlali S, Korgel B A, Barois P, Drisko G L and Aymonier C 2018 Silicon-based dielectric metamaterials: focus on the current synthetic challenges *Angew. Chem. Int. Ed.* **57** 4478–98

[118] Sun K *et al* 2018 An overview of metamaterials and their achievements in wireless power transfer *J. Mater. Chem.* C **6** 2925

[119] Zhang Z, Zhang B, Deng B, Wei X and Wang J 2018 Opportunities and challenges of metamaterial-based wireless power transfer for electric vehicles *Wirel. Power Transf.* **5** 9–19

[120] Palma G, Mao H, Burghignoli L, Göransson P and Iemma U 2018 Acoustic metamaterials in aeronautics *Appl. Sci.* **8** 971

[121] Wu X, Su Y and Shi J 2019 Perspective of additive manufacturing for metamaterials development *Smart Mater. Struct.* **28** 093001

[122] Fischer S C L, Hillen L and Eber C 2020 Mechanical metamaterials on the way from laboratory scale to industrial applications: challenges for characterization and scalability *Materials* **13** 3605

[123] Lvov V A, Senatov F S, Veveris A A, Skrybykina V A and Lantada A D 2022 Auxetic metamaterials for biomedical devices: current situation, main challenges, and research trends *Materials* **15** 1439

[124] Sinha P and Mukhopadhyay T 2023 Programmable multi-physical mechanics of mechanical metamaterials *Mater. Sci. Eng.* R **155** 100745

[125] Zhou X, Ren L, Song Z, Li G, Zhang J, Li B, Wu Q, Li W, Ren L and Liu Q 2023 Advances in 3D/4D printing of mechanical metamaterials: from manufacturing to applications *Compos. Part* B **254** 110585

[126] Aijaz U and Ruchi 2023 Metamaterial inspired antennas for gain enhancement—a review *Proc. of the 8th Int. Conf. on Communication and Electronics Systems (ICCES 2023)* pp 438–44

Chapter 2

Microwave metamaterial sensors

Mehmet Bakir

Symbols

ε	Permittivity
μ	Permeability
κ	Dielectric constant
λ	Wavelength (m)

2.1 Introduction

Metamaterial sensors are a type of sensor that utilizes the unique properties of metamaterials to detect changes in their environment. Metamaterials are artificial materials that are designed to have properties not found in natural materials, such as negative refractive index, which allows them to bend light in unusual ways. Metamaterial sensors work by detecting changes in the way that electromagnetic waves interact with the metamaterial. For example, a metamaterial sensor might be designed to detect changes in the frequency or polarization of light, or changes in the magnetic or electric fields around it.

Metamaterial-based sensors can be used in a variety of sensing applications, including microfluidic sensors, terahertz sensors, absorber-based sensors, 3D metamaterial sensors, and machine learning based sensors.

Microfluidic sensors: Metamaterials can be used in microfluidic sensors to detect changes in the refractive index or concentration of fluids flowing through microchannels. These sensors can be used for a wide range of applications, including point-of-care diagnostics, environmental monitoring, and chemical analysis.

Terahertz sensors: Terahertz radiation is a type of electromagnetic radiation that has potential applications in imaging, spectroscopy, and communication. Metamaterials can be used to create terahertz sensors that are more sensitive and selective than conventional sensors. For example, metamaterials can be designed to

doi:10.1088/978-0-7503-5754-8ch2

detect specific molecules or materials based on their unique terahertz absorption spectra.

Absorber-based sensors: Metamaterials can be used as absorbers to detect changes in the surrounding environment. These sensors can be used to detect changes in temperature, humidity, pressure, or other physical parameters. Absorber-based sensors can be designed to be highly sensitive and can be used in a wide range of applications, including structural health monitoring and environmental sensing.

3D metamaterial sensors: 3D metamaterials are structures that have unique electromagnetic properties in all three dimensions. 3D metamaterial sensors can be used to detect changes in the surrounding environment, such as changes in temperature or pressure. These sensors can be designed to be highly sensitive and selective and can be used in a wide range of applications, including biomedical sensing and environmental monitoring.

Machine learning based sensors: Metamaterial sensors can be integrated with machine learning algorithms to create intelligent sensing systems. These systems can learn from past data and make predictions about future events based on real-time sensor data. Machine learning based sensors have potential applications in a wide range of fields, including robotics, smart cities, and environmental monitoring.

Since most of the metamaterial-based sensor applications have been developed in four headlines, in this part of the study we decided to give examples of them.

2.2 Microfluidic sensors

The microfluidic sensors' operation is based on an establishment of connection between the dielectric properties of a liquid sample and resonance frequency. High sensitivity and accurate identification of chemical and biological liquid samples using microwave dielectric and cylindrical resonators have been studied and demonstrated in [1–5]. After these improvements, complementary split-ring resonators (CSSRs) and split-ring resonators (SRRs) have been widely implemented in microfluidic sensors both in transmission line inspired metamaterials and in waveguides. Determination of the complex permittivity of liquid based on changes in the resonance frequency and peak attenuation of the transmission response have been widely used by different scientists and work groups. A water–ethanol mixture was used to validate the proposed sensor for both simulations and experiments. Microstrip coupled CSSR used the advantage of a strong electric field that is produced in a small area, which results with high sensitivity to complex permittivity changes.

Another microfluidic sensor was characterized by Rawat *et al* in 2014 [6]. Their proposal is also CSSR-shaped and validated by water, petrol, and ethanol. To increase the sensitivity and decrease the losses due to substrates, the researchers generally employed low loss dielectric substrates such as Rogers 5870 and its derivatives. They successfully showed that the split-ring resonator based metamaterials can be used as petrol sensors as well as water–ethanol sensors, as shown in figure 2.1.

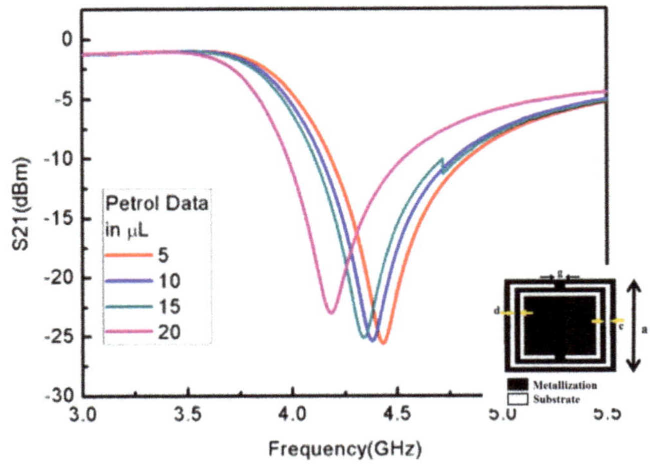

Figure 2.1. Rawat *et al*'s proposed design and reflection coefficient simulation for petrol. (Reprinted from [6], with the permission of AIP Publishing.)

Following this work, several fuel and fuel adulteration sensors have been designed and proposed [7–14, 18]. In 2015, to show the miniaturized quasi-lumped resonators for dielectric characterization of liquid mixtures, Benkhaoua *et al* proposed a double-sided spiral split-ring resonator (DSS-SRR) shaped metamaterial sensor for characterizing a water–ethanol mixture in a practical application. The proposed sensor has high sensitivity, short assay time, small volume, low cost, and can be produced via lab chips since its operating frequency is around 150 MHz [15].

Another CSSR-based microfluidic sensor was proposed by Salim and Lim in 2016 [16]. Two tightly coupled concentric CSRRs were loaded on a patch realized on a Rogers RT/Duroid 5870 substrate. Then, a microfluidic channel was engraved on polydimethylsiloxane which was integrated for chemical sensor applications. While the resonant frequency of the structure before loading the microfluidic channel was 4.72 GHz, after loading the channel, a 550 MHz shift in the resonant frequency was achieved when the ethanol concentration is changed from 0% to 100%. While their proposal is seen in figure 2.2, the measurement results are given in figure 2.3.

There are also THz microfluidics sensor applications in the current literature. For example, Geng *et al* presented a unique characteristic for biosensing applications due to their accurately tunable resonance frequency, which is in accord with the vibrational frequency of some important biomolecules such as cancer biomarkers [17]. Therefore, they numerically designed two kinds of THz metamaterials biosensors integrated with microfluidics and experimentally fabricated and tested for detection of the Alpha fetoprotein (AFP) and Glutamine transferase isozymes II (GGT-II) of the liver cancer biomarker at an early stage. They obtained a 19 and 14.2 GHz resonance shift for GGT-II and AFPi, respectively. Also, their experimental study results are in a good agreement with the simulation. Their design is presented in figure 2.4 and the measurement results are presented in figure 2.5.

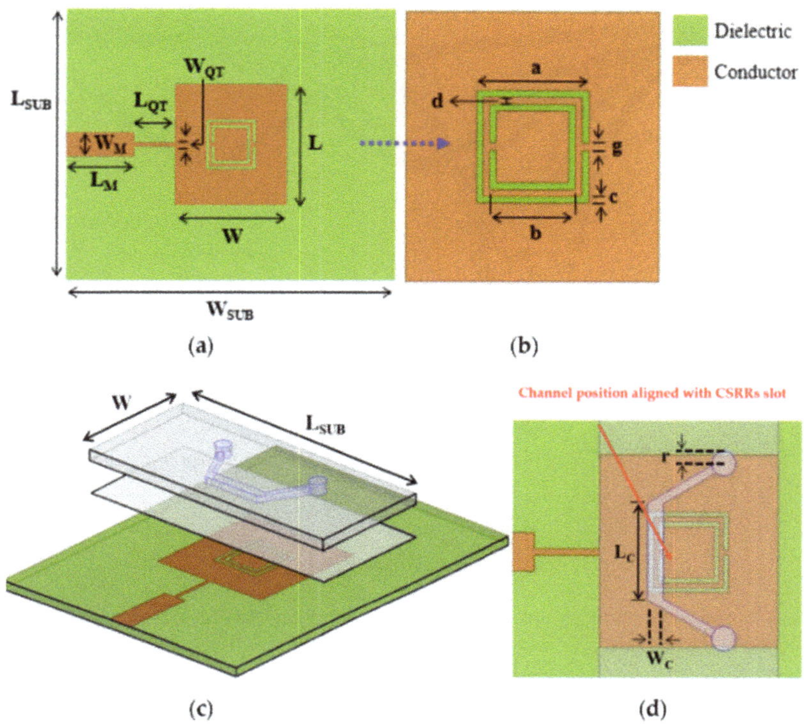

Figure 2.2. (a) Top view of the proposed sensor (b) zoomed-in view (c) bird's-eye view with microfluidic channel; (d) top view of channel alignment with CSRRs slot. Reprinted from [16] CC BY 4.0.

Microfluidic sensor applications have formed the most important parts in metamaterial sensor applications for over a decade. Sometimes, different scientists show different microfluidic applications in a single sensitive design. To achieve this, the geometry of the proposed design must be created carefully to respond to different complex permittivity values. For example, Bakır *et al*, designed a chiral-shaped SRR to reveal two different types of sensor applications both experimentally and numerically. These applications are different types of alcohol derivatives and fuel adulterations, respectively [18]. Researchers both in this and other studies generally measure the complex permittivity in each frequency band and design their metamaterials according to the measured values. In this work, the researchers preferred to validate their proposals in a waveguide to increase sensitivity as shown in figure 2.6. This work also has a higher Q factor when we look at the simulation and experimental study plot as given in figure 2.7.

Microfluidic sensor applications have a wide variety of application types, such as the glucose monitoring work by Govind and Akhtar in Aqueous Solutions published in 2019 [19]. The unique SRR design of the proposed sensor includes an interdigital capacitor (IDC) in the cavity of the resonator. To increase the electric field concentration over a larger surface area, this IDC is used to provide higher sensitivity. To perform experimental validation of lossy fluids, the microfluidic

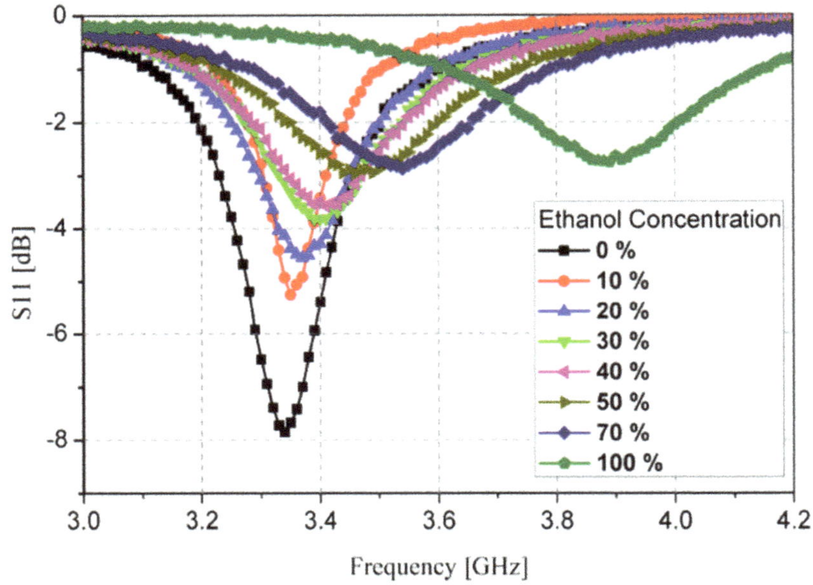

Figure 2.3. Simulated and measured S-parameters comparison. Reprinted from [16] CC BY 4.0.

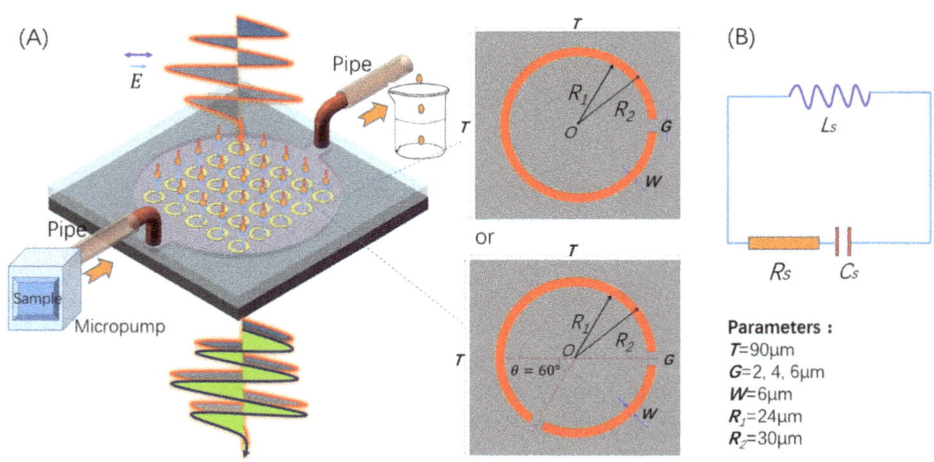

Figure 2.4. (A) A sketch of the THz metamaterials biosensor chip integrated with microfluidics. (B) Equivalent circuit with {RLC}s for the SRRs. Reprinted from [17] CC BY 4.0.

channel is conveniently positioned over the IDC region of the SRR through which glucose solutions are allowed to flow [19]. The measurement sensitivity of the polydimethylsiloxane integrated SRR sensor recommended for glucose testing has been found to be 2.60 E−02 MHz/mg dl^{-1}.

Another interesting microfluidic sensor was proposed by Abdulkarim *et al* in 2020 [20]. The interesting part of the proposed work is that the design is realized for

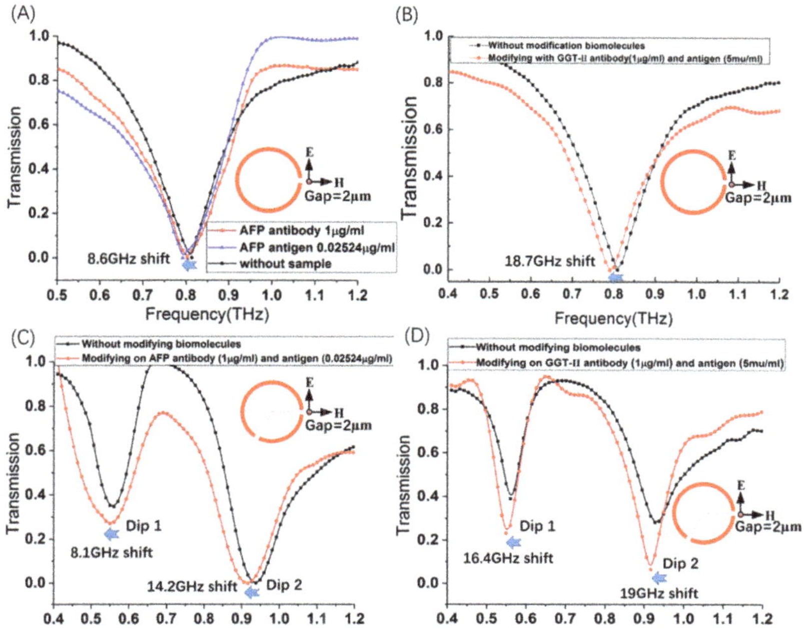

Figure 2.5. Test results of a liver cancer biomarker: (A) SRRs with a 2 μm gap; (B) transmission coefficient of a GGT-II marker; (C) the AFP test results; (D) GGT-II liver cancer marker transmission coefficient of SRRs with two gaps. Reprinted from [17] CC BY 4.0.

transformer oil condition sensing and FR4 is selected as the substrate in contrast to the low loss substrates. The researchers proposed a liquid sensor integrating an omega-shaped resonator with a microstrip transmission line. Strong electric and magnetic coupling has been observed on the proposed work in the operation band. The proposed sensor was not only simulated but also experimentally tested for transformer oil. Also, this sensor structure has been examined by methanol–water, ethanol–water mixtures, PEG 300, isopropyl alcohol, PEG1500, ammonia, and water mixtures too. The proposed design is given in figure 2.8 and the simulated and measured transformer oil transmission coefficient is presented in figure 2.9 for this work.

Up to now, we have only shown microfluidic sensor applications; however, it is possible to design a microfluidic sensor as a solid sensor as realized by Velez *et al* [21]. They combined a composite resonator with a shunt connected step impedance resonator (SIR) and a CSRR etched in the ground plane. In contrast to similar works, this work has demonstrated sensor applications by feeding the sensor with an envelope detector. By designing the CSRR-loaded SIR resonator with closely spaced pole and transmission zeros, highly sensitive sensors can be obtained. This microfluidic sensor proposal can discriminate volume fractions of 5% isopropanol in DI water, with a maximum measured sensitivity of 4 mV/%, which is a suitable value for defining the percentage of alcohol in the wine fermentation process [21]; therefore, the proposed sensor could be adapted to such industrial processes.

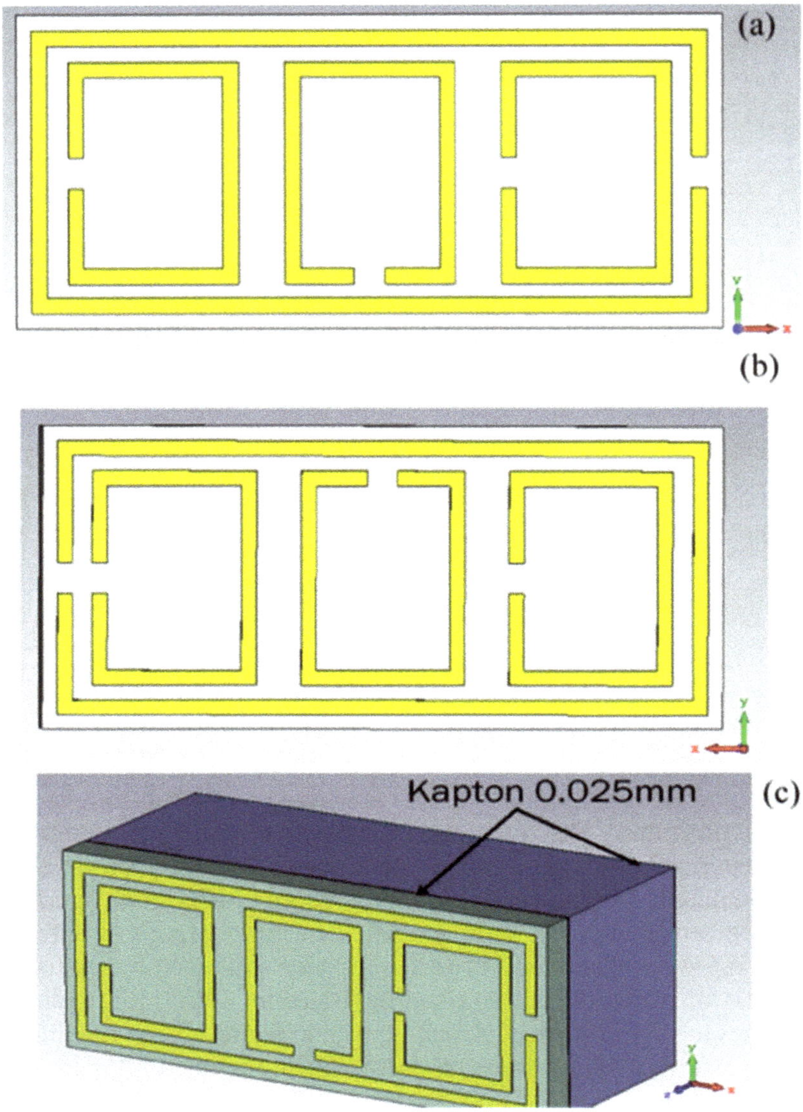

Figure 2.6. Proposed structure: (a) front view, (b) back view, (c) profile view. Reproduced from [18]. Copyright IOP Publishing Ltd. All rights reserved.

It is possible to realize liquid sensors for the food industry. For example, Islam *et al* suggested that a metamaterial sensor has a good quality factor and sensitivity in both frequency shifting and amplitude changing [26]. The proposed work's resonance frequency shifted to 100 MHz between olive and corn oils and 70 MHz between sunflower and palm oil. This work also employed SRR shapes to work within the WR90 waveguide by using an FR4 substrate with a thickness of 1.5 mm. The authors obtained this design by trying different SRR shapes and exploiting parametric studies of CST microwave studio programs. Their proposal has a

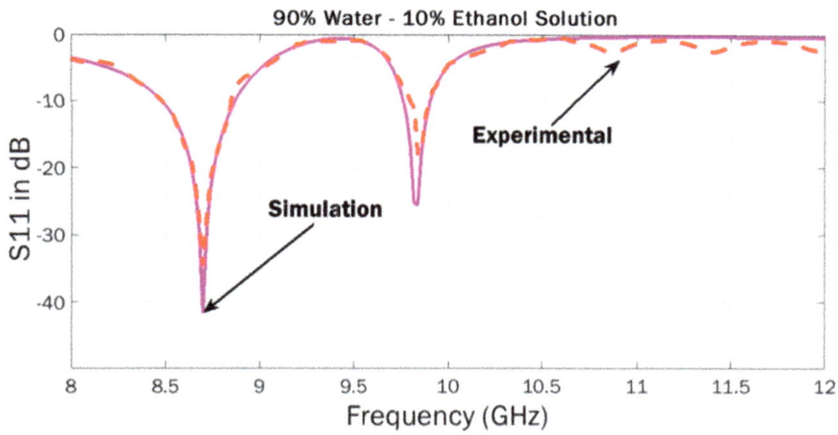

Figure 2.7. Simulated and tested reflection coefficient for 10%, volume fraction of ethanol–water mixture. Reproduced from [18]. Copyright IOP Publishing Ltd. All rights reserved.

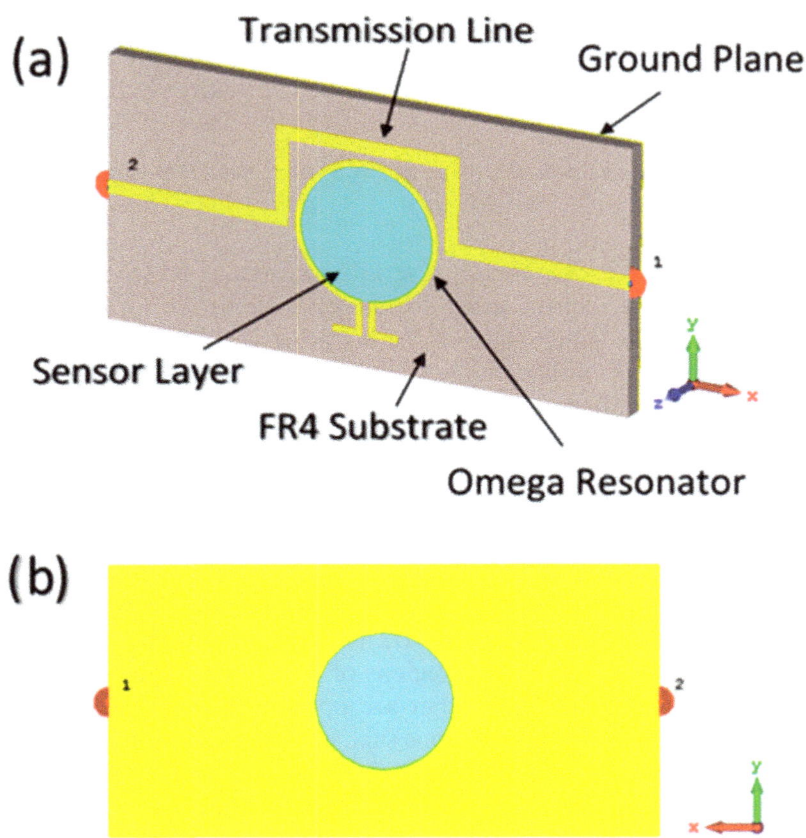

Figure 2.8. Proposed metamaterial-based liquid sensor design: (a) perspective view; (b) back view. Reproduced from [20] CC BY 4.0.

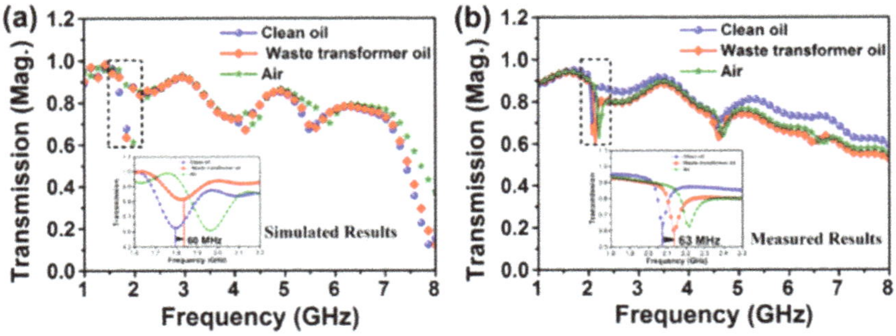

Figure 2.9. Transmission coefficient for the clean and waste transformer oils at frequency 1–8 GHz: (a) simulated result; (b) fabricated result. Reproduced from [20] CC BY 4.0.

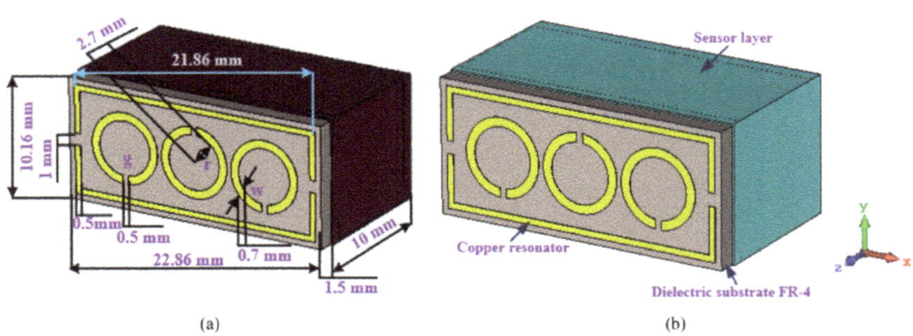

Figure 2.10. (a) Main construction with sizes, (b) perspective position of the MTM-based sensor. Reproduced from [26] CC BY 4.0.

Q factor of 135, which is obtained by a sensitive design as presented in figure 2.10 and the simulations are realized using a special sample holder for testing [26].

In this section of the study, some of the literature is referenced for emphasizing the microfluidic sensor applications of metamaterial sensors. There are also important studies for different sensing applications [22–28]. It would be useful for the reader to understand the operation and design principles of microfluid sensors as well.

2.3 THz metamaterial sensors

Terahertz (THz) radiation (also known as T-rays) refers to electromagnetic waves with frequencies between 0.1 and 10 THz (1 THz = 10^{12} Hz). These waves lie between the microwave and infrared regions of the electromagnetic spectrum, and they can penetrate many materials that are opaque to visible light. This makes terahertz radiation useful for applications such as medical imaging, security screening, and quality control in manufacturing. Terahertz metamaterial sensors (TMSs) typically consist of a thin film of metamaterial that is placed on a substrate. When terahertz radiation is directed at the sensor, it interacts with the metamaterial structures and produces a characteristic spectral response. By analyzing the changes in this spectral response, researchers can detect the presence of specific chemicals or biomolecules that are presented on the sensor surface [29–33, 37]. For example, terahertz metamaterial sensors have been used to detect trace amounts of explosives, drugs, and biomolecules such as glucose and DNA. THz waves are non-ionizing that differentiates them from X-rays and microwaves and makes them safe for wave–material interaction [34, 35]. THz waves are not only sensitive to the vibrational and rotational modes of molecules [36], but also, they can penetrate nonpolar substances and nonmetallic materials such as paper, leather, etc [37].

The advantages of terahertz metamaterial sensors include high sensitivity, selectivity, and speed, as well as their ability to operate in ambient conditions and penetrate many materials. However, they also have some limitations, such as their relatively high cost and the need for specialized equipment to generate and detect terahertz radiation. Nevertheless, research in this area is ongoing, and terahertz metamaterial sensors have the potential to revolutionize many fields of science and engineering [38, 39]. A TMS consists of an array of subwavelength resonant structures that are designed to interact with terahertz radiation in a specific way. When terahertz radiation is incident on the metamaterial, it excites collective oscillations of the free electrons in the resonant structures, which in turn leads to the generation of resonant modes. The resonant modes produced by the metamaterial sensor are highly sensitive to changes in the local environment. For example, if the sensor is placed in the vicinity of a material with a different dielectric constant or conductivity than the surrounding medium, the resonant modes will shift in frequency and/or amplitude [40]. By measuring the changes in the resonant modes, the terahertz metamaterial sensor can detect and quantify the presence of the target material. The sensitivity and selectivity of the sensor can be further enhanced by optimizing the design of the resonant structures and the configuration of the sensor array. Overall, the sensing mechanism of a terahertz metamaterial sensor relies on the interaction between terahertz radiation and the resonant structures in the sensor array, and the ability of these structures is to detect and respond to changes in the local environment [41, 85].

To increase the performance of the TMS, it is necessary to understand the sensing mechanism. Low and high frequency resonance can be observed on the TMS, which are caused by material capacitance and inductance, respectively. The following equations can be used to express them [42–44]. While d depends on the geometric parameters of metamaterial, ε_0 is the permittivity of vacuum, v space variable, $\varepsilon(v)$ is the material permittivity, $E(v)$ is the electric field, and ε_{eff} is the effective dielectric constant of the test environment. λ is the volume ratio of air in the surrounding media. According to these formulas, it can be understood that as the target substance and metamaterial composition change, the material parameters change accordingly. Furthermore, structure design and metamaterial composition directly affect the resonance frequency of the TMS.

$$W_{\text{LC}} = \frac{1}{\sqrt{\text{LC}}} = \frac{1}{\sqrt{L}\sqrt{\varepsilon_0 \int_0^V \varepsilon(v)E(v)dv}} \tag{2.1}$$

$$W_d \approx \frac{1}{2d\sqrt{\varepsilon_{\text{eff}}}} \tag{2.2}$$

$$\varepsilon_{\text{eff}} = \varepsilon_{\text{sub}} + \lambda\varepsilon_{\text{air}} + (1 - \lambda)\lambda\varepsilon_a \tag{2.3}$$

The performance metrics of a TMS can be listed as sensitivity, quality factor, and figure of merit (FoM). Sensitivity can be formulated as resonance frequency change/refractive index change. The quality factor can be expressed as resonance frequency/full width at the half maximum of the resonance peak. The FoM value is the multiplication of both [45].

Fungi and bacteria are a sub-type of microorganisms which may cause human disease and should be identified accurately for effective treatment. For this purpose, Park *et al* proposed a TMS study for the detection of microorganisms by resonance frequency shift. This shift is dependent on the number density and dielectric constants of the microorganisms [46]. The proposed resonator is an SRR type and is shown in figure 2.11. The authors verified their proposal with both a simulation technique and experimentally. The amount of shift is found to be 9 GHz (6% of the FWHM) for the density of penicillia, which can easily be sensed by a vector network analyzer.

Another study was conducted by Yang *et al* to sense four species of bacteria (*Escherichia coli*, *Staphylococcus aureus*, *Pseudomonas aeruginosa*, and *Acinetobacter baumannii*) according to their dielectric constant in [47]. In the work, the authors only employ time domain spectroscopy. For this reason, discrimination of them is only possible according to the absorption coefficient, which is not clear as the resonance frequency shift as shown in figure 2.12.

Animal derived food due to incorrect use of veterinary medicine endangers human life. Therefore, using antibiotics in this type of food is limited. Qin *et al* proposed a TMS that enables one to sense tetracycline hydrochloride (TCH) [37]. Circular ring-shaped metamaterials were prepared by employing beam lithography on a silicon substrate. They used metamaterials in the experiments consisting of an array. GHz/μm sensitivity is obtained with their proposal as shown in figure 2.13.

Xu *et al* prepared a TMS for the detection of chlorpyrifos-methyl (CM) because of its broad spectrum [48]. The results demonstrate that sensitivity is much higher than the limit given by the World Health Organization. In their work, first, CM powder is placed into pellets and then electromagnetic energy is squeezed into small volumes near a metal dielectric interface, which verifies that the limit of detection of CM reaches 0.204 mg l^{-1} as seen in figure 2.14.

Kim *et al* developed hybrid humidity sensing methods by employing silk fibroin protein onto metamaterials that operate in the terahertz (THz) frequencies. They observed additional resonance shifts when they applied humid air to silk-coated metamaterials, due to the increased water molecule numbers on the film. Their proposal in THz spectroscopy measurement reveals that the time response to the humid air affects the decaying process [49]. The sensitivity in terms of Δf/RH reaches 0.22 GHz/%, and they can identify an RH change of ~2.3%. The proposed design of Kim *et al* and obtained results are presented in figure 2.15.

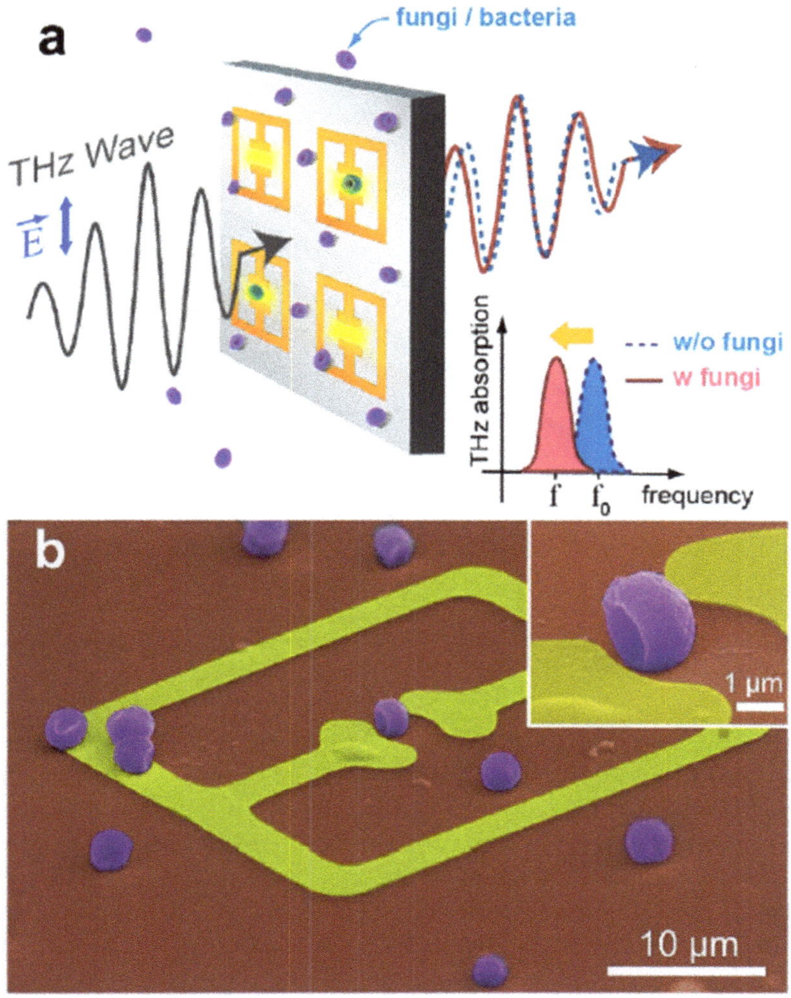

Figure 2.11. Park *et al*'s TMS proposal. (a) A schematic presentation of THz metamaterials sensing of microorganisms. (b) A color-enhanced SEM image of metamaterials coated by penicillia. (inset) Magnified image of the fungi located in the micro-gap. Reproduced from [46] CC BY 3.0.

Zhong *et al* investigated both theoretically and experimentally a metasurface-enhanced aqueous solution sensor based on attenuated total reflection (ATR) [50]. In summary, a metasurface-enhanced ATR sensor for aqueous solution in the THz range is designed. The metasurface in their proposal is composed of periodic CSRRs, which are lithographed directly on the ATR prism. If the basic operation principle is given, the incident *p*-polarized THz waves act on the CSRRs to excite LC resonance. This resonance can be used to measure the concentration of sucrose solution from 1 mol l^{-1} to 0.031 25 mol l^{-1} and the detection sensitivity of the metasurface-enhanced ATR is four times higher than that of the conventional ATR as illustrated in figure 2.16 [50].

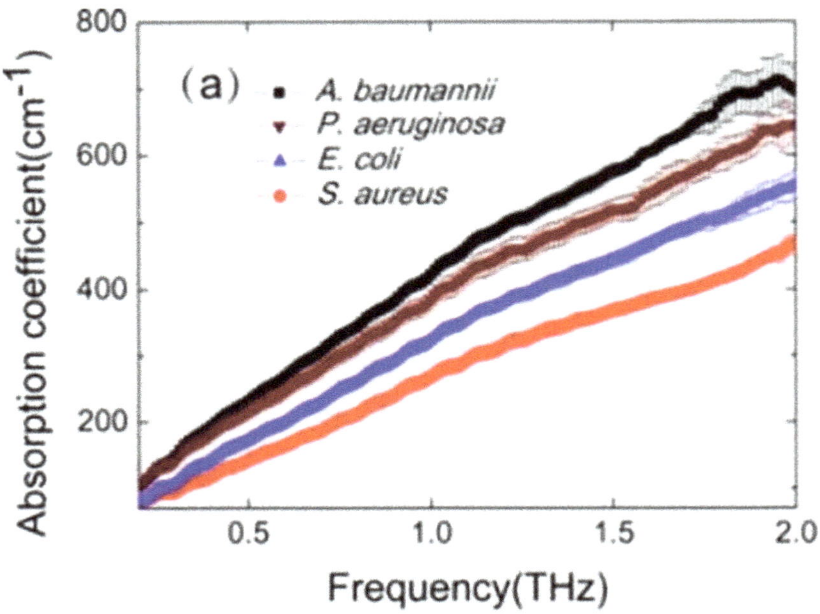

Figure 2.12. Bacterial species-specific frequency (THz) vs absorption. Reprinted from [47], copyright (2016) John Wiley & Sons.

Yang *et al* proposed a TMS for distinguishing normal and abnormal concentration of glucose in 2021 [51]. Their proposal comprises two asymmetric thick metals on a silicon dioxide substrate which showed transmission dips at 0.26 and 0.4 THz. Four glucose samples with concentrations from 4 mmol l^{-1} to 10 mmol l^{-1} with the step of 2 mmol l^{-1} feed on to the sensor shows that this design yields the sensitivity of 1.22 THz (mol l^{-1}) and 0.52 THz (mol l^{-1}). The proposed sensor and obtained results are presented in figure 2.17.

Citrate salt has been widely used in the food and pharmaceutical industries. To sense the citrate concentration in a solution, Deng *et al* proposed a TMS sensor in 2022 [39]. The proposed TMS is an asymmetric double-opening ring-shaped design, which is validated by simulation and experiment. To increase the sensitivity, the design and opening position are carefully chosen and six different citrate solutions with low concentrations are prepared. They chose a 1.5–2.5 THz operation band for validation and obtained 402 GHz/RIU. The obtained results are given in figure 2.18.

Another glucose sensing study was proposed by Lee *et al* in 2023 [52]. While previous studies calculate the sensitivity of metamaterial by considering the RI of the analyte, this study developed a modified Lorentz model. SRR-based metamaterials were designed, fabricated, and measured for verification of sensing ranging from 0 to 500 mg dL^{-1}. Finally, based on the modified Lorentz model, the resonance frequency and transmittance changes were calculated and compared with the measurement results, which are briefly given in figure 2.19. They fabricated a THz-based SRR pattern on a polyimide substrate with a resonance frequency of 1.455 THz and Q factor of approximately 22. After the design, simulation, and

(a)

(b)

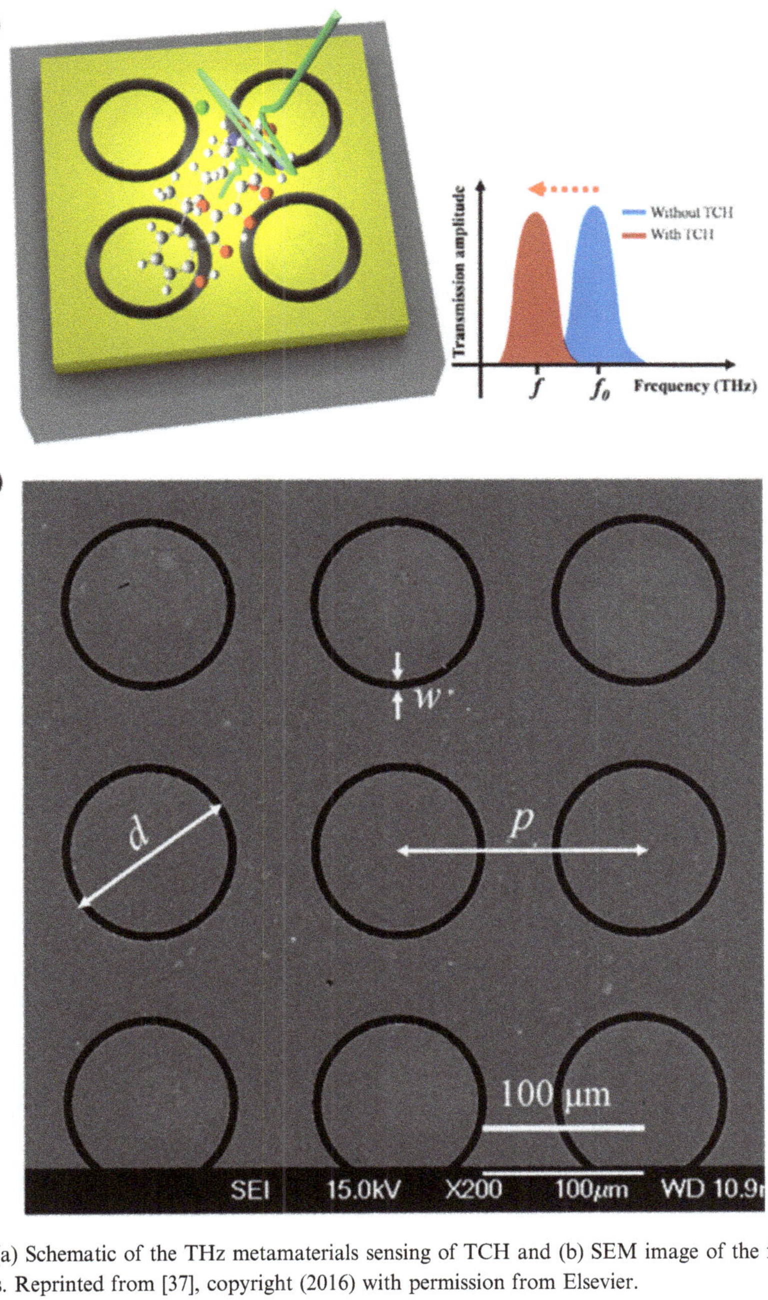

Figure 2.13. (a) Schematic of the THz metamaterials sensing of TCH and (b) SEM image of the fabricated metamaterials. Reprinted from [37], copyright (2016) with permission from Elsevier.

measurement process, a modified Lorentz dispersion model is used to obtain a better congruency between the measurement and simulation results.

TMSs have attracted great attention in biomedical sensing because of their abilities to predict the vibration modes of molecules in the terahertz operation band [30]. TMSs are employed for accurate discrimination in biomedical applications

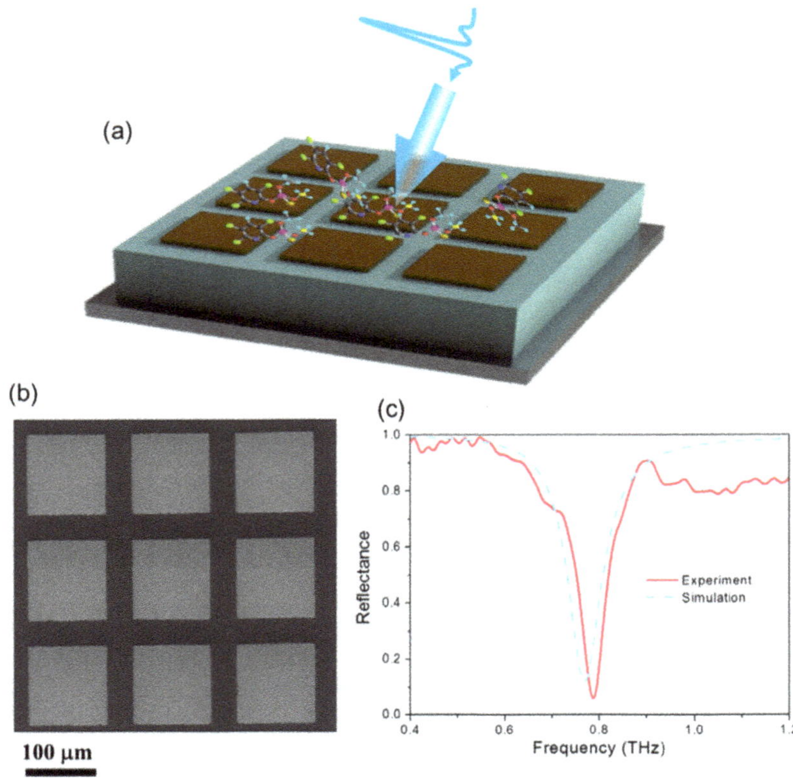

Figure 2.14. (a) Schematic diagram of terahertz metamaterials' detection of CM. (b) A scanning electron microscope (SEM) image of the terahertz metamaterials. (c) Reflectance curves of the simulation and experimental results of terahertz metamaterials. Reprinted from [48], Copyright (2017), with permission from Elsevier.

such as protein detection [52, 53], cancer cells [31, 54], DNA molecules [55, 56], viruses [57, 58], and microorganisms [59, 60]. Some examples of the literature will be given in the following sections to outline TMS applications in biomedicine.

Zhu *et al* designed a single graphene layer which is integrated into the structure of a simple metamaterial absorber. The absorber enables a relative change in the reflectance of about 361% at infrared frequencies [61]. In their work, they designed a high-speed optical modulator which is a key device for the construction of on-chip optical interconnects and optical communication systems as shown in figure 2.20.

Abdolrazzaghi *et al* proposed a TMS for thin DNA strands with high sensitivity in 2015 [62]. Their proposal is based on a simulation used for the identification of very thin analytes according to their dielectric constant. According to their obtained results, for the unit change in dielectric constant of load, 3.8% shift in resonance is obtained after introduction of the structure to 10 nm of DNA strands (r = 3.2).

El-Naib designed a conductively coupled TMS by using two SRRs and connected them for biomedical application in 2016 [63]. He employed conductively joined split-ring resonators for sensing DNA and RNA. He investigated the sensitivity level

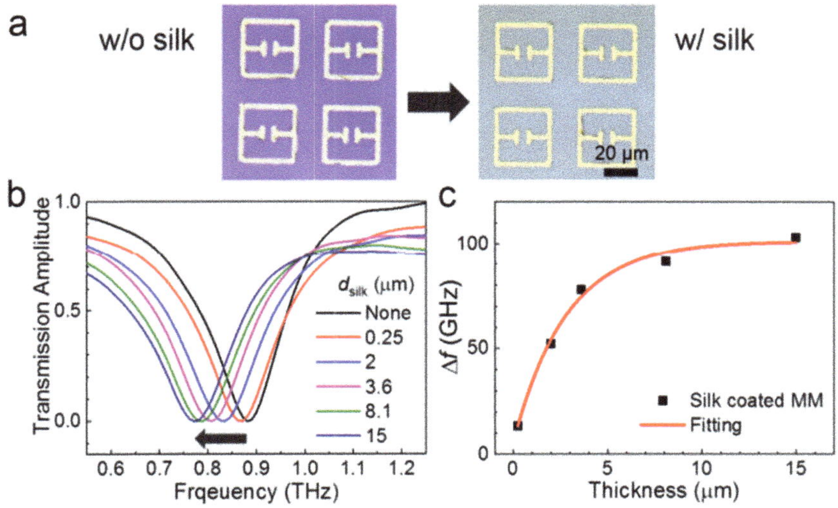

Figure 2.15. (a) Microscopy image of the metamaterials before and after the deposition of a silk fibroin film with silk $d = 8.1$ μm. (b) Transmission amplitudes for different silk film thickness from silk $d = 0$–15 μm. (c) Plot of the frequency shift as a function of silk d extracted from (b). Reproduced from [49]. CC BY 4.0.

when the whole area of the sensor is covered with different thicknesses and a range of refractive index values of an analyte. Due to DNA and RNA, the refractive index varies from 1.4 and 2. He designed the sensor to be operable between these values with a maximum Q factor of 41. He also investigated the sensor performance with different coating thickness. Interestingly, there is a shift of 9, 14, 20, 28, and 38 GHz for the analyte refractive index of 1.2, 1.4, 1.6, 1.8, and 2.0, respectively. In the b section the shifts are observed at 3, 5, 8,10, and 12 GHz, which shows that the small change in the refractive index can induce a measurable shift in the frequency response for differentiating analytes with different refractive indices.

Chen and Fan developed a planar terahertz metamaterial sensor by using rectangular groove-shaped stripes in 2017 [64]. By employing Fabry–Perot resonance of plasmons, they obtained a sharp resonance for biomedical applications. They obtained av frequency sensitivity of 1.966 THz per refractive index unit and a figure of merit of 19.86 together with a thickness sensitivity higher than 52.5GHz μm^{-1}. The thickness of the analyte directly affects the sensor performance as in previous studies. A schematic illustration of their proposal is given in figure 2.21. The proposed design is periodically arranged in x and y directions and polyimide (PI) with a dielectric constant of 3 and loss tangent of 0.005 33 has been used as a substrate in this work.

Keshavarz and Vafapour presented a TMS for skin cancer detection using a water-based metamaterial and a semiconductor film in 2018 [65]. They applied terahertz pulsed imaging in reflection geometry to study skin tissue and related cancers. Their recommendations are based on the refractive index change by incorporating normal skin and basal cell carcinoma (BCC) into their design.

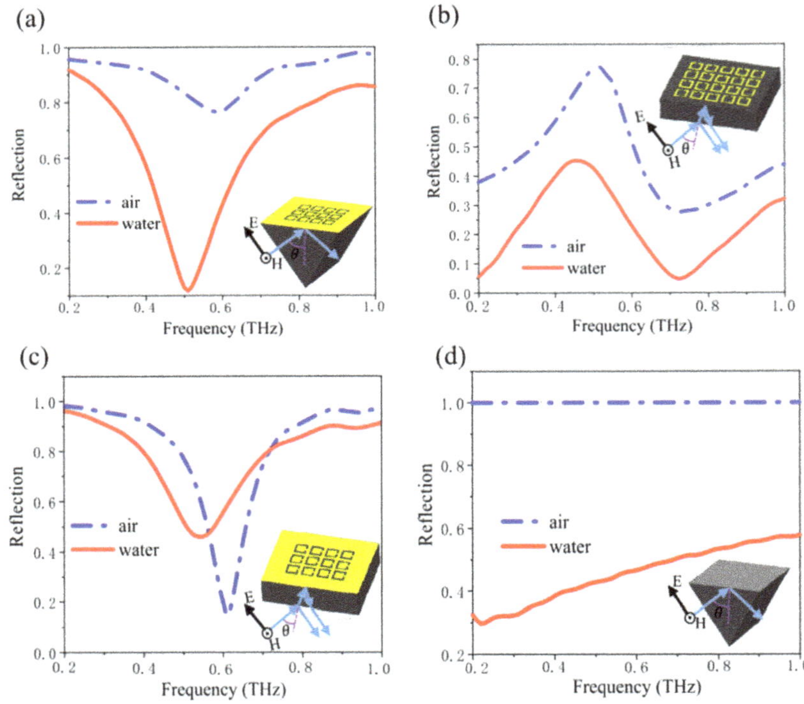

Figure 2.16. Reflection simulation results: (a) metasurface-enhanced ATR structure with and without water. (b) Split-ring structure with and without water. (c) CSRR structure with and without water. (d) Conventional ATR with and without water. Reprinted from [50], copyright (2020), with permission from Elsevier.

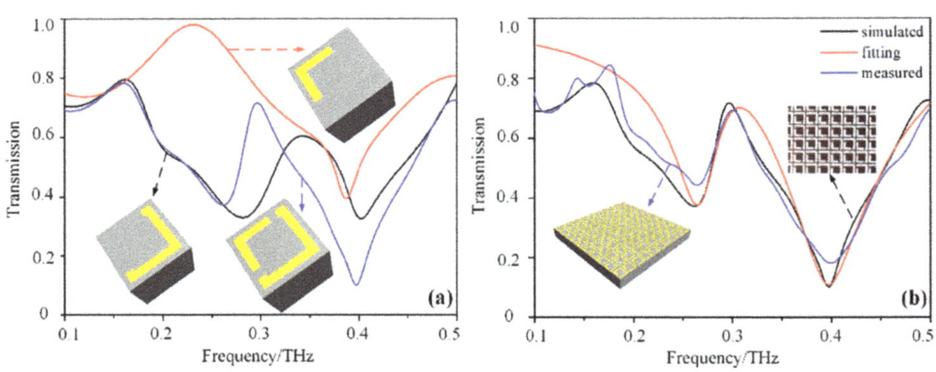

Figure 2.17. (a) Simulated transmission spectra of the metamaterial: the combined shape (blue curve), the large L-shaped (black curve), and the small L-shaped (red curve) of the MM structure, and (b) the simulated, fitting, and measured transmission spectra for the sensor. Reprinted from [51], copyright (2020), with permission from Elsevier.

The sensitivity value of their design reaches 117 μm/RIU. They also suggest that this water-based metamaterial could be used in the control of gene expression. They placed the proposed water-based metamaterial on the skin and simulated the

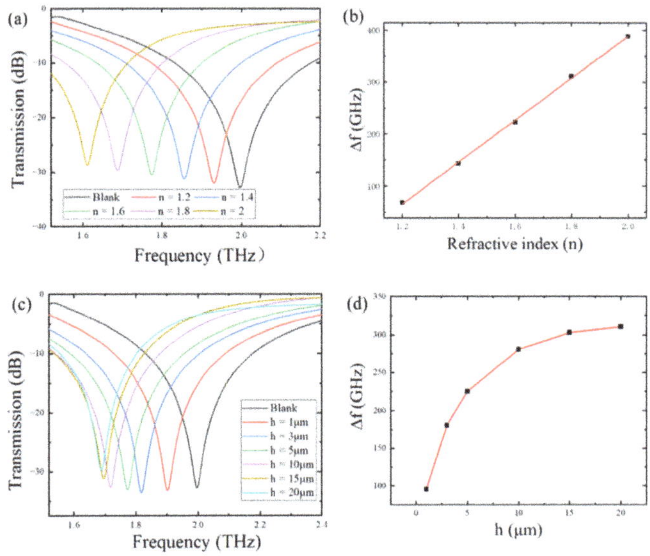

Figure 2.18. (a) Simulated transmission coefficient by using different refractive index materials at h = 20 μm; (b) variation in the resonant peak according to the refractive indexes of analytes; (c) transmission coefficients with different thickness analytes at n = 1.8; (d) variation in the resonant peak. Reproduced from [39] CC BY 4.0.

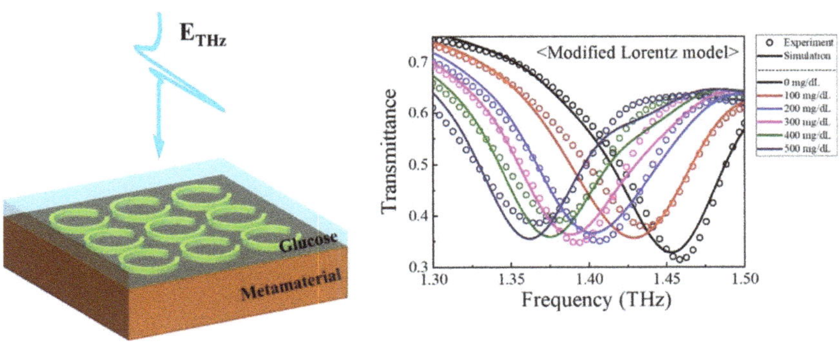

Figure 2.19. Lee *et al*'s proposal for a TMS-based glucose sensor and modified Lorentz model transmittance plot between 1.3 and 1.5 THz. Reprinted from [52], copyright (2023), with permission from Elsevier.

metamaterial's reflective light between 1–1.5 THz for healthy and BCC cases. In the simulated result, the resonance frequency is 1.398 and 1.382 for healthy and BCC cases, respectively [65].

Saadeldin *et al* proposed a perfect metamaterial absorber based TMS in 2019 [66]. Although perfect metamaterial absorber (PMA) based sensors are discussed in the next section, this work is summarized here due to its high sensitivity in the biomedical field. The proposed work has a sensitivity of 300 GHz/RIU and a figure of merit (FoM) of 2.94 when analyte thickness is 1.0 μm. They also realized an experimental study and compared this with simulation in their work. Since the

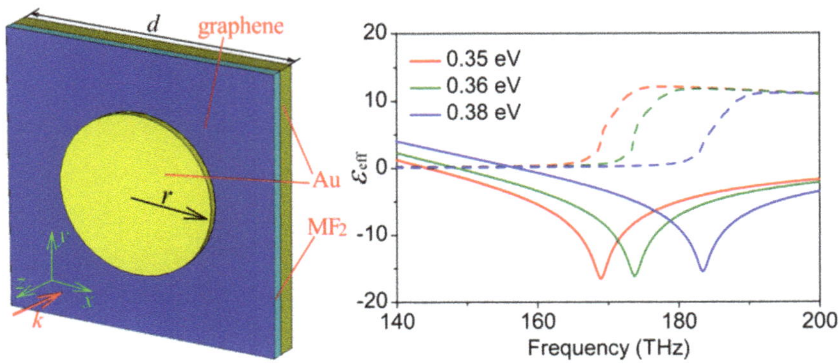

Figure 2.20. Zhu *et al*'s proposal. (a) Unit cell of a graphene-coupled metamaterial absorber. (b) Effective permittivity of graphene for Fermi-level energies of 0.35, 0.36, and 0.38 eV 5 THz. Reprinted from [61], with the permission of AIP Publishing.

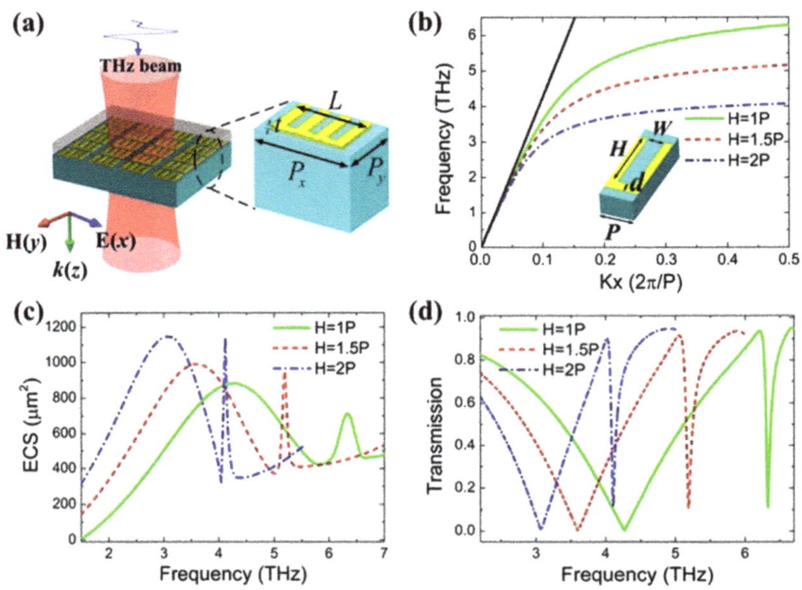

Figure 2.21. (a) Schematic diagram. (b) The calculated dispersion curves, (c) and transmission spectra, (d) transmission spectra of the structure with different groove depth. Reproduced from [64] CC BY 4.0

refractive index of biomaterials changes as blood varies, analyte thickness was kept at 1 μm and the proposed work was simulated as a refractive index sensor.

A sickle-shaped metamaterial (SSM) based biochemical sensor with multiple resonances was investigated by Ou *et al* in 2020 [67]. Dipolar, quadrupolar, octupolar, and hexadecapolar resonances are generated from SSM and perpendicularly incident terahertz waves. The sensing performances of SSM-based biochemical sensors are evaluated by changing ambient environments and analyte varieties. A sensitivity and FoM value of 471 GHz/RIU (refraction index unit) and 94 RIU^{-1}

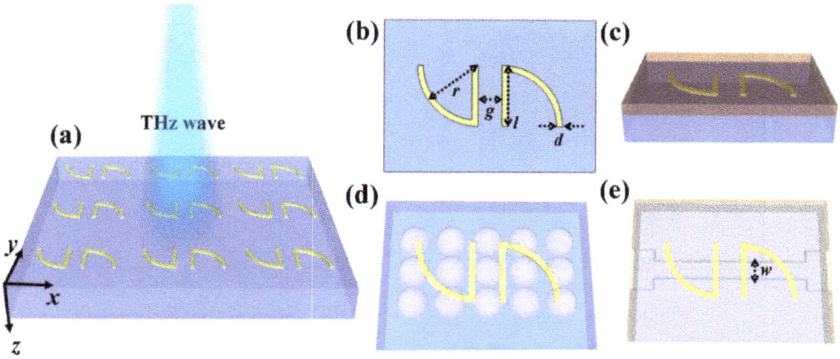

Figure 2.22. (a) Schematic drawing of the sickle-shaped metamaterial (SSM). (b) Geometrical denotations for the corresponding SSM unit cell. (c–e) Schematic drawings of SSM covered with (c) analyte thin film, (d) dielectric hemispheres, and (e) microfluidic channels. Reproduced from [67] CC BY 4.0.

was obtained, respectively. The authors also adapted their design to imitate dry and hydrous biological specimens. When their design is employed as a microfluidic chip, they obtained a sensitivity of 405 GHz/RIU. After that, they simulated three models for an illustration of different situations, and the corresponding highest sensitivity is found to be 502 GHz/RIU. While their proposal is given in figure 2.22, the obtained results are briefly given in figure 2.23.

Sun *et al* proposed a TMS with high sensitivity which is accomplished by using different modes of coupling, low loss substrate–high conductivity metal usage [44]. Electromagnetically induced transparent (EIT)-like resonance showed good results in this work. They analyzed double split-ring resonators (DSRRs). Their reports showed that the width of the DSRR had the most significant effect on the EIT-like resonance. The effect of four spatially arranged macrocell metamaterials consisting of multi-DSRRs is compared with diagonally arranged metamaterials in their work. While their designs are presented in figure 2.24, a comparison of the results is presented in figure 2.25.

Patel *et al* proposed several metasurface-based TMS designs that include and exclude gaps for the detection of several types of cancer affected cells [68]. Normal and cancer contaminated cells have different refractive indices. They successfully designed their unit cells to respond to these changes. They also investigated the maximum Q factor and sensitivity as 13.11, 207.14 GHz/RIU. Their operation band is between 0.1–1 THz. While their design is given in figure 2.26, the obtained results are presented in figure 2.27.

Huang *et al* proposed another biomolecular sensor by employing a concave metallic copper single-layer metamaterial structure to achieve improved detection sensitivity in 2023 [69]. Their anisotropic and asymmetric metasurface design reacts to the incident polarization angle by giving different responses. To analyze the material under the test carefully, they extracted the frequency shift data of different polarization angles to construct a multi-dimensional hexagonal radar map. They also did an experimental validation of their design by producing it with

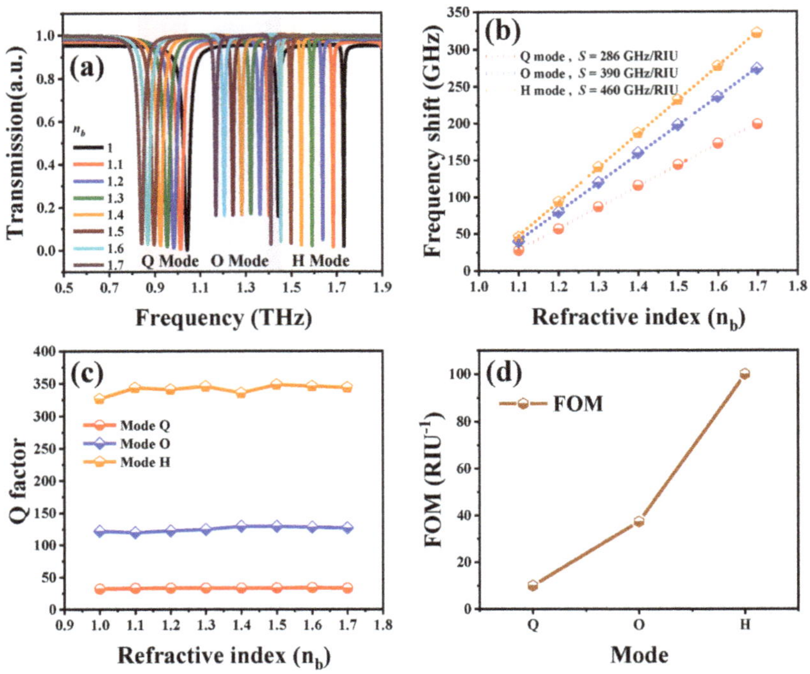

Figure 2.23. (a) Transmission spectra of SSM with different nb values. (b) Frequency shifts and (c) Q factors plotted against the nb values. (d) The average figure of merit (FoM) values of Q, O, and H modes. Reproduced from [67] CC BY 4.0.

photolithography and tested it on a bovine serum albumin biomolecule. Simulation studies show that the proposed metasurface-based TMS's sensitivity can be improved by using a double layer which produces different resonance peaks. These peaks can be used to provide comparison for subsequent detection which reduces costs and produces label-free detection of biomolecules [69]. The transmission spectra and resonance frequency shift of their proposal is given in figure 2.28.

Since most of the developed sensors are for biomedical and food science, a brief discussion of the literature is given for these application areas. There also other application areas such as environmental monitoring [70–74] in TMS applications.

2.4 The metamaterial absorber based sensors

Metamaterial absorbers can be designed as a type of sensor that can be used to detect and measure the presence of different substances or materials. These sensors are made up of a series of metamaterial layers, which are designed to interact with electromagnetic waves in specific ways. When electromagnetic waves propagate through a metamaterial absorber, the waves are absorbed and converted into heat energy. This absorption process causes a change in the temperature of the metamaterial layers, which can be detected and measured using various techniques such as infrared cameras or thermocouples [75–79]. Metamaterial absorbers can be

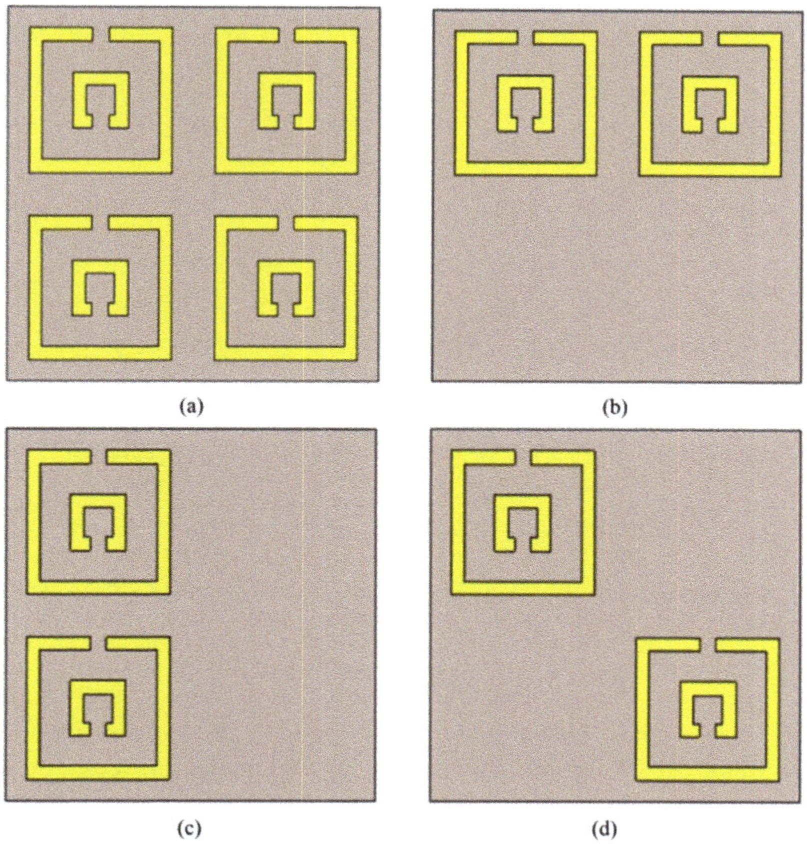

Figure 2.24. Different proposed designs by Sun *et al.* Reprinted from [44], Copyright (2020), with permission from Elsevier.

used in a variety of sensing applications, such as detecting the presence of specific gases or chemicals. The absorption properties of the metamaterial can be tuned to target specific wavelengths of light, allowing for selective detection of certain materials. One advantage of the metamaterial absorber sensors is high sensitivity and the ability to detect very small quantities of a material. Additionally, because they rely on electromagnetic wave absorption rather than chemical reactions, they can be used in a wide range of environments and do not require any sample preparation [80–82]. Overall, metamaterial absorber sensors have the potential to be a powerful tool in a variety of sensing applications, from environmental monitoring to medical diagnostics. For this reason, in this section, we decided to give more examples from current state-of-the-art sensors.

One of the first studies in this field was realized by Kearney *et al* in 2013 [83] to increase the sensitivity of uncooled thermal sensors in the terahertz (THz) spectral range (1 to 10 THz). They investigated thin metamaterial layers that are composed of a periodic array of aluminum and silicon oxide substrate showing resonant absorption in this field. They also replaced silicon oxide with silicon dioxide to show

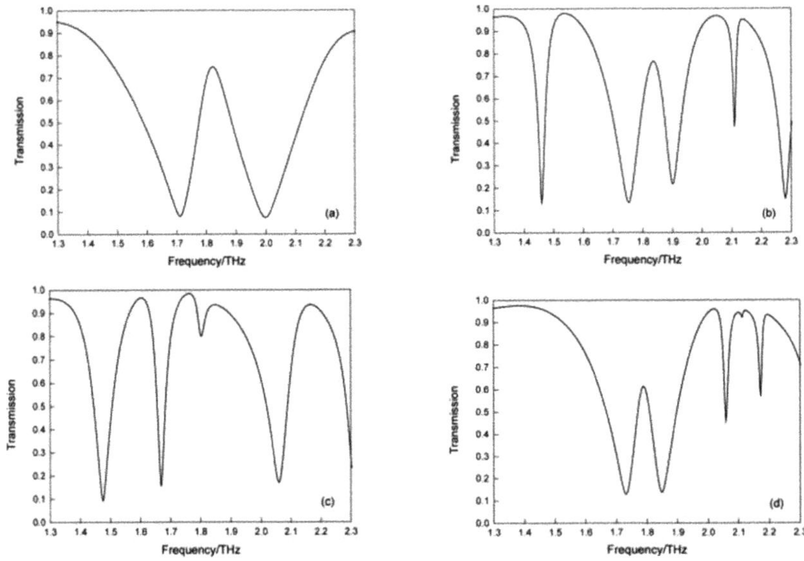

Figure 2.25. Simulated transmission of the proposed designs. Reprinted from [44], Copyright (2020), with permission from Elsevier.

the effects on resonance. They tested the absorption to show the sensor application using eight different square dimensions. They also simulated the dielectric substrate thickness on resonance frequency by using eight various thicknesses. The results are shown in figure 2.29.

Another study in this field realized in the X band waveguide was by Sabah *et al* in 2014 [84]. They proposed a metamaterial absorber based sensor which is integrated in the X band waveguide for high resolution measurements of the dielectric constant and thickness of the superstrate layer. They offered this study to show the effects of parameters which are caused by physical, chemical, or biological factors. Their proposed design corresponds to any parameter change by shifting resonance frequency. A 1.1 μm change in thickness or 0.023 absolute change in relative permittivity of the sensing layer cause a 10 MHz resonance frequency shift. To consolidate the simulation works, they carried out an experimental verification. A simulation study of different permittivity versus absorption plot is presented in figure 2.30 [84].

Wang *et al* presented a dual-band terahertz metamaterial absorber formed by a patterned metallic strip and a dielectric layer on top of a metallic ground plane which operates between 1–3.5 THz [85]. They also claimed that their proposal has a high Q factor and FoM value since their values are 8.4 and 22.7 times higher than that of the fundamental resonance, respectively. Due to the proposed frequency band, their design is promising for solar cells, detection, and imaging applications. For sensor application, refractive index sensing simulations have been done between 1 and 1.1 with a step of 0.02. The obtained plot is presented in figure 2.31.

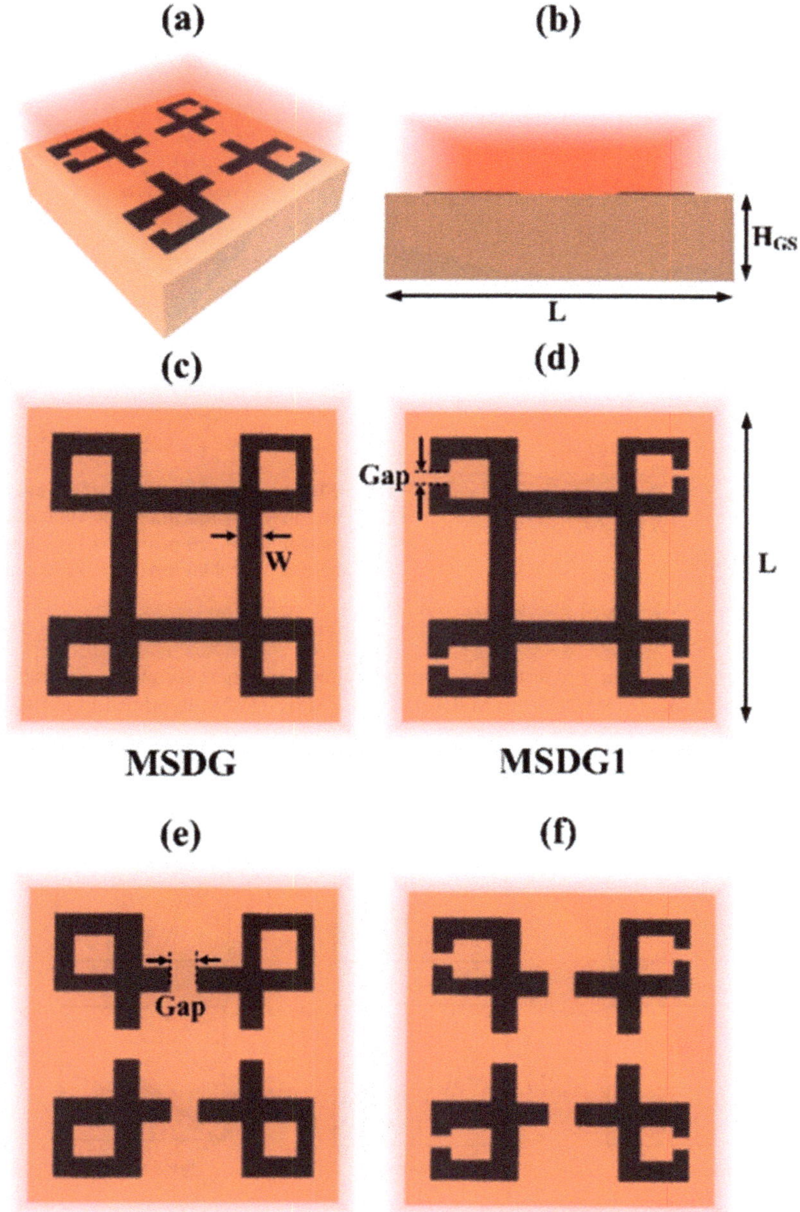

Figure 2.26. Different views of the developed cancer sensor: (a) 3D view, (b) front view, (c) overhead view, (d) overhead view with gaps at the outer boundary, (e) overhead view with gaps at the inner boundary, (f) overhead view of the design with gaps at both the outer and inner boundary. Reprinted from [68], Copyright (2022), with permission from Elsevier.

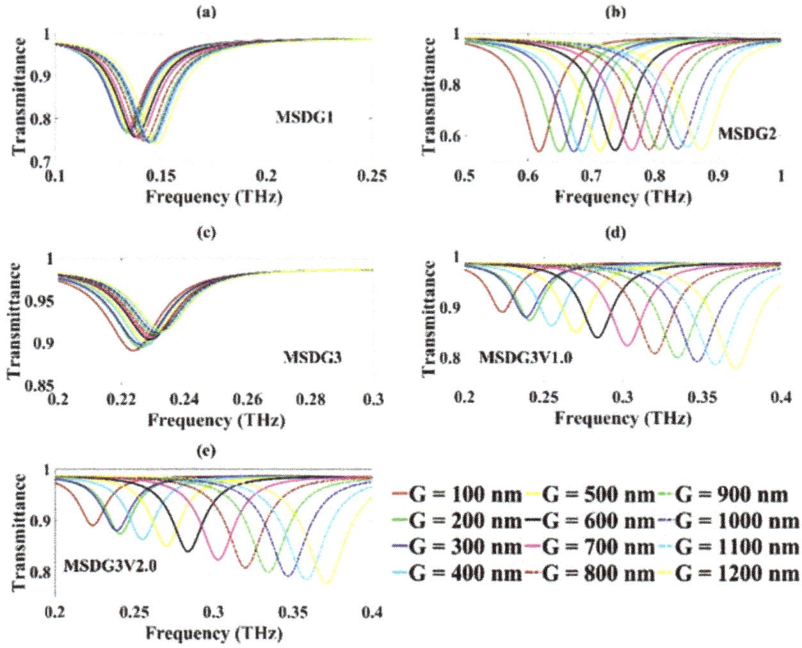

Figure 2.27. Effect of increasing gap distance on the transmittance response for proposed design of sensors: (a) MSDG1, (b) MSDG2, (c) MSDG3, (d) MSDG3V1.0, (e) MSDG3V2.0. Reprinted from [68], Copyright (2022), with permission from Elsevier.

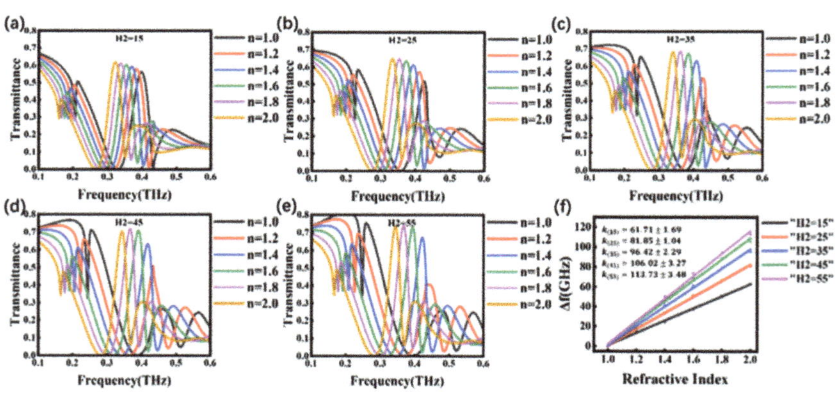

Figure 2.28. (a–e) Transmission coefficients simulation changes according to the refractive index values and silicon depths. (f) The relationship between the frequency shift and refractive index for various silicon depths and its linear fitting results. Reprinted from [69], Copyright (2023), with permission from Elsevier.

Bakır *et al* gave an example of a density sensor application by using the design given in figure 2.32. Furthermore, this application shows the relation between lumped element power and different substrate materials. Since the dielectric constant of the substrate directly affects the resonance frequency of the system and lumped element power level, the simulation will show the whole effect, as presented in

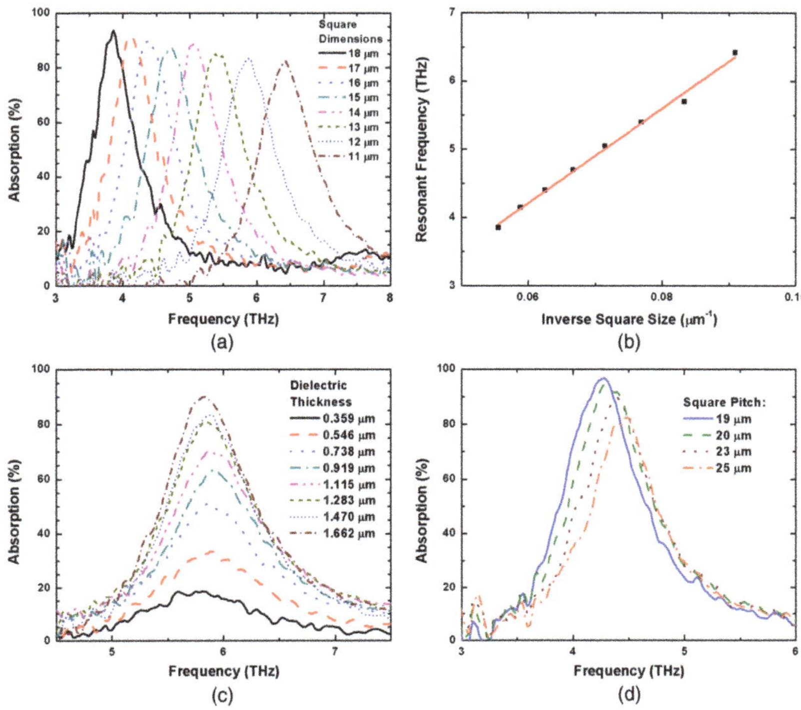

Figure 2.29. Measurement plots for (a) varying square dimensions, (b) inverse square size (1/s) to resonant absorption frequency, (c) dielectric thickness versus absorption, (d) square pitch plot. Reproduced from [83] CC BY 4.0.

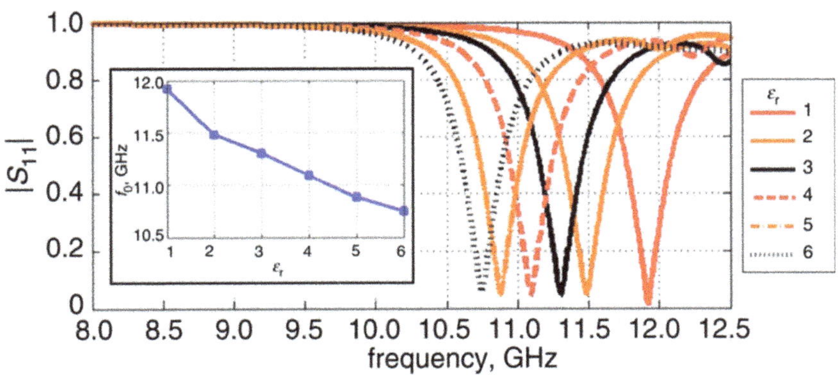

Figure 2.30. Sabah *et al*'s designed sensor for different dielectric permittivity values. Reprinted from [84], copyright (2014) John Wiley & Sons.

figure 2.33. They showed a resonance frequency shift from 3.5 to 5.8 GHz, which is a strong shift for a given frequency band [86].

One of the advantages of metamaterial absorber based energy harvesting is that it can be designed to absorb specific wavelengths of electromagnetic radiation, which makes it more efficient than traditional energy harvesting technologies.

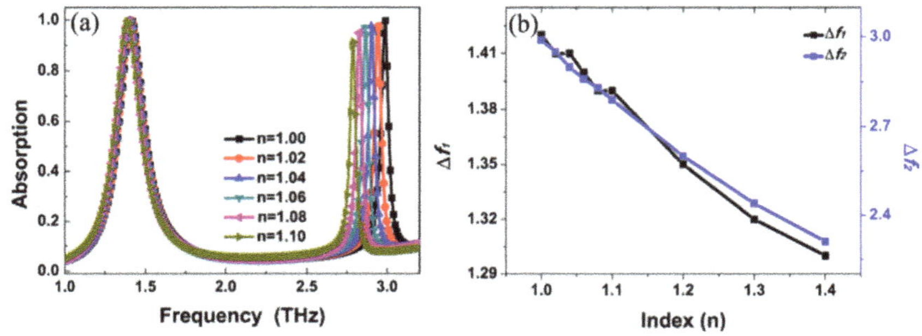

Figure 2.31. (a) Absorption plot under a low RI environment. (b) Absorption peaks (f1 and f2) as a function of the RI. Reprinted from [85], with the permission of AIP Publishing.

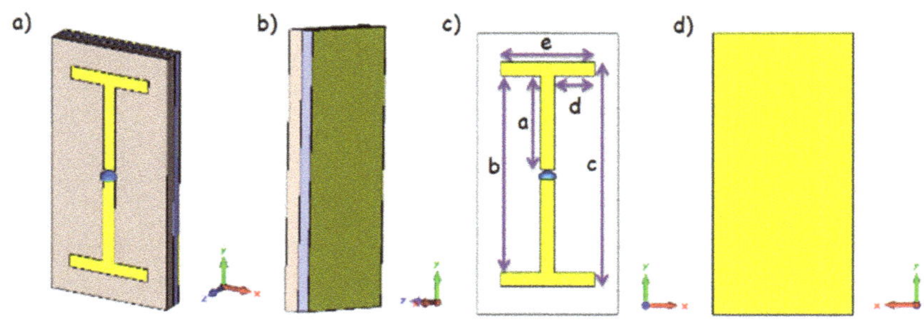

Figure 2.32. Bakır *et al*'s proposed design for metamaterial absorber based energy harvesting and sensor application. Reproduced with permission from [85].

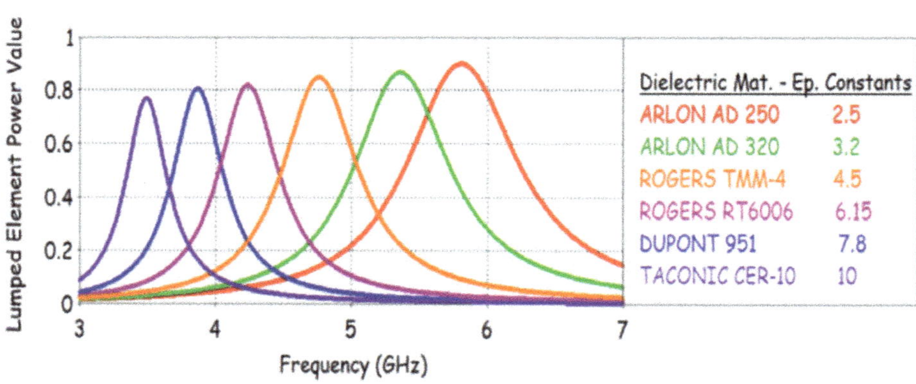

Figure 2.33. Bakır *et al*'s lumped element power value according to different substrate material. Reproduced with permission from [85].

Additionally, the use of metamaterials allows for the absorber to be made thinner and lighter than traditional absorbers, making it easier to integrate into electronic devices. Adding sensor applications to absorber designs enables them to be used in a variety of applications where low-power sensing is required. Here are some examples:

Environmental monitoring: Metamaterial absorber based sensors can be used to monitor the environment for various parameters such as temperature, humidity, and air quality. The energy harvested from the environment can be used to power the sensors, eliminating the requirement for batteries or wires.

Structural health monitoring: Metamaterial absorber based sensors can be used to monitor the health of structures such as bridges and buildings. The sensors can detect vibration, strain, and other parameters and use the harvested energy to transmit the data wirelessly to a monitoring station.

Smart agriculture: Metamaterial absorber based sensors can be used in agriculture to monitor soil moisture, temperature, and other parameters. The harvested energy can be used to power the sensors, eliminating the need for batteries or wired connections.

Wearable technology: Metamaterial absorber based sensors can be used in wearable technology such as smart watches and fitness trackers. The sensors can monitor various parameters such as heart rate, temperature, and activity levels, and use the harvested energy to power the device.

Industrial monitoring: Metamaterial absorber based sensors can be used to monitor various parameters in industrial settings such as temperature, pressure, and vibration. The harvested energy can be used to power the sensors, eliminating the need for batteries or wired connections. Overall, metamaterial absorber based energy harvesting sensors have the potential to revolutionize the way we monitor and sense the environment around us and offer a more sustainable and efficient alternative to traditional sensing technologies [87–91].

Deng *et al* proposed a tunable metamaterial absorber based on a liquid crystal intended for the F frequency band, which was presented in 2017 [92]. The analysis showed that the determined design bias affects the orientation of the nematic liquid crystal embedded in the mesh cavities and causes the resonance frequency to shift from 110.9 to 103.8 GHz with a frequency tunability of 6.4%. Their proposal is important when tuning is necessary in sensor applications.

Hasan and Lee proposed a hybrid metamaterial absorber platform for sensing of CO_2 gas at mid-IR in 2018 [93]. Their design utilized a metal–oxide semiconductor with a smart, gas-selective-trapping polymer for high sensitivity. They obtained a minimum 40 ppm detection limit with a considerably small form factor as shown in figure 2.34 with a fast response time.

Qi *et al* presented a dual-band perfect metamaterial absorber for sensor applications to be operated between 500 and 900 nm in 2019 [94]. MoS_2 dielectric metamaterial is used as a substrate to obtain two absorption bands with maximum absorption up to 99.9% in the visible range. Their proposal shows polarization-insensitive and high scalability in the frequency range by tuning the structural parameters. Furthermore, the design shows good sensing properties as a refractive index sensor around physical property change sensing.

Another tunable dual-band absorber-based refractive index sensor on monolayer graphene array was realized by Li *et al* in 2020 [95]. The proposed work numerically investigated the tunable dual-band perfect metamaterial absorber in the terahertz region. A perforated graphene disk array placed on top of a gold ground plate is

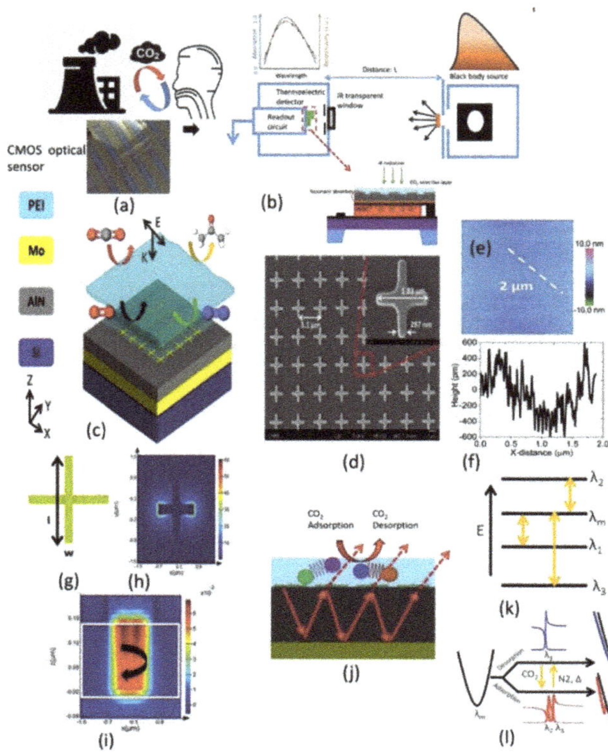

Figure 2.34. (a) CMOS platform. (b) Conceptual diagram of state-of-the-art nondispersive infrared (NDIR) sensing platform. (c) Proposed 'hybrid metamaterial' absorber, (d) field emission scanning electron microscopy image, (e) atomic force microscopy (AFM) image, (f) AFM height profile of the PEI surface. (g) Crosswire metamaterial pattern (h), (i) 3D simulated electric field and magnetic field intensity distribution. (j) Coupling of metamaterial absorption resonance. (k) and (l) Energy diagram representation and effect of coupling. Reproduced from [93]. CC BY 4.0.

spaced by a SiO_2 dielectric substrate. The proposed study achieved almost 100% absorption peaks at 4.7 and 10.7 THz with an independence of incident angle. They also showed a controlling option by graphene conductivity and structural parameters with a maximum sensitivity of 2.33 THz/RIU [95].

As is briefly shown, metamaterial absorber based sensor applications are realized generally in THz frequencies. Some researchers have attempted to develop metamaterials which have a high quality factor and refractive index sensing, such as Mohonty *et al* [96]. They employed two identical metallic patches, a dielectric spacer, and a ground metal plane in their work. Furthermore, they achieved 100% absorbance at 4.5 THz frequency with a quality factor of 225.

Singh *et al* designed and analyzed an ultrathin metamaterial absorber for w band biomedical sensing application in 2022 [97]. They designed a metamaterial absorber that consisted of a key resonator and had a low peak of 99.7% at 105.7 GHz. The proposed design has full width half maxima and Q factor of 5.4 GHz and 19.57, respectively. Since most of the biological samples had refractive indices in extremely close values and sensing of them is a substantial challenge, their proposed design is tuned for obtaining high sensitivity. The obtained average sensitivity and FoM value are 14.81 GHz/RIU and 3.48, respectively. They also simulated their work at different refractive index values of 1 to 2 at 85 and 110 GHz.

2.5 New approaches in metamaterial sensors by using machine learning or a three-dimensional (3D) metamaterial-based sensor

Three-dimensional (3D) metamaterial-based sensors have attracted significant attention due to their ability to manipulate electromagnetic waves in three dimensions, providing unprecedented control over light–matter interactions. The design of 3D metamaterial-based sensors involves engineering the geometry, size, and arrangement of the metamaterial structures to manipulate the interaction between the electromagnetic field and the surrounding environment. This manipulation allows the sensors to detect and measure changes in the environment with multiple advantages. One of the key advantages of 3D metamaterial-based sensors is their ability to provide highly selective and sensitive measurements. This is achieved by tailoring the geometry and size of the metamaterial structures to interact with specific wavelengths of light or specific molecules or chemicals. Additionally, the 3D design of these sensors allows for precise control over the directionality of the electromagnetic field, which can enhance their sensitivity and selectivity [98–102].

Examples of applications of 3D metamaterial-based sensors include environmental monitoring, medical diagnostics, and chemical sensing. For example, 3D metamaterial-based sensors can be used to detect specific biomolecules in blood samples, or to monitor the concentration of pollutants in the environment. Overall, the unique properties of metamaterials make 3D metamaterial-based sensors highly promising for a wide range of sensing applications. In this part of the work, some examples are chosen to present briefly the application of 3D metamaterial-based sensors.

Aristov *et al* presented a 3D metamaterial-based biosensor which is composed from a woodpile structure with high spectral sensitivity (>2600 nm per refractive index unit (RIU) change) and prominent phase-sensitive response (>3 × 104 deg. of phase per RIU) [103]. Their proposed metamaterial is given in figure 2.35.

In the same year, Sreekanth *et al* designed a biosensor with a grating coupled hyperbolic metamaterial, which is a bulk 3D subwavelength structure that enhances the angular sensitivity of a 3D metamaterial-based sensor [104]. They successfully showed cowpea mosaic virus detection with the proposed 3D metamaterial with a maximum sensitivity of 7000 deg per RIU. Their proposed design and simulation result are given in figure 2.36.

An optical fiber surface plasmon resonance (SPR) biosensor was developed for measuring time and concentration-dependent DNA hybridization kinetics by Li *et al* in 2021 [105].

A localized surface plasmon resonance optical fiber biosensor based on a 3D gold nanoparticle and multilayer graphene films was developed by Li *et al* in 2020 [105]. A U-shaped multi-mode fiber (U-MMF) is utilized in this work for DNA sensing.

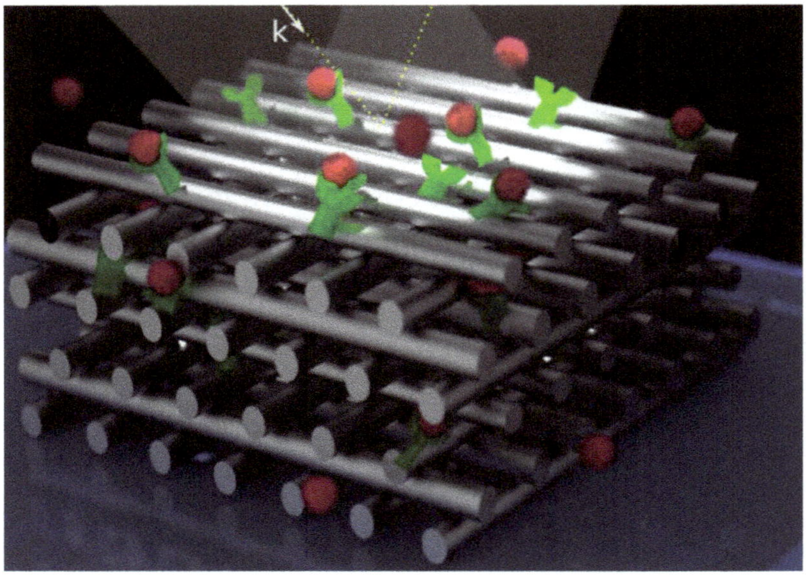

Figure 2.35. 3D schematic illustration of an Ag-coated woodpile crystal developed by Aristov *et al.* Reproduced from [103]. CC BY 4.0.

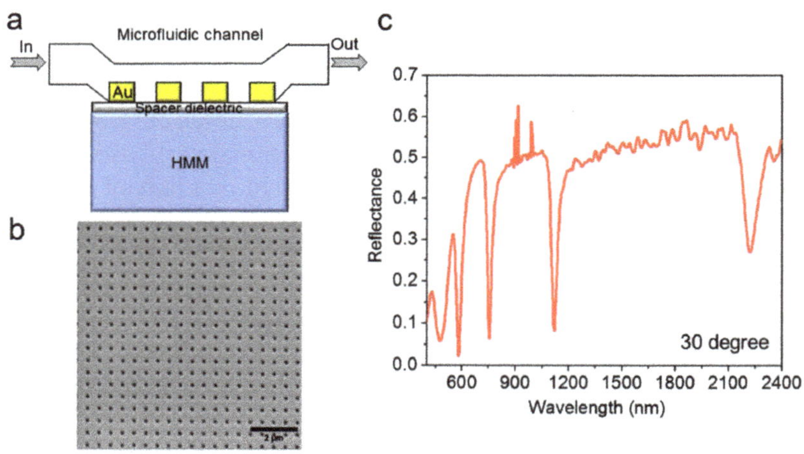

Figure 2.36. (a) Schematic representation of a microfluidic channel integrated with a grating coupled hyperbolic metamaterial biosensor. (b) SEM image of a 2D gold hole array. (c) Reflectance spectrum of a grating coupled Au/Al2O3 HMM at a 30° angle of incidence. Reprinted from [104], copyright (2016) John Wiley & Sons.

The gold nanoparticles are synthesized, and graphene films are formed. Finally, the number of gold nanoparticles are found to be critical for this sensor application. The sensor, after these design stages, shows a sensitivity of 1251.44 nm/RIU, as well as high stability and repeatability. The authors also verified the experimental study of the proposed work as briefly given in figure 2.37, which shows good performance

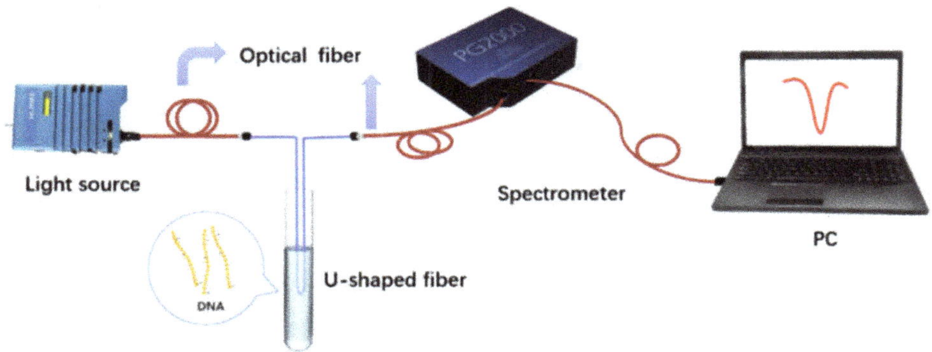

Figure 2.37. Schematic of the experimental set-up used for the U-MMF LSPR sensor. Reproduced from [105] CC BY 4.0.

after preparation of DNA samples ranging from 0.1–100 nM. One year later, a similar work was realized by Li *et al* by using D-shaped metamaterials [106].

Machine learning (ML) is a type of artificial intelligence that involves the use of algorithms and statistical models to enable computers to learn from data and improve their performance on a given task over time. ML has been increasingly applied in the development of metamaterial-based sensors, particularly in the areas of signal processing, data analysis, and pattern recognition.

In metamaterial-based sensors, ML algorithms can be used to analyze the data collected by the sensors and extract useful information about the environment being measured. For example, in chemical sensing applications, ML can be used to identify and classify different chemicals based on their unique spectral fingerprints. ML can also be used to enhance the sensitivity and selectivity of the sensors by optimizing the design of the metamaterial structures.

One approach to using ML in metamaterial-based sensors involves training the algorithm using data collected from the sensors under different environmental conditions. This can enable the algorithm to identify patterns in the data and learn to recognize specific features that are associated with environmental conditions or analytes. Once the algorithm has been trained, it can be used to predict the presence of specific environmental conditions or analytes based on the data collected by the sensors.

Another approach involves using ML algorithms to optimize the design of the metamaterial structures for specific sensing applications. By using the algorithm to simulate the performance of different designs and evaluate their sensitivity and selectivity, an optimal design can be identified and then fabricated for use in the sensor.

Overall, the use of ML in metamaterial-based sensors has the potential to enhance their performance and expand their applications by enabling more efficient and accurate data analysis and design optimization.

One of the first studies in this field is realized by Moon *et al* in 2020 [107]. In the work, plasmonic biosensor designs are integrated with metamaterials and machine learning algorithms. Meta-plasmonic biosensors are designed for optimized

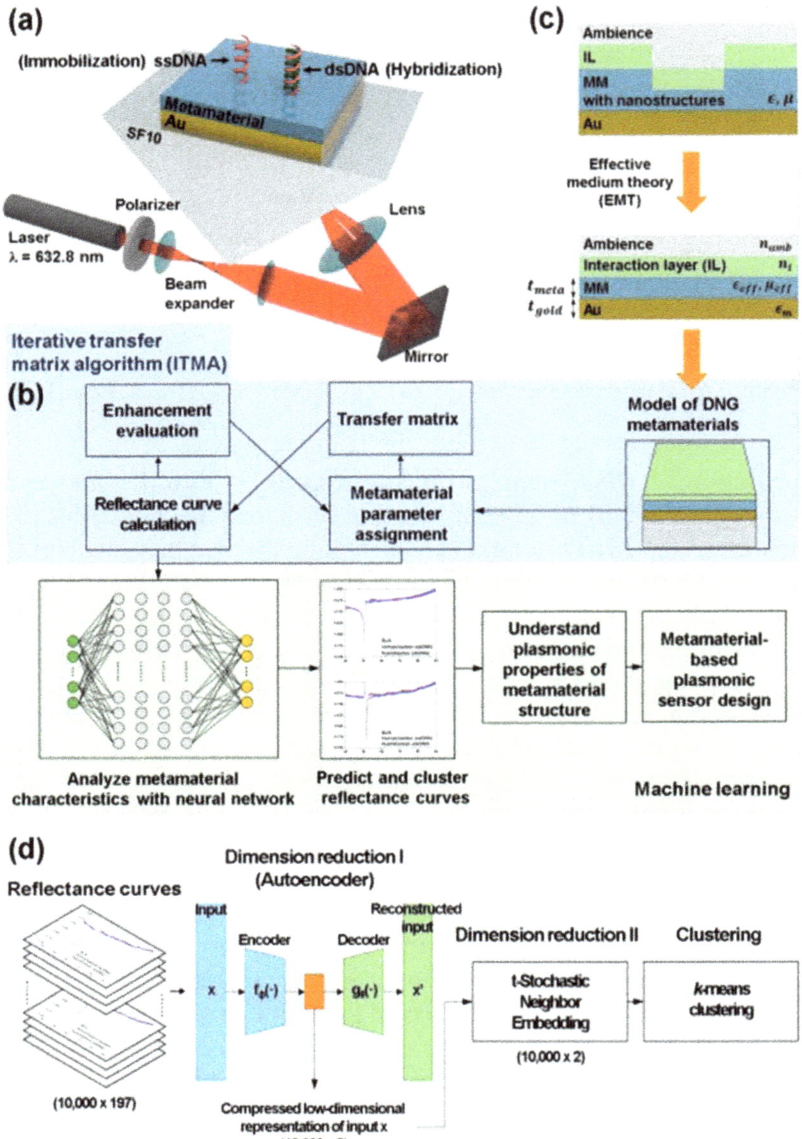

Figure 2.38. (a) Schematics of metamaterial-based DNA biosensing. (b) Workflow of pre-processing classical calculation of metamaterial-based plasmonic detection. (c) Cross section of metamaterial-based plasmonic detection. (d) Machine learning algorithm that consists of three modalities. Reprinted from [107], Copyright (2020), with permission from Elsevier.

detection of DNA. A transfer matrix approach is used for training and testing. Optical characteristics are used to detect optical responses of the proposed sensor. The authors show that it is possible to enhance the sensitivity of the proposed work by using machine learning algorithms. A block diagram is given in figure 2.38.

Another similar work in this field is realized by Jain *et al* in 2023 for biomedical applications by employing hepta band metamaterial-based absorbers with T shaped resonators placed on polyamide for sensing applications [108]. They obtained a quality factor of 117 and sensitivity of 4.72 THz/RIU for glucose detection application. An extreme randomized tree (ERT) model is utilized to predict absorption for intermediate frequencies with physical parameters as unit cell dimensions, substrate thickness, etc. The authors showed that simulation time can be saved by up to 60% in their work. Their proposed sensor shape is presented in figure 2.39.

2.6 Future challenges and future works

If future challenges are to be explained, design complexity is the first challenge. The theoretical modeling and design of metamaterial sensors can be complex due to the intricate physics involved, and the impact of environmental factors. As we move toward more complex and multi-functional metamaterial sensors, the design complexity may increase. The second challenge to be solved is the limitation of fabrication. The fabrication of metamaterial-based sensors at the micro and nano-scale levels can be challenging due to material processing and assembly techniques. Achieving high precision and accuracy in fabrication is necessary to maintain the unique properties of the metamaterials. The third challenge is real-time sensing. Many metamaterial-based sensors still face challenges in real-time sensing and quick response time, which is crucial for practical applications. The fourth challenge is scaling up, which is related to the production of metamaterial sensors for industrial uses. This is a significant challenge, mainly due to cost and technical issues. The fifth challenge is related to stability and durability, which is necessary for the use of metamaterial sensors under different environmental conditions.

The future works in metamaterial-based sensors can be summarized as follows. Advanced theoretical models need to be developed to guide the design and optimization of new classes of metamaterial sensors. Machine learning and artificial intelligence could be used to predict the performance of metamaterials and guide design processes. Further work would be in improving the fabrication techniques. There is a need to develop more precise and cost-effective fabrication techniques to

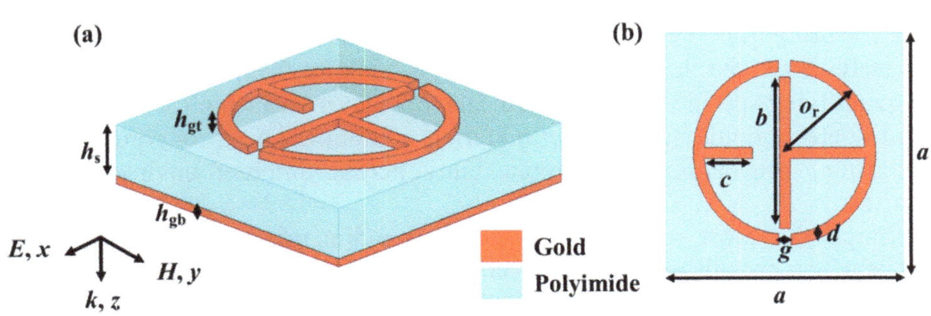

Figure 2.39. (a) Perspective, (b) top view of the proposed sensor. Reproduced from [108] CC BY 4.0.

make metamaterial-based sensors more readily available for wide-ranging applications. Integration with other technologies can be defined as another future work. The integration of metamaterial sensors with other technologies, like Internet-of-Things (IoT) devices and wearable electronics, could open up new applications and use-cases. Furthermore, research into new metamaterials with unique sensing properties could expand the capabilities and applications of metamaterial-based sensors. This includes studying metamaterials that operate at different ranges of the electromagnetic spectrum.

Real-world testing and application work needs to be done to validate the performance of these sensors in real-world conditions and to explore their practical applications across different sectors, such as healthcare, security, and environmental monitoring. Finally, future work could be in the field of environmental and health impact studies. As we move toward wider application of metamaterials, an understanding of their impact on the environment and health becomes crucial. Future work could focus on conducting comprehensive studies to understand these impacts and propose mitigation measures.

2.7 Conclusion

In conclusion, metamaterial-based sensors offer a fascinating and versatile approach to sensing applications, harnessing the unique properties of artificial materials to detect changes in the environment. Through the utilization of the negative refractive index and other extraordinary characteristics, these sensors can manipulate electromagnetic waves in ways that natural materials cannot achieve. As discussed, metamaterial sensors have been applied in various sensing applications, each presenting distinct advantages and capabilities.

Microfluidic sensors capitalize on the connection between the dielectric properties of liquid samples and resonance frequency, enabling high sensitivity and accurate identification of chemical and biological samples. Terahertz sensors, operating in the THz frequency range, demonstrate remarkable potential for medical imaging, security screening, and quality control. The ability of THz waves to penetrate various materials safely and interact with molecular vibrations provides valuable insights into detecting explosives, drugs, and biomolecules.

Metamaterial absorbers, on the other hand, offer a novel approach to detecting and measuring the presence of substances or materials. These absorbers can selectively target specific wavelengths of light, making them valuable in detecting and quantifying specific gases or chemicals. Moreover, their high sensitivity allows them to detect minute quantities of materials without the need for complex sample preparation, broadening their applicability across different environments and scenarios.

Perhaps the most intriguing are the three-dimensional metamaterial-based sensors, which provide unprecedented control over light–matter interactions in three dimensions. This unique design enables highly selective and sensitive measurements by tailoring the metamaterial structures to interact with specific wavelengths or target molecules or chemicals. Precise control over the electromagnetic field direction enhances their sensitivity and selectivity, making them valuable in a wide range of sensing applications.

In summary, metamaterial-based sensors represent a promising frontier in the world of sensing technology. As research and development continues, we can anticipate further breakthroughs and innovations in this field, unlocking new possibilities in biomedical sensing, environmental monitoring, robotics, and smart city applications. The integration of machine learning algorithms with metamaterial sensors also holds great promise for creating intelligent and adaptive sensing systems, paving the way for more sophisticated and efficient applications in various industries. As we move forward, it is imperative to explore and harness the full potential of metamaterial sensors to address emerging challenges and contribute to a more connected and informed world.

References

[1] Lee H J, Lee J H, Moon H S, Jang I S, Choi J S, Yook J G and Jung H I 2012 A planar split-ring resonator-based microwave biosensor for label-free detection of biomolecules *Sensors Actuators* B **169** 26–31

[2] Gennarelli G, Romeo S, Scarfì M R and Soldovieri F 2013 A microwave resonant sensor for concentration measurements of liquid solutions *IEEE Sens. J.* **13** 1857–64

[3] Kapilevich B and Litvak B 2011 Optimized microwave sensor for online concentration measurements of binary liquid mixtures *IEEE Sens. J.* **11** 2611–6

[4] Chretiennot T, Dubuc D and Grenier K 2012 A microwave and microfluidic planar resonator for efficient and accurate complex permittivity characterization of aqueous solutions *IEEE Trans. Microwave Theory Tech.* **61** 972–8

[5] Ebrahimi A, Withayachumnankul W, Al-Sarawi S and Abbott D 2013 High-sensitivity metamaterial-inspired sensor for microfluidic dielectric characterization *IEEE Sens. J.* **14** 1345–51

[6] Rawat V, Dhobale S and Kale S N 2014 Ultra-fast selective sensing of ethanol and petrol using microwave-range metamaterial complementary split-ring resonators *J. Appl. Phys.* **116** 164106

[7] Agarwal S, Prajapati Y K and Mishra V 2015 Thinned fibre Bragg grating as a fuel adulteration sensor: simulation and experimental study *Opto-Electron. Rev.* **23** 231–8

[8] Bakır M, Karaaslan M, Dinçer F, Delihacioglu K and Sabah C 2016 Tunable perfect metamaterial absorber and sensor applications *J. Mater. Sci., Mater. Electron.* **27** 12091–9

[9] Tümkaya M A, Dinçer F, Karaaslan M and Sabah C 2017 Sensitive metamaterial sensor for distinction of authentic and inauthentic fuel samples *J. Electron. Mater.* **46** 4955–62

[10] Bakır M, Dalgaç Ş, Karaaslan M, Karadağ F, Akgol O, Unal E and Sabah C 2019 A comprehensive study on fuel adulteration sensing by using triple ring resonator type metamaterial *J. Electrochem. Soc.* **166** B1044

[11] Tamer A, Karadağ F, Ünal E, Abdulkarim Y I, Deng L, Altintas O and Karaaslan M 2020 Metamaterial based sensor integrating transmission line for detection of branded and unbranded diesel fuel *Chem. Phys. Lett.* **742** 137169

[12] Dalgac S, Akdogan V, Kiris S, Incesu A, Akgol O, Unal E and Karaaslan M 2021 Investigation of methanol contaminated local spirit using metamaterial based transmission line sensor *Measurement* **178** 109360

[13] Tabassum S, Nayemuzzaman S K, Kala M, Kumar Mishra A and Mishra S K 2022 Metasurfaces for sensing applications: gas, bio and chemical *Sensors* **22** 6896

[14] Islam M R, Islam M T, Hoque A, Alshammari A S, Alzamil A, Alsaif H and Soliman M S 2023 Star enclosed circle split ring resonator-based metamaterial sensor for fuel and oil adulteration detection *Alex. Eng. J.* **67** 547–63

[15] Benkhaoua L, Benhabiles M T, Mouissat S and Riabi M L 2015 Miniaturized quasi-lumped resonator for dielectric characterization of liquid mixtures *IEEE Sens. J.* **16** 1603–10

[16] Salim A and Lim S 2016 Complementary split-ring resonator-loaded microfluidic ethanol chemical sensor *Sensors* **16** 1802

[17] Geng Z, Zhang X, Fan Z, Lv X and Chen H 2017 A route to terahertz metamaterial biosensor integrated with microfluidics for liver cancer biomarker testing in early stage *Sci. Rep.* **7** 16378

[18] Bakır M, Karaaslan M, Unal E, Karadag F, Alkurt F Ö, Altıntaş O and Sabah C 2018 Microfluidic and fuel adulteration sensing by using chiral metamaterial sensor *J. Electrochem. Soc.* **165** B475

[19] Govind G and Akhtar M J 2019 Metamaterial-inspired microwave microfluidic sensor for glucose monitoring in aqueous solutions *IEEE Sens. J.* **19** 11900–7

[20] Abdulkarim Y I, Deng L, Karaaslan M, Altıntaş O, Awl H N, Muhammadsharif F F and Luo H 2020 Novel metamaterials-based hypersensitized liquid sensor integrating omega-shaped resonator with microstrip transmission line *Sensors* **20** 943

[21] Velez P, Munoz-Enano J, Ebrahimi A, Herrojo C, Paredes F, Scott J and Martin F 2021 Single-frequency amplitude-modulation sensor for dielectric characterization of solids and microfluidics *IEEE Sens. J.* **21** 12189–201

[22] Qiu Z, Xia Y, Wang B, Hua Y, Li W, Tang W and Cui T J 2022 A metamaterial based microfluidic sensor for permittivity detection of liquid *J. Phys. D: Appl. Phys.* **55** 435001

[23] Salim A and Lim S 2018 Review of recent metamaterial microfluidic sensors *Sensors* **18** 232

[24] Zhou H, Hu D, Yang C, Chen C, Ji J, Chen M and Mu X 2018 Multi-band sensing for dielectric property of chemicals using metamaterial integrated microfluidic sensor *Sci. Rep.* **8** 14801

[25] Jing X, Qin G and Zhang P 2022 Broadband silicon-based tunable metamaterial microfluidic sensor *Photonics Res.* **10** 2876–85

[26] Islam M R, Islam M T, Bais B, Almalki S H, Alsaif H and Islam M S 2022 Metamaterial sensor based on rectangular enclosed adjacent triple circle split ring resonator with good quality factor for microwave sensing application *Sci. Rep.* **12** 6792

[27] Sharafadinzadeh N, Abdolrazzaghi M and Daneshmand M 2020 Investigation on planar microwave sensors with enhanced sensitivity from microfluidic integration *Sensors Actuators* A **301** 111752

[28] Sreekanth K V, Mahalakshmi P, Han S, Mani Rajan M S, Choudhury P K and Singh R 2019 Brewster mode-enhanced sensing with hyperbolic metamaterial *Adv. Opt. Mater.* **7** 1900680

[29] Xu W, Xie L and Ying Y 2017 Mechanisms and applications of terahertz metamaterial sensing: a review *Nanoscale* **9** 13864–78

[30] Shen S, Liu X, Shen Y, Qu J, Pickwell-MacPherson E, Wei X and Sun Y 2022 Recent advances in the development of materials for terahertz metamaterial sensing *Adv. Opt. Mater.* **10** 2101008

[31] Zhou H, Yang C, Hu D, Li D, Hui X, Zhang F and Mu X 2019 Terahertz biosensing based on bi-layer metamaterial absorbers toward ultra-high sensitivity and simple fabrication *Appl. Phys. Lett.* **115** 143507

[32] Beruete M and Jáuregui-López I 2020 Terahertz sensing based on metasurfaces *Adv. Opt. Mater.* **8** 1900721

[33] Ng B, Wu J, Hanham S M, Fernández-Domínguez A I, Klein N, Liew Y F and Maier S A 2013 Spoof plasmon surfaces: a novel platform for THz sensing *Adv. Opt. Mater.* **1** 543–8

[34] Houard A, Liu Y, Prade B, Tikhonchuk V T and Mysyrowicz A 2008 Strong enhancement of terahertz radiation from laser filaments in air by a static electric field *Phys. Rev. Lett.* **100** 255006
Chen H T, Padilla W J, Cich M J, Azad A K, Averitt R D and Taylor A J 2009 A metamaterial solid-state terahertz phase modulator *Nat. Photon.* **3** 148–51

[35] Sun Y, Du P, Lu X, Xie P, Qian Z, Fan S and Zhu Z 2018 Quantitative characterization of bovine serum albumin thin-films using terahertz spectroscopy and machine learning methods *Biomed. Opt. Express* **9** 2917–29

[36] Qin J, Xie L and Ying Y 2015 Determination of tetracycline hydrochloride by terahertz spectroscopy with PLSR model *Food Chem.* **170** 415–22

[37] Qin J, Xie L and Ying Y 2016 A high-sensitivity terahertz spectroscopy technology for tetracycline hydrochloride detection using metamaterials *Food Chem.* **211** 300–5

[38] Wang W, Yan F, Tan S, Zhou H and Hou Y 2017 Ultrasensitive terahertz metamaterial sensor based on vertical split ring resonators *Photon. Res.* **5** 571–7

[39] Deng X, Shen Y, Liu B, Song Z, He X, Zhang Q and Wei D 2022 Terahertz metamaterial sensor for sensitive detection of citrate salt solutions *Biosensors* **12** 408

[40] Park S J, Jun S W, Kim A R and Ahn Y H 2015 Terahertz metamaterial sensing on polystyrene microbeads: shape dependence *Opt. Mater. Express* **5** 2150–5

[41] Cong L, Tan S, Yahiaoui R, Yan F, Zhang W and Singh R 2015 Experimental demonstration of ultrasensitive sensing with terahertz metamaterial absorbers: a comparison with the metasurfaces *Appl. Phys. Lett.* **106** 031107

[42] Driscoll T, Andreev G O, Basov D N, Palit S, Cho S Y, Jokerst N M and Smith D R 2007 Tuned permeability in terahertz split-ring resonators for devices and sensors *Appl. Phys. Lett.* **91** 062511

[43] Padilla W J, Taylor A J, Highstrete C, Lee M and Averitt R D 2006 Dynamical electric and magnetic metamaterial response at terahertz frequencies *Phys. Rev. Lett.* **96** 107401

[44] Sun R, Li W, Meng T and Zhao G 2021 Design and optimization of terahertz metamaterial sensor with high sensing performance *Opt. Commun.* **494** 127051

[45] Hossain M B, Faruque M R I, Alshammari A S and Islam M T 2022 Wide bandwidth enriched symmetric hexagonal split ring resonator based triple band negative permittivity metamaterial for satellite and Wi-Fi applications *Res. Phys.* **37** 105511

[46] Park S J, Hong J T, Choi S J, Kim H S, Park W K, Han S T and Ahn Y H 2014 Detection of microorganisms using terahertz metamaterials *Sci. Rep.* **4** 1–7

[47] Yang X, Wei D, Yan S, Liu Y, Yu S, Zhang M and Fu W 2016 Rapid and label-free detection and assessment of bacteria by terahertz time-domain spectroscopy *J. Biophoton.* **9** 1050–8

[48] Xu W, Xie L, Zhu J, Wang W, Ye Z, Ma Y and Ying Y 2017 Terahertz sensing of chlorpyrifos-methyl using metamaterials *Food Chem.* **218** 330–4

[49] Kim H S, Cha S H, Roy B, Kim S and Ahn Y H 2018 Humidity sensing using THz metamaterial with silk protein fibroin *Opt. Express* **26** 33575–81
[50] Zhong Y, Du L, Liu Q, Zhu L and Zhang B 2020 Metasurface-enhanced ATR sensor for aqueous solution in the terahertz range *Opt. Commun.* **465** 125508
[51] Yang J, Qi L, Li B, Wu L, Shi D, Uqaili J A and Tao X 2021 A terahertz metamaterial sensor used for distinguishing glucose concentration *Res. Phys.* **26** 104332
[52] Lee G, Cho Y and Ok G 2023 Improved analysis of THz metamaterials for glucose sensing based on modified Lorentz dispersion model *Spectrochim. Acta, Part* A **293** 122519
[53] Ahi K, Jessurun N, Hosseini M P and Asadizanjani N 2020 Survey of terahertz photonics and biophotonics *Opt. Eng.* **59** 061629
[54] Zhan X, Yang S, Huang G, Yang L, Zhang Y, Tian H and Fu W 2021 Streptavidin-functionalized terahertz metamaterials for attomolar exosomal microRNA assay in pancreatic cancer based on duplex-specific nuclease-triggered rolling circle amplification *Biosens. Bioelectron.* **188** 113314
[55] Park S J, Cha S H, Shin G A and Ahn Y H 2017 Sensing viruses using terahertz nano-gap metamaterials *Biomed. Opt. Express* **8** 3551–8
[56] Alkurt F O, Altintas O, Ozakturk M, Karaaslan M, Akgol O, Unal E and Sabah C 2020 Enhancement of image quality by using metamaterial inspired energy harvester *Phys. Lett.* A **384** 126041
[57] Yang Y, Xu D and Zhang W 2018 High-sensitivity and label-free identification of a transgenic genome using a terahertz meta-biosensor *Opt. Express* **26** 31589–98
[58] Hou X, Chen X, Li T, Li Y, Tian Z and Wang M 2021 Highly sensitive terahertz metamaterial biosensor for bovine serum albumin (BSA) detection *Opt. Mater. Express* **11** 2268–77
[59] Baqir M A, Farmani A, Fatima T, Raza M R, Shaukat S F and Mir A 2018 Nanoscale, tunable, and highly sensitive biosensor utilizing hyperbolic metamaterials in the near-infrared range *Appl. Opt.* **57** 9447–54
[60] Lee S H, Lee Y K, Lee S H, Kwak J, Song H S and Seo M 2022 Detection and discrimination of SARS-CoV-2 spike protein-derived peptides using THz metamaterials *Biosens. Bioelectron.* **202** 113981
[61] Zhu W, Rukhlenko I D and Premaratne M 2013 Graphene metamaterial for optical reflection modulation *Appl. Phys. Lett.* **102** 241914
[62] Abdolrazzaghi M, Abdolali A and Daneshmand M 2015 Highly sensitive miniaturized bio-sensor using 2-layer double split ring resonators *2015 IEEE Int. Symp. on Antennas and Propagation and USNC/URSI National Radio Science Meeting* (Piscataway, NJ: IEEE) pp 736–7
[63] Al-Naib I 2016 Biomedical sensing with conductively coupled terahertz metamaterial resonators *IEEE J. Sel. Top. Quantum Electron.* **23** 1–5
[64] Chen X and Fan W 2017 Ultrasensitive terahertz metamaterial sensor based on spoof surface plasmon *Sci. Rep.* **7** 2092
[65] Keshavarz A and Vafapour Z 2018 Water-based terahertz metamaterial for skin cancer detection application *IEEE Sens. J.* **19** 1519–24
[66] Saadeldin A S, Hameed M F O, Elkaramany E M and Obayya S S 2019 Highly sensitive terahertz metamaterial sensor *IEEE Sens. J.* **19** 7993–9
[67] Ou H, Lu F, Xu Z and Lin Y S 2020 Terahertz metamaterial with multiple resonances for biosensing application *Nanomaterials* **10** 1038

[68] Patel S K, Surve J and Parmar J 2022 Detection of cancer with graphene metasurface-based highly efficient sensors *Diam. Relat. Mater.* **129** 109367

[69] Huang X, Li C, Fang B, Lu J, Ma X, Hong Z and Jing X 2023 Improved biomolecular detection sensitivity based on concave metamaterial structure *Infrared Phys. Technol.* **128** 104524

[70] Purkayastha A, Srivastava T and Jha R 2016 Ultrasensitive THz–plasmonics gaseous sensor using doped graphene *Sensors Actuators* B **227** 291–5

[71] Srivastava T, Purkayastha A and Jha R 2016 Graphene based surface plasmon resonance gas sensor for terahertz *Opt. Quantum Electron.* **48** 1–11

[72] Xiang Y, Zhu J, Wu L, You Q, Ruan B and Dai X 2017 Highly sensitive terahertz gas sensor based on surface plasmon resonance with graphene *IEEE Photonics J.* **10** 1–7

[73] Lai W H, Li B, Fu S H and Lin Y S 2023 Tunable MEMS-based terahertz metamaterial for pressure sensing application *Micromachines* **14** 169

[74] Wang Y, Qiu Y, Zhang Y, Lang T and Zhu F 2023 High-sensitivity temperature sensor based on the perfect metamaterial absorber in the terahertz band *Photonics* (Multidisciplinary Digital Publishing Institute) Vol 10 p 92

[75] Abdulkarim Y I, Mohanty A, Acharya O P, Appasani B, Khan M S, Mohapatra S K and Dong J 2022 A review on metamaterial absorbers: microwave to optical *Front. Phys.* **359**

[76] Mazare A G, Abdulkarim Y I, Karim A S, Bakır M, Taouzari M, Muhammadsharif F F and Bizon N 2022 Enhanced sensing capacity of terahertz triple-band metamaterials absorber based on pythagorean fractal geometry *Materials* **15** 6364

[77] Hakim M L, Islam M T, Alam T, Abdul Rahim S K, Bais B, Islam M S and Soliman M S 2023 Triple-band square split-ring resonator metamaterial absorber design with high effective medium ratio for 5G sub-6 GHz applications *Nanomaterials* **13** 222

[78] Banerjee S, Dutta P, Basu S, Mishra S K, Appasani B, Nanda S and Thounthong P 2023 A new design of a terahertz metamaterial absorber for gas sensing applications *Symmetry* **15** 24

[79] Abdulkarim Y I, Muhammadsharif F F, Appasani B, Bakır M, Karaaslan M, Rezaei P and Dong J 2023 A novel terahertz metamaterial absorber based on split-relief design for the application of refractive index sensing

[80] Prakash D and Gupta N 2022 Applications of metamaterial sensors: a review *Int. J. Microwave Wireless Technol.* **14** 19–33

[81] Hedayati M K, Faupel F and Elbahri M 2014 Review of plasmonic nanocomposite metamaterial absorber *Materials* **7** 1221–48

[82] Yu P, Besteiro L V, Huang Y, Wu J, Fu L, Tan H H and Wang Z 2019 Broadband metamaterial absorbers *Adv. Opt. Mater.* **7** 1800995

[83] Kearney B, Alves F, Grbovic D and Karunasiri G 2013 Al/SiO_x/Al single and multiband metamaterial absorbers for terahertz sensor applications *Opt. Eng.* **52** 013801

[84] Sabah C, Turkmen-Kucuksari O Z N U R and Turhan-Sayan G 2014 Metamaterial absorber-based sensor embedded into X-band waveguide *Electron. Lett.* **50** 1074–6

[85] Wang B X, Zhai X, Wang G Z, Huang W Q and Wang L L 2015 A novel dual-band terahertz metamaterial absorber for a sensor application *J. Appl. Phys.* **117** 014504

[86] Bakir M, Karaaslan M, Dincer F, Akgol O and Sabah C 2016 Electromagnetic energy harvesting and density sensor application based on perfect metamaterial absorber *Int. J. Mod. Phys.* B **30** 1650133

[87] Vivek A, Shambavi K and Alex Z C 2019 A review: metamaterial sensors for material characterization *Sensor Rev.* **39** 417–32

[88] Butt M A, Khonina S N, Kazanskiy N L and Piramidowicz R 2022 Hybrid metasurface perfect absorbers for temperature and biosensing applications *Opt. Mater.* **123** 111906

[89] Altintas O *et al* 2019 High sensitive metamaterial sensor for water treatment centres *Water, Air, Soil Pollut.* **230** 1–9

[90] Khan Y, Butt M A, Khonina S N and Kazanskiy N L 2022 Thermal sensor based on polydimethylsiloxane polymer deposited on low-index-contrast dielectric photonic crystal structure *Photonics* (MDPI) Vol. 9 p 770

[91] Du John H V, Jose T and Pandiaraj S 2022 Design of 5.8 GHz single band metamaterial unit cell absorber for wi-fi applications *2022 6th Int. Conf. on Devices, Circuits and Systems (ICDCS)* (Piscataway, NJ: IEEE) pp 310–2

[92] Deng G, Xia T, Jing S, Yang J, Lu H and Yin Z 2017 A tunable metamaterial absorber based on liquid crystal intended for F frequency band *IEEE Antennas Wirel. Propag. Lett.* **16** 2062–5

[93] Hasan D and Lee C 2018 Hybrid metamaterial absorber platform for sensing of CO2 gas at Mid-IR *Adv. Sci.* **5** 1700581

[94] Qi B, Zhao Y, Niu T and Mei Z 2019 Ultra-broadband metamaterial absorber based on all-metal nanostructures *J. Phys. D: Appl. Phys.* **52** 425304

[95] Li J, Liao Q, Li H, Liu W, Yu T and Wang T 2020 Tunable dual-band perfect metamaterial absorber based on monolayer graphene arrays as refractive index sensor *Jpn. J. Appl. Phys.* **59** 095002

[96] Mohanty A, Acharya O P, Appasani B, Mohapatra S K and Khan M S 2021 Design of a novel terahertz metamaterial absorber for sensing applications *IEEE Sens. J.* **21** 22688–94

[97] Singh H, Gupta A, Kaler R S, Singh S and Gill A S 2022 Designing and analysis of ultrathin metamaterial absorber for W band biomedical sensing application *IEEE Sens. J.* **22** 10524–31

[98] Navau C, Mach-Batlle R, Parra A, Prat-Camps J, Laut S, Del-Valle N and Sanchez A 2017 Enhancing the sensitivity of magnetic sensors by 3D metamaterial shells *Sci. Rep.* **7** 1–9

[99] Chen J, Qi S, Hong X, Gu P, Wei R, Tang C and Zhao C 2019 Highly sensitive 3D metamaterial sensor based on diffraction coupling of magnetic plasmon resonances *Res. Phys.* **15** 102791

[100] Kadic M, Milton G W, van Hecke M and Wegener M 2019 3D metamaterials *Nat. Rev. Phys.* **1** 198–210

[101] Hassan M M, Sium F S, Islam F and Choudhury S M 2021 A review on plasmonic and metamaterial based biosensing platforms for virus detection *Sens. Bio-Sens. Res.* **33** 100429

[102] Norouzi M, Jarchi S, Ghaffari-Miab M, Esfandiari M, Lalbakhsh A, Koziel S and Moloudian G 2023 3D metamaterial ultra-wideband absorber for curved surface *Sci. Rep.* **13** 1043

[103] Aristov A I, Manousidaki M, Danilov A, Terzaki K, Fotakis C, Farsari M and Kabashin A V 2016 3D plasmonic crystal metamaterials for ultra-sensitive biosensing *Sci. Rep.* **6** 1–8

[104] Sreekanth K V, Alapan Y, ElKabbash M, Wen A M, Ilker E, Hinczewski M and Strangi G 2016 Enhancing the angular sensitivity of plasmonic sensors using hyperbolic metamaterials *Adv. Opt. Mater.* **4** 1767–72

[105] Li C, Li Z, Li S, Zhang Y, Sun B, Yu Y and Yue W 2020 LSPR optical fiber biosensor based on a 3D composite structure of gold nanoparticles and multilayer graphene films *Opt. Express* **28** 6071–83

[106] Li C, Gao J, Shafi M, Liu R, Zha Z, Feng D and Jiang S 2021 Optical fiber SPR biosensor complying with a 3D composite hyperbolic metamaterial and a graphene film *Photon. Res.* **9** 379–88

[107] Moon G, Choi J R, Lee C, Oh Y, Kim K H and Kim D 2020 Machine learning-based design of meta-plasmonic biosensors with negative index metamaterials *Biosens. Bioelectron.* **164** 112335

[108] Jain P, Chhabra H, Chauhan U, Prakash K, Gupta A, Soliman M S and Islam M T 2023 Machine learning assisted hepta band THz metamaterial absorber for biomedical applications *Sci. Rep.* **13** 1792

IOP Publishing

Advanced Metamaterials for Engineers

Lulu Wang and Muharrem Karaaslan

Chapter 3

Metamaterial absorbers in the microwave range

Fatih Özkan Alkurt and Emin Ünal

3.1 Introduction

Metamaterials (MTMs) are a class of artificially engineered materials with various exotic properties that natural materials do not have, including negative permittivity, negative permeability, and negative refractive index [1–3]. Due to their unique attributes, MTMs are used in an extensive application area from the microwave to terahertz region of the electromagnetic spectrum [4–7]. These materials are often arranged in a repeating pattern with a side length smaller than the operating wavelength and they are known as subwavelength materials with negative constitutive parameters [8, 9]. With their geometry, design shape and size, MTMs offer diverse capabilities in manipulating electromagnetic waves, encompassing absorption, reflection, transmission, filtering, and polarization conversion. Consequently, MTMs have become instrumental in numerous technological sectors, spanning antennas [10–12], sensors [13–15], absorber designs [16, 17], energy harvesters [18, 19], medical applications [20–22], imaging systems [23–25], military purposes [26, 27], radomes [28, 29], lasers [30, 31], aerospace [32, 33], and telecommunication [34–36], among others.

In the context of absorbers, MTMs play a crucial role in enabling efficient absorption of incident electromagnetic waves at specific operating frequencies. Absorbers are specially designed to convert absorbed energy into heat within the dielectric substrate layer, typically resulting in irreversible processes. However, in certain applications, known as microwave energy harvesting, the absorbed energy can be converted into DC electrical power.

Some parameters such as reflection magnitude, reflection phase, transmission magnitude, and transmission phase are the important parameters of the absorption mechanism. Another important criterion for the absorption mechanism is the loss characteristics in the substrate of the absorber. Each criterion should be defined according to the application area. In some cases, broadband absorbers can be preferred, and the design criteria should be regulated to the broadband purposes.

In the pursuit of enhanced absorption characteristics, MTMs are frequently integrated into microwave absorbers, as they bolster absorption efficiency within the subwavelength regime. Moreover, tunable absorbers based on MTM technology have emerged, offering versatile absorption control. This chapter presents the design procedure and the operating mechanism of the MTM absorbers in the microwave region. On the other hand, MTMs can also improve the bandwidth of the absorption band. By harnessing the unique properties of MTMs, microwave absorbers have the potential to revolutionize various industries, offering innovative solutions for absorbing, controlling, and manipulating electromagnetic waves across diverse applications.

3.2 Microwave region of the electromagnetic spectrum

In the vast expanse of the electromagnetic (EM) spectrum, electromagnetic waves manifest as the oscillation of both electric and magnetic field components emanating from energy sources. By capturing these waves across a wide spectrum, we unveil the entire electromagnetic spectrum, revealing a plethora of EM waves distinguished by their unique frequencies. This comprehensive spectrum encompasses an array of waves, ranging from radio waves to microwaves, infrared, visible light, ultraviolet, X-rays, and even gamma rays [37]. In this section, we will delve into the significance and utilization of the microwave region within the electromagnetic spectrum, focusing on frequencies that span from 300 MHz to 300 GHz. Notably, the microwave region assumes a major role as one of the most prominent and widely employed frequency ranges, particularly in the fields of medical and telecommunication technologies. The term 'microwaves' refers to centimeter and millimeter waves that fall within this frequency range, encompassing various bands such as L, S, C, X, Ku, K, Ka, Q, U, V, W, F, and D. The remarkable diversity and versatility of microwave frequencies hold paramount importance in diverse domains, underpinning the efficiency and effectiveness of medical and telecommunication systems on a global scale.

The microwave region, nestled within the broader EM spectrum, covers frequencies ranging from 300 MHz to 300 GHz. This frequency range has garnered significant attention and widespread application in various cutting-edge technologies due to its unique properties and advantageous characteristics. Frequencies falling within this range facilitate efficient transmission and manipulation of electromagnetic signals, making it an indispensable frequency band for numerous practical applications. The microwave region forms the backbone of modern telecommunication technologies, serving as an essential frequency range for wireless communication systems. Microwave frequencies enable high data transfer rates and long-distance communication links, making them ideal for various applications, including satellite communication, wireless broadband, and microwave point-to-point links. Moreover, the microwave bands are extensively used in radar systems, where they offer precise and rapid detection capabilities for navigation, weather monitoring, and military surveillance.

Microwave metamaterial absorbers are an intriguing class of engineered structures designed to efficiently absorb electromagnetic waves within the microwave frequency range. These absorbers leverage the unique properties of metamaterials, which possess negative refractive indices, to achieve exceptional absorption characteristics. By meticulously tailoring the geometry and arrangement of subwavelength unit cells, microwave metamaterial absorbers can achieve near-total absorption of incident waves across a specific frequency band. This remarkable absorption efficiency stems from the ability of metamaterial absorbers to create artificial impedance matching with the surrounding environment, thereby minimizing reflection and maximizing energy dissipation. As a result, these absorbers find versatile applications in various domains, including antenna performance enhancement, stealth technology, electromagnetic interference (EMI) shielding, and radar cross-section reduction. The continuous advancements in metamaterial design and fabrication techniques hold great promise for further enhancing the performance and expanding the applications of microwave metamaterial absorbers in wireless communication systems and beyond.

3.3 Microwave absorption mechanism

The absorption mechanism's primary purpose is to effectively block both reflection and transmission of the incident electromagnetic (EM) wave across all frequencies within the desired operating band. To achieve this, a typical unit cell is designed with three distinct layers: a metallic back layer, a dielectric substrate layer, and a resonator layer at the top.

The metallic back layer serves as a barrier, preventing the transmission of the incident EM wave. The dielectric substrate layer, positioned between the back layer and the resonator, can be composed of lossy material, allowing it to convert absorbed EM energy into heat energy.

The resonator layer holds great significance in the design, as it not only determines the operating frequency but also plays a vital role in the absorption mechanism. It effectively confines the incident EM energy within the substrate layer, ensuring maximum absorption. Additionally, the resonator layer facilitates impedance matching between the incident region and the absorber, allowing for efficient penetration of the EM energy.

The absorption mechanism aims to eliminate reflection and transmission of EM waves across the operating band by using a carefully designed unit cell comprising metallic, dielectric, and resonator layers. This approach maximizes energy absorption and efficiency while maintaining impedance matching for optimal EM energy penetration.

The mathematical background of the absorption mechanism is as follows [38],

$$A = 1 - |S_{11}|^2 - |S_{21}|^2 \tag{3.1}$$

where A is the absorption magnitude, S_{11} is the reflection from the designed layer, and S_{21} is the transmission to the back side of the designed layer.

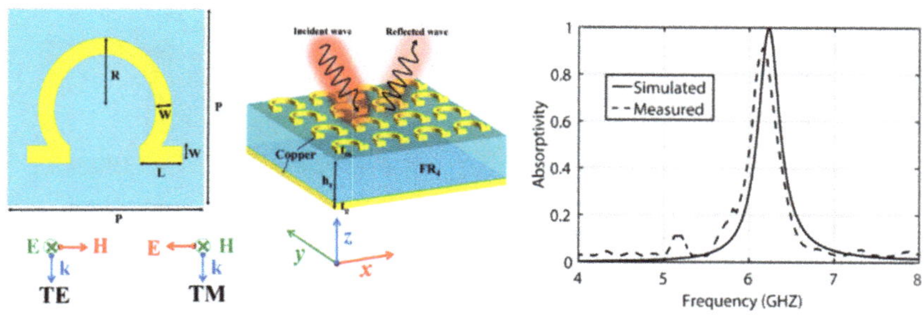

Figure 3.1. Omega-shaped MTM absorber design. Reprinted from [38], Copyright (2021), with permission from Elsevier.

Figure 3.1 shows the omega-shaped design of a metamaterial-based absorption structure, which aims to prevent reflection and transmission at the operating frequency of 6.2 GHz, which is in the wireless communication bands. As shown in the figure, the back layer is covered by metallic copper in full arrangement that blocks the EM wave transmission to the back side which means S_{21} goes to zero (0). In this case, equation (3.1) can be written as,

$$A = 1 - | S_{11} |^2 \tag{3.2}$$

Considering equation (3.2), it is easy to analyze the absorption characteristics because the absorption parameter depends only on the reflection magnitude. As shown in the omega-shaped design, FR4 dielectric substrate is used with an acceptable loss characteristic for the respective frequency bands. There is also a periodic arrangement of unit cell in both transverse electric (TE) and transverse magnetic (TM) incident wave modes with an excellent absorption rate. Moreover, this design can work as a polarization independent perfect metamaterial absorber, with nearly unity absorption characteristics in ISM bands both in simulation and experiment [38].

3.4 Absorber design processes

There are infinite ways to design a microwave absorber; the main point is to propose a novel and effective resonator layer design. One of these designs has been proposed by Hannan *et al* as shown in figure 3.2 [39]. In this design process, four identical and symmetric resonators are placed next to each adjacent cell on the top layer. This design contains split-ring resonator combinations as shown on each microwave section and each cell corresponds to a resonance point due to its geometric dimensions. This multiband configuration covers the operating bands for C, X, and Ku band absorber applications. The back layer is covered by a metallic plate aiming to provide zero transmission as specified in equation (1.2).

As seen in the figure 3.2, the proposed design has multi and wide band absorption characteristics at the frequencies of 4.192, 5.24, 8.632, 9.264, and 10.152 GHz with nearly perfect absorption. This absorption characteristic changes under different polarization conditions. Due to the specified features, this structure can be used for

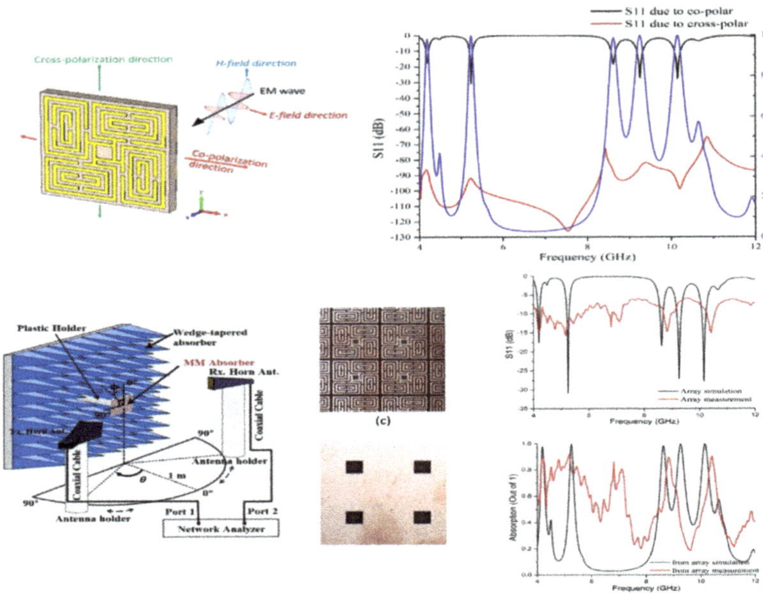

Figure 3.2. Multiband metamaterial absorber for C and X band applications. Reprinted from [39], Copyright (2022), with permission from Elsevier.

radar, defense, and sensing applications. The designed absorber structure is fabricated in the microwave laboratory, its measurements are carried out experimentally, and the measured results have been compared with the results obtained from the simulation, and as a result, it has been clearly shown that the results are in good agreement [39].

In another design process, Zhang *et al* have modeled a multi-layered absorber design which is based on the metamaterial subwavelength approach [40]. The frequency band aimed at in this study is the low-frequency region between 2 and 4 GHz in the S band. The substrate layer is chosen as the active magnetic material, which creates a broadband characteristic at low frequencies, especially in the S band. Another important feature of this multi-layered design is impedance matching. When the impedance of the designed periodic absorber and the medium of the incident wave match, the absorption characteristic can be significantly improved.

Figure 3.3 shows periodic patches in each resonator layer and aluminum back layer to prevent transmission [40]. This sandwiched structure has a different sized unit cell resonator in each layer and this approach has significantly increased the operating bandwidth between 2 and 10 GHz. In the substrate layer, FR4 dielectric material is used with the permittivity and thickness of 4.3 and 0.2 mm, respectively. The use of magnetic coating in each layer creates a perfect impedance matching in the low-frequency region. This designed multi-layered structure is fabricated and investigated experimentally, the results obtained are a strong match with the simulations.

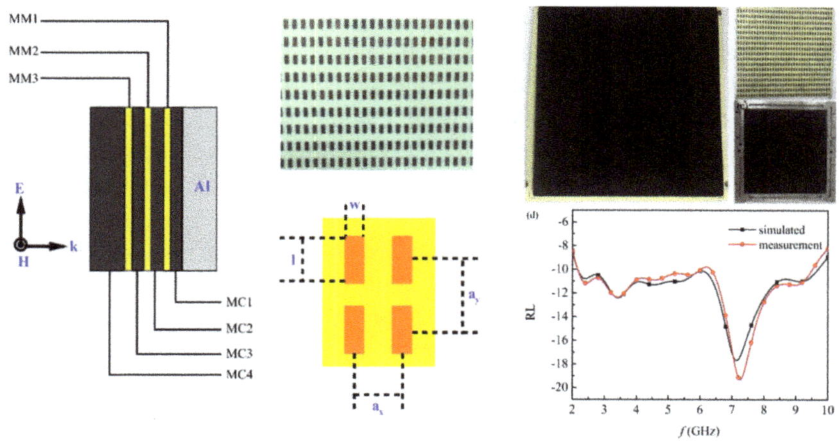

Figure 3.3. Multi-layered MTM absorber design. Reprinted from [40], Copyright (2020), with permission from Elsevier.

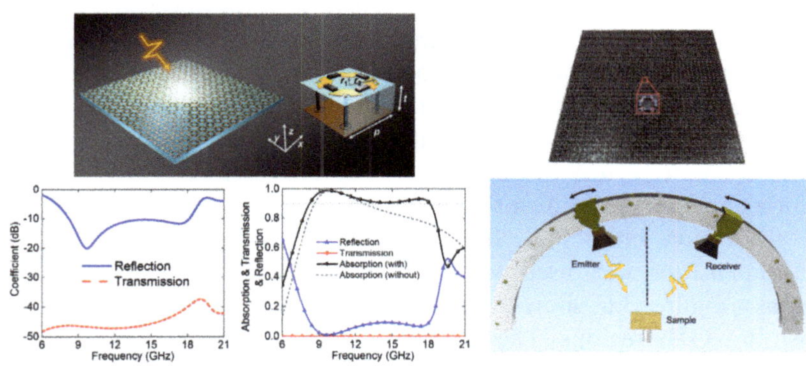

Figure 3.4. Lumped element loaded MTM absorber design. Reproduced from [41]. CC BY 4.0.

Microwave resonators can be assumed as electronic circuits at microwave frequencies. For this reason, microwave absorbers can be represented by an equivalent circuit model. Electronic lumped element components can be used in a unit cell of a metamaterial-based absorber design to modify the absorption mechanism electronically.

Chen *et al* have designed a broadband absorber structure based on a metamaterial design approach [41]. The aimed operating frequency band is the X and Ku band region, and the thickness of the designed layer is less than 1/10 of the related frequency. The thickness modification shows the subwavelength approach which is the basis of the metamaterial design procedure.

Figure 3.4 shows the designed unit cell; each cell contains four symmetrical lumped elements with a chip resistor set to 110 ohms. The purpose of using four lumped elements is to create a polarization independent symmetry. As can be clearly seen from figure 1.5, absorption is improved by using appropriate lumped elements in each cell. The fabricated sample and measurement setup are also given in figure 3.4, and two horn antennas are used as the emitter–receiver scenario in the measurements and the absorption has been calculated with equation (3.2) since the back side is covered by metallic copper that prevents transmission [41].

In another absorber study, Wang and Cheng have proposed a metamaterial absorber with a meander line approach [42]. This absorber is compact and has a low-frequency absorption characteristic up to 6 GHz. Double-negative constitutive parameters have been observed within the operating frequencies as shown in figure 3.5. The real part of the permittivity of the designed meander line absorber is negative up to 4 GHz and the real part of the permeability is negative at frequencies greater than about 3.6 GHz. This structure can be considered as having a good absorption characteristic between the frequencies of 2 and 4 GHz. Furthermore, the authors have analyzed this meander line by a parametric approach and obtained the optimum frequency response.

To carry out the measurements, this metamaterial absorber structure has been fabricated by using lossy FR4 dielectric material. The array size is 20 × 20 in the fabrication, which ensures the minimum measurement criteria. The fabricated metamaterial absorber sample and the measured absorption characteristics are given in figure 3.5. When the measurement results and simulation results are compared, it is seen that they are in good agreement, so this design can be used for 1.84 GH–5.96 GHz communication bands. Moreover, this metamaterial absorber can be used in many electromagnetic applications such as energy harvesting, stealth technology, and so on [42].

In another metamaterial absorber approach, a multi-layered design is used to increase absorption ratios [43]. In this design, a '+' shaped unit cell with two different dimensions was used to create absorptions for both GHz and THz applications. Figure 3.6 shows the designed cell structure and its absorption mechanism for GHz regions. The smaller '+' shaped unit cells correspond to THz frequencies. As shown, this structure absorbs microwave EM energy at three frequencies as 110, 161, and 185 GHz, with nearly unity absorptions. Moreover, this absorber has been analyzed under different TE and TM modes and the absorption ratios are nearly unity [43].

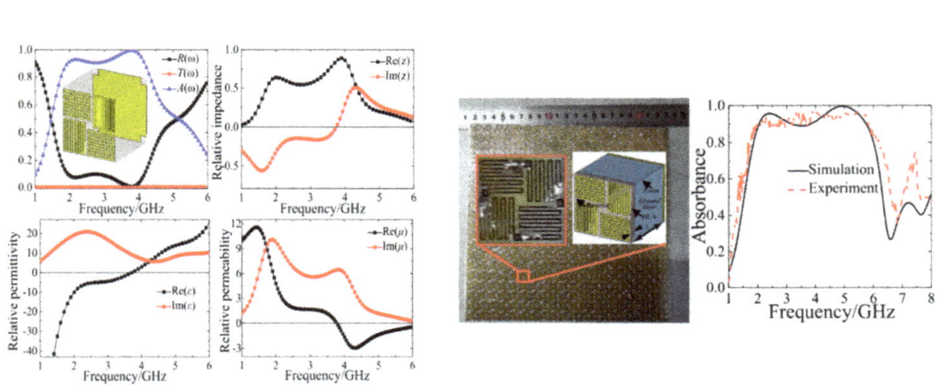

Figure 3.5. Meander wire loaded MTM absorber design. Reprinted from [42], Copyright (2020), with permission from Elsevier.

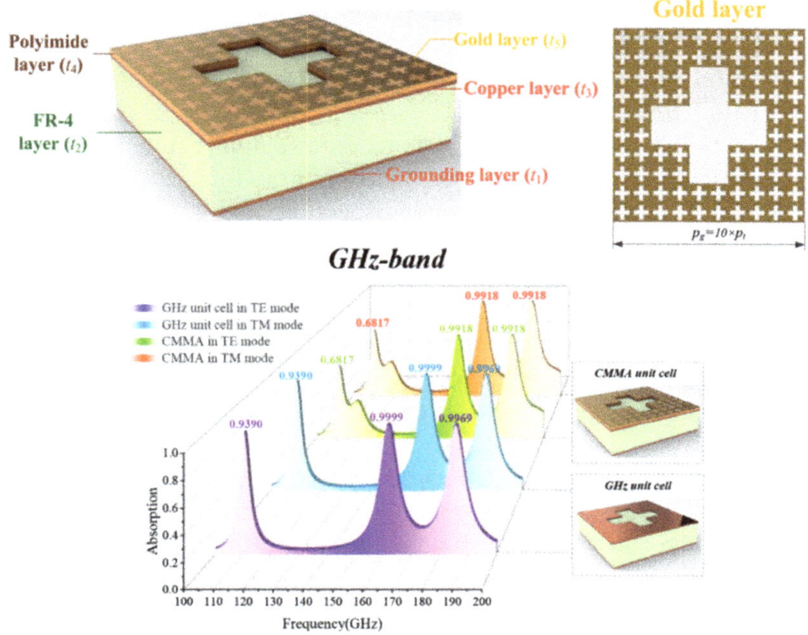

Figure 3.6. Multiband metamaterial absorber. Reprinted from [43], Copyright (2021), with permission from Elsevier.

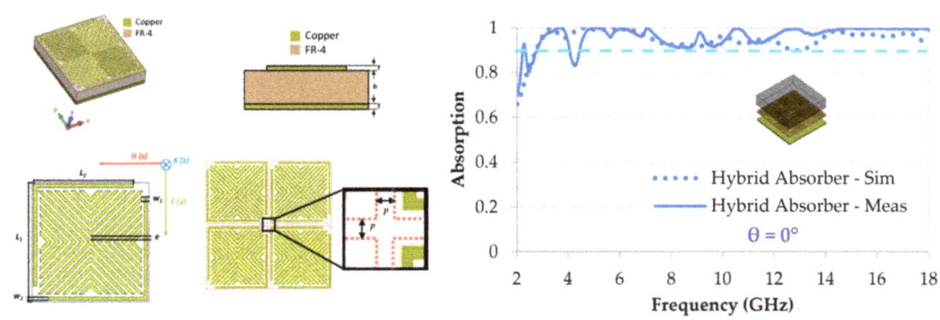

Figure 3.7. Hybrid MTM absorber design. Reproduced from [44]. CC BY 4.0.

In another metamaterial-based absorber design, Assal *et al* proposed an ultra-wide-band hybrid metamaterial absorber structure for 2–18 GHz frequency bands which covers the main communication bands [44]. This hybrid approach includes two different layers, a metallic metamaterial layer and a lossy dielectric foam layer. The foam is an epoxy composite with low weight percentage (0.075 wt.%) of 12 mm length carbon fibers. The absorber cell is composed of FR4 dielectric material and a metallic copper layer as illustrated in figure 3.7. This absorber is designed for incident EM wave absorption applications in the TEM mode. The interesting point of this study is the integratation of dielectric foam into a metallic absorber resonator. This designed absorber structure has multiband absorption characteristics with different absorption levels without dielectric foam integrations. The simulated and measured absorption characteristics are given in figure 3.7. It can be seen that a very good absorption is obtained between 2 and 20 GHz band for the transverse electromagnetic (TEM) mode wave.

In many microwave applications, systems require multiband absorption characteristics. Thummaluru *et al* proposed a symmetric four element metamaterial absorber structure to obtain a multiband absorption mechanism [45]. The designed novel metamaterial absorber integrates two different resonators as shown in figure 3.8. As seen, the absorption peaks reach 91%, 98.9%, and 99.5% absorptivity at 4.2, 7, and 7.4 GHz, respectively. In this design, an equivalent circuit design is also modeled and presented as shown in figure 3.8. The designed absorber structure and the absorption characteristics are also given in this figure. The interesting point of this design is that it can be adapted to other frequencies such as THz applications [45].

In another literature study, a 3D absorber approach has been examined by Chikhi *et al* [46]. As shown in figure 3.9, the pyramidal metamaterial absorber structure has been proposed. The aim of this approach is to demonstrate mode suppression in a real pillbox cavity by using a pyramidal metamaterial that acts as a lightweight, small volume damper at the operating frequencies between 3 and 4 GHz.

The unit cell design of the absorber structure, as mentioned earlier, is of great importance because the absorption mechanism mostly depends on the resonator design. As concluded, the geometric parameters of each resonator element provide an absorption for different frequency points [47]. The authors have explained that the advantage of the proposed design is the capability to absorb incident microwaves with a polarization independent characteristic. There are four band absorption peaks with a metamaterial absorber that has the absorption peaks at 3.46, 6.44, 7.89,

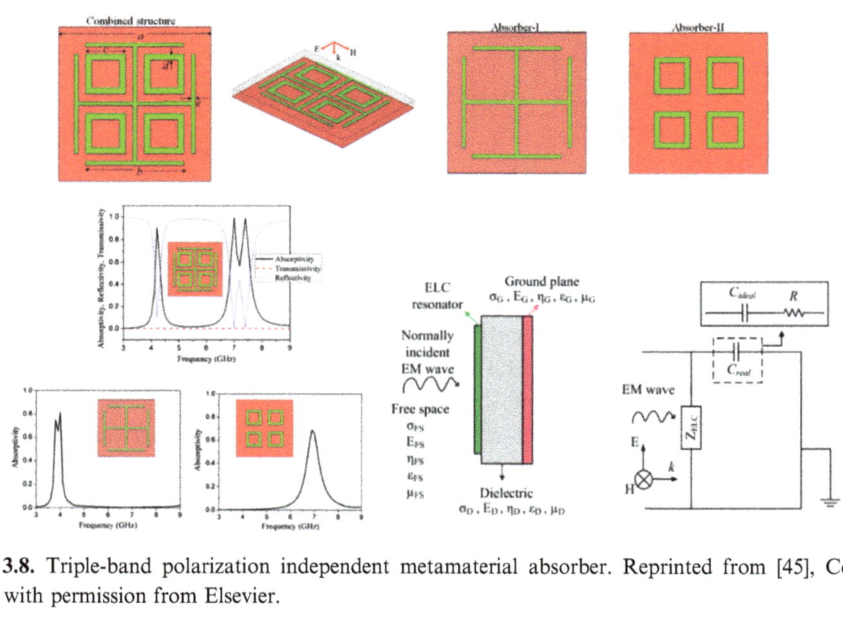

Figure 3.8. Triple-band polarization independent metamaterial absorber. Reprinted from [45], Copyright (2017), with permission from Elsevier.

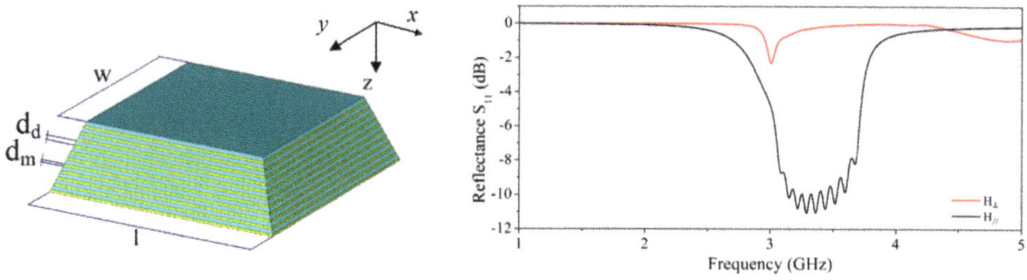

Figure 3.9. Ultra-thin broadband MTM absorber design. Reproduced from [46]. CC BY 4.0.

and 12.44 GHz with the absorption rates of 97%, 99.9%, 99.9%, and 99.6%, respectively. This absorber is designed for EMI shielding applications in TE mode. Each resonator section has splits to create a capacitive effect under resonance and the metallic lines correspond to inductive and resistive effects at resonant frequencies. The authors also examined the parametric analysis of the designed unit cell to see how the geometric parameters affect the absorption frequency and magnitude. In this study, it is also shown how to calculate the shielding effectiveness (SE) using S-parameters.

$$\mathrm{SE(dB)} = \mathrm{SE}_R + \mathrm{SE}_A \tag{3.3}$$

$$\mathrm{SE}_R = 10\log\left(\frac{1}{1 - S_{11}^2}\right) \tag{3.4}$$

$$\mathrm{SE}_A = 10\log\left(\frac{1 - S_{11}^2}{S_{12}^2}\right) \tag{3.5}$$

$$\mathrm{SE}(dB) = 10\log\left(\frac{1}{1 - S_{11}^2}\right) + 10\log\left(\frac{1 - S_{11}^2}{S_{12}^2}\right) \tag{3.6}$$

where SE_R and SE_A are the shielding effectiveness corresponding to reflection and absorption, respectively [47].

To provide optical transparency and flexibility, a cheap and easily available material has been chosen, 'water' [48]. As concluded, the water-based absorber structure has a periodic arrangement, and each unit cell is composed of a metallic back cover, water dielectric resonator, and resin top cover. The absorption characteristic has been calculated by using equation (3.2) because of the metallic back cover blocking the transmission. The permittivity and the loss tangent of the used resin material is 3.0 and 0.02, respectively. Furthermore, the dielectric constant of the water used in the simulation is calculated by the Debye formula. The unit cell has circular geometry that can create polarization independency at the resonance frequencies.

After the parametric analysis, the designed water-based absorber structure has been fabricated and the 3D fabrication process and water injection are realized during the fabrication process. The measured results are mostly the same as the simulated, with excellent absorption characteristics from 7.7 to 36.8 GHz with a minimum absorption rate of 90%. The aim of this structure can be focused to microwave energy harvesting, stealth technology, anti-radar systems, etc [48].

In another study, Stephen *et al* have proposed a bidirectional and bandwidth enhanced metamaterial absorber structure as illustrated in figure 3.10 [49]. The aim of this design is to absorb incident electromagnetic microwaves between the front and back sides. The four-fold rotational symmetry approach creates polarization independency for each incident wave from the front and back sides. Equation (3.1) is used to obtain the absorption characteristic because there is no fully metallic cover to block transmission to other side. As clearly shown in the absorption graph, the proposed structure has a good absorption rate between 13.6 and 14.2 GHz.

Bilal *et al* have proposed a split-ring resonator combined E-shaped fractal unit cell metamaterial for absorption applications, especially for mm wave applications including K and Ka bands [50]. All simulations and experiments are carried out for both TE and TM mode incident waves. In the design part, the back side is completely covered with a metallic copper layer that blocks the transmissions and ensures that the transmission is zero in equation (3.2). The simulated and measured absorption graphs with the designed absorber structure are given in figure 3.11.

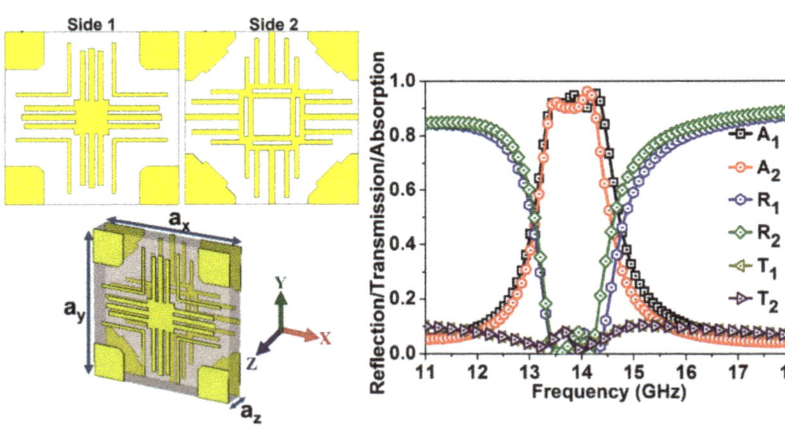

Figure 3.10. Ultra-thin broadband MTM absorber design. Reproduced from [49]. CC BY 4.0.

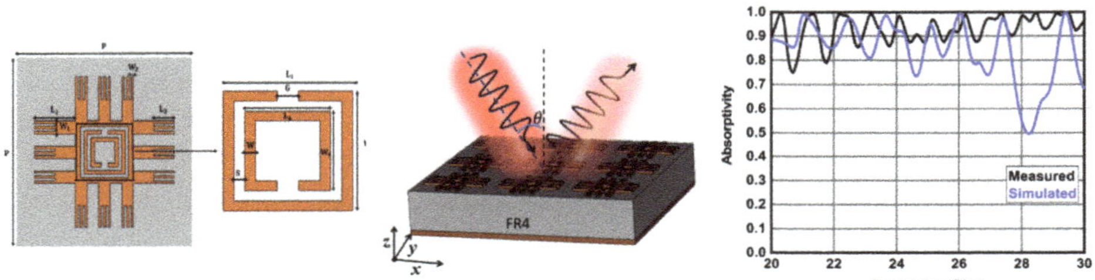

Figure 3.11. Ultra-thin broadband MTM absorber design. Reproduced from [50]. CC BY 4.0.

There is a good agreement between the simulated and measured results. Finally, this structure is a good candidate for 5G applications such as filtering, isolation, and attenuation applications [50].

There are many ways to design microwave absorbers, one of which is the Yagi–Uda approach, used by Bilal *et al* [51]. Figure 3.12 describes the designed unit cell of the multiband metamaterial absorber for X and Ku band applications. This periodic absorber has nearly unity absorption performances at 10.64, 12.08, and 14.09 GHz bands. This structure has wide incident angle stability and polarization controllability behavior. The interesting point of this design is the capability of polarization detection.

The absorption characteristic has been obtained with the reflection coefficient as S11 because of the fully metallic background. Equation (3.2) has been satisfied to obtain absorption as shown in figure 3.12. Moreover, the surface current distributions for TE mode polarization have been investigated to see the magnetic resonance between the adjacent arms of the designed Yagi–Uda structure. This structure is a good candidate for airborne and radar communication applications [51].

Moniruzzaman *et al* have proposed a quad band metamaterial absorber which is based on asymmetric circular split-ring resonator unit cells [52]. The designed unit cell and boundary conditions are given in figure 3.13, periodic boundary conditions are assigned for both x and y directions, and waveguide ports are set in the z direction as seen. This absorber has four resonances located in the C, X, and Ku bands. The FR4 dielectric material is used as substrate layer and copper metal is used for both the resonator and metallic ground layer. This absorber has absorption ratios of 97.9%, 99.1%, 99.5%, and 99.95% for 4.1, 6.86, 11.3, and 13.45 GHz, respectively, as shown in the figure. The unit cell structure has single negative behaviour with a Q factor greater than 20. The designed absorber structure has been fabricated as in the figure and measured in the microwave laboratory. As clearly shown in the figure, the measured and simulated results are in good agreement [52].

Hossain *et al* have proposed a double E-shaped symmetric split-ring resonator based metamaterial absorber for microwave EMI shielding applications at C and X band frequencies [53]. The designed unit cell of the proposed absorber is illustrated in figure 3.14. The design process is similar to previous studies, in line with equation (3.2), because of the fully copper back layer. As shown in the figure, this absorber

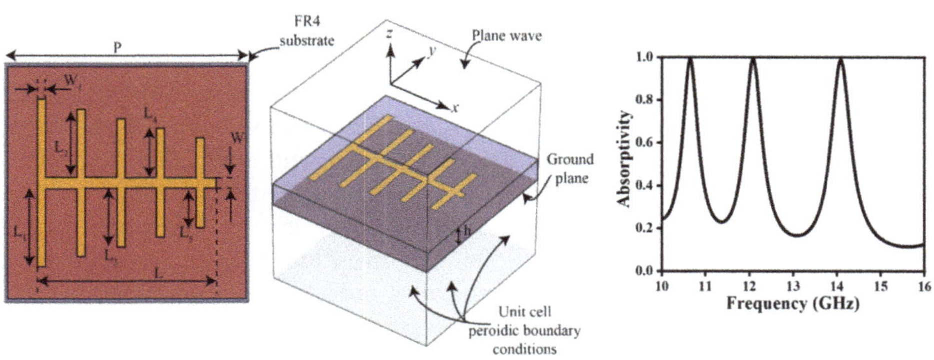

Figure 3.12. Ultra-thin broadband MTM absorber design. Reproduced from [51]. CC BY 4.0.

Figure 3.13. Ultra-thin broadband MTM absorber design. Reproduced from [52]. CC BY 4.0.

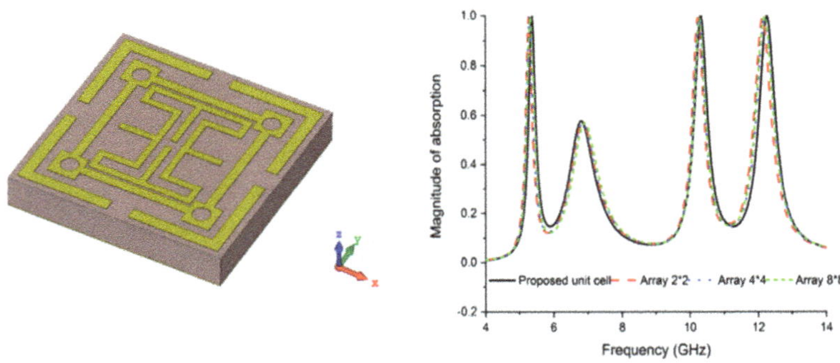

Figure 3.14. Ultra-thin broadband MTM absorber design. Reproduced from [53]. CC BY 4.0.

has three absorption peaks at 5.376, 10.32, and 12.25 GHz with the absorption rates of 99.9%, 99.9%, and 99.7%, respectively. This structure has been analyzed under four cases as a unit cell, 2×2, 4×4, 8×8 array. Each trial has the same absorption results as given in the figure ensuring that the absorption mechanism is not dependent on the array size. This metamaterial absorber structure has a triple band shielding capability with shielding effectiveness greater than 40 dB, which is a good ratio to reduce electromagnetic interference [53].

3.5 Flexible metamaterial absorber designs

Flexible metamaterial absorbers have emerged as a groundbreaking technology with immense potential in various applications. These absorber types are designed to efficiently mitigate microwave reflection and transmission across a wide range of frequencies while maintaining flexibility and adaptability. They have proven to be particularly useful in modern communication systems, stealth technology, radar systems, and even biomedical devices. The core concept behind these flexible metamaterial absorbers lies in their unique structure and composition. Typically, they consist of a flexible substrate, such as a thin polymer or elastomer sheet, incorporated with specially engineered metamaterial elements. These metamaterial elements are meticulously designed to exhibit extraordinary electromagnetic properties that can manipulate and control microwave radiation.

One of the key advantages of flexible metamaterial absorbers is their ability to conform to various surfaces and shapes. This feature makes them highly versatile for applications where traditional rigid materials would be impractical or limited. Their flexibility enables seamless integration into complex geometries, curved surfaces, or even wearable devices. Furthermore, these absorbers offer excellent performance and customization options. By adjusting the metamaterial's geometry, dimensions, and arrangement on the substrate, their absorption characteristics can be precisely tailored to match specific frequency bands or operational requirements. Flexible metamaterial absorbers are also known for their lightweight and low-profile nature, making them ideal for reducing the overall size and weight of devices and systems. Additionally, their thin and compact design ensures minimal interference with the surrounding environment. Moreover, these absorbers boast impressive broadband absorption capabilities, effectively tackling multiple frequencies within a given range. This broadband response is especially advantageous in scenarios where a wide spectrum of microwave frequencies needs to be absorbed efficiently.

In 2018, Kim *et al* developed a flexible ultra-thin metamaterial absorber which has wideband absorption characteristics [54]. The aim of this study is to design and develop a flexible absorber to cover a wide band characteristic between 0.87 and 6.11 GHz. The designed and fabricated metamaterial absorber structure is shown in figure 3.15 with absorption characteristics. Each unit cell contains different sized patch and microstrip elements. Finally, different bending angle configurations have been analyzed and presented as shown [54].

In another study, Gao *et al* proposed an optically transparent broadband metamaterial absorber structure for mm wave bands, especially radar-infrared applications [55]. The unit cell design steps and TE-TM mode results are given in figure 3.16. In the resonator layer, indium tin oxide (ITO) and frequency selective surface (FSS) combination is used. The first substrate layer is PET dielectric material with high optical transparency and the third one is an air layer. The fourth layer is a water layer which has high lossy factor to increase absorption and the ground layer is an ITO backplane which blocks the transmission. As shown in the simulation absorption results, the proposed metamaterial absorber has absorption over 90% in the 14.4 to 39.4 GHz range. The advantage of using a water layer is to avoid infrared radiation because of its high loss factor, which ensures infrared invisibility.

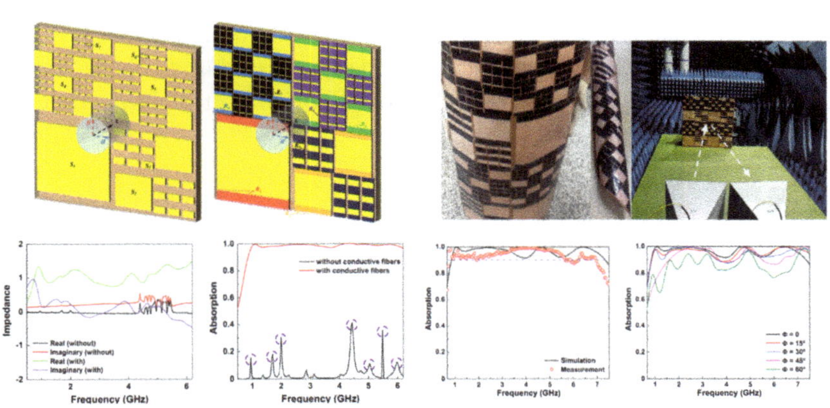

Figure 3.15. Flexible ultra-thin metamaterial absorber design. Reproduced from [54]. CC BY 4.0.

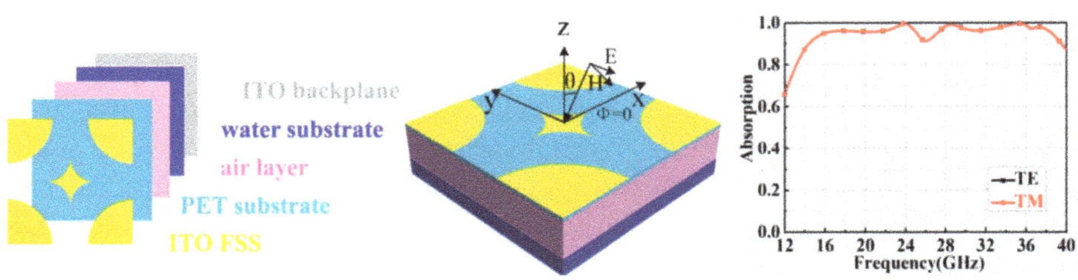

Figure 3.16. Ultra-thin broadband MTM absorber design. Reprinted from [55], Copyright (2021), with permission from Elsevier.

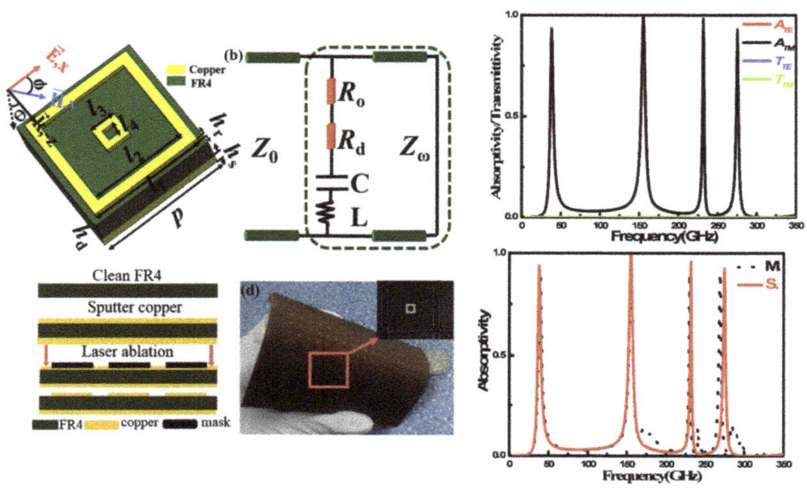

Figure 3.17. Four band metamaterial absorber. Reprinted from [56], Copyright (2017), with permission from Elsevier.

Another study presents a multiband flexible metamaterial absorber structure with its theory, design, simulation, fabrication, and performance analysis [56]. Figure 3.17 shows a designed unit cell and corresponding equivalent circuit. As shown in figure 3.17, the unit cell has two square rings with different sizes that creates four resonance peaks at 93.887%, 99.479%, 98.296%, and 93.035% at 38.165, 155.48, 231.71, and 275.27 GHz, respectively. Multiband characteristics are obtained by using only two resonator layers; this is the most important factor of this study [56].

3.6 Discussions

In this book chapter, we have explored the fascinating world of metamaterial absorbers in the microwave range, delving into their principles, design considerations, and diverse applications. Metamaterial absorbers have garnered significant attention in recent years due to their unique ability to manipulate electromagnetic waves, offering unprecedented control over absorption and reflection properties. As we conclude our exploration, we summarize the key findings and implications of this research. We have discussed the fundamental principles underpinning metamaterial absorbers, which leverage engineered subwavelength structures to achieve remarkable control over the interaction between incident electromagnetic waves and the absorber. By carefully designing the unit cell geometry and material properties, we can tailor the resonant behavior of the metamaterial absorber to match specific frequencies of interest, thereby enabling efficient absorption of electromagnetic energy.

The success of metamaterial absorbers hinges on meticulous design and optimization processes. We have examined various design parameters, such as unit cell geometry, thickness, and material properties, and their impact on the absorption characteristics. Techniques such as full-wave electromagnetic simulations and optimization algorithms play a crucial role in fine-tuning the absorber's performance, ensuring high absorption efficiency over a broad frequency range.

Throughout the chapter, we have showcased the diverse applications of metamaterial absorbers in microwave engineering. These absorbers find utility in radar systems, stealth technology, electromagnetic interference (EMI) shielding, and antenna performance enhancement. By effectively suppressing undesired reflections and reducing cross-polarization, metamaterial absorbers contribute to improved system performance and increased efficiency in a wide range of microwave applications. We have explored the emerging trend of multiband and broadband metamaterial absorbers, where unique designs enable simultaneous absorption at multiple frequencies or extended bandwidths. These absorbers open up new possibilities for complex applications, where a single device can efficiently absorb electromagnetic energy over a broad frequency spectrum, simplifying system design and reducing complexity. As metamaterial technology continues to evolve, novel concepts and materials emerge, offering even greater opportunities for absorber performance enhancement. However, challenges remain, such as fabrication techniques, material losses, and scalability for large-scale applications. Further research and collaboration across disciplines will be essential to address these challenges and unlock the full potential of metamaterial absorbers.

3.7 Future works

As we venture into the realm of metamaterial absorbers in the microwave range, the potential for groundbreaking research and technological advancements becomes evident. This book chapter has laid a solid foundation by exploring the principles, design considerations, and applications of metamaterial absorbers. Looking ahead, numerous exciting opportunities for future works arise, promising to further enrich this fascinating field of study. The following are key areas where future research and development can significantly contribute to the advancement and proliferation of metamaterial absorbers in the microwave range:

- Multi-functionality and tunability:

  One promising avenue for future research lies in exploring the potential for multi-functionality and tunability of metamaterial absorbers. Investigating ways to incorporate additional functionalities, such as polarization conversion or beam steering, into the absorber's design can lead to versatile and adaptable devices. Furthermore, dynamic tuning mechanisms that allow real-time adjustments to the absorption characteristics can enhance the absorber's performance across varying conditions and frequencies.

- Advanced material and fabrication techniques:
  Exploring novel materials and fabrication techniques is paramount to unlocking the full potential of metamaterial absorbers. Advanced manufacturing processes, such as additive manufacturing or nano-fabrication, hold promise in achieving greater precision and scalability, enabling the creation of complex and tailored absorber structures. Additionally, research into low-loss and high-temperature-resistant materials can enhance the overall efficiency and durability of metamaterial absorbers.
- Broadband and ultra-wide-band absorbers:
  Efforts should be directed towards designing and optimizing metamaterial absorbers capable of achieving broadband and ultra-wide-band absorption. Developing absorbers that can efficiently operate over an extensive frequency range will find applications in various wireless communication systems and electronic devices. Expanding the absorber's operational bandwidth will contribute to greater versatility and applicability in real-world scenarios.
- Metamaterial absorber arrays and configurations:
  Investigating the potential of metamaterial absorber arrays and configurations is a fruitful area for future exploration. By integrating multiple metamaterial absorber elements, researchers can achieve enhanced absorption efficiency and novel functionalities, such as spatial beam shaping and steering. This could lead to applications in adaptive antenna systems, efficient energy harvesting, and high-resolution imaging.
- Integrated metamaterial absorber systems:
  Looking ahead, integrating metamaterial absorbers with other components and systems holds great promise. For instance, combining absorbers with antennas, waveguides, or metamaterial lenses can create compact and efficient microwave devices. Such integrated systems can revolutionize microwave engineering, leading to miniaturized and highly functional devices for various applications.
- Practical applications and industry implementation:
  For metamaterial absorbers to have a widespread impact, further efforts should be dedicated to demonstrating their practicality and feasibility in real-world applications. Collaborations with industry partners can facilitate the transition of metamaterial absorbers from theoretical concepts to commercially viable products, fostering adoption in sectors such as aerospace, telecommunications, and defense.

Finally, the future of metamaterial absorbers in the microwave range holds immense promise, with a myriad of exciting research directions and practical applications on the horizon. By embracing multi-functionality, advanced materials, and integration with other microwave components, researchers can unlock new possibilities for these absorbers. As the field continues to evolve, interdisciplinary collaborations and a focus on practical implementation will be key to realizing the full potential of metamaterial absorbers in revolutionizing microwave engineering and shaping the future of wireless technologies.

3.8 Conclusions

This chapter presents metamaterial absorbers and their working mechanisms for microwave applications. The exploration of microwave frequency bands, which encompass a wide range from 2 to 300 GHz, has led us to a dedicated investigation into the fascinating world of metamaterial absorbers and their profound implications for microwave applications. At the heart of an absorber's functionality lies its primary objective: to efficiently absorb a substantial portion of the incident wave, exceeding 90%, while simultaneously inhibiting transmission and reflection at the specified operating frequency. Achieving such remarkable capabilities hinges on harnessing the power of metamaterial structures endowed with unique negative constitutive parameters, such as negative permittivity and negative permeability, effectively creating what is known as a double-negative (DNG) medium. This metamaterial integration significantly improves absorption ratios at the designated operating frequency, culminating in nearly unity absorption. A typical absorber cell configuration comprises a resonator layer positioned on top, a dielectric material substrate layer, and a fully metallic back cover, strategically engineered to quell transmission across most scenarios.

In the unit cell design, the absorber cell has a top layer as resonator, a substrate layer as dielectric material, and a fully metallic back cover to block transmission in most studies. The most important part of the design step is to model the resonator layer because each metallic element corresponds to an electrically small component. To further optimize the absorber's performance, each resonator cell is represented as an RLC microwave circuit and subjected to thorough analysis within a circuit simulator, facilitating the extraction of crucial S-parameters. Additionally, achieving polarization-insensitive absorption characteristics is a critical design criterion, which can be successfully realized through the implementation of symmetric geometry. In addition, some studies present optically transparent and flexible absorbers that use PET dielectric material as a substrate layer. Finally, a metamaterial absorber structure can be used in microwave regions in many applications such as medical, telecommunication, radar stealth, defense, EMI isolation, and beyond.

References

[1] Engheta N and Ziolkowski R W (ed) 2006 *Metamaterials: Physics and Engineering Explorations* (New York: Wiley)

[2] Ziolkowski R W 2003 Design, fabrication, and testing of double negative metamaterials *IEEE Trans. Antennas Propag.* **51** 1516–29

[3] Ozbay E, Aydin K, Cubukcu E and Bayindir M 2003 Transmission and reflection properties of composite double negative metamaterials in free space *IEEE Trans. Antennas Propag.* **51** 2592–5

[4] Chen T, Tang W, Mu J and Cui T J 2019 Microwave metamaterials *Ann. Phys.* **531** 1800445

[5] Bakır M, Karaaslan M, Unal E, Akgol O and Sabah C 2017 Microwave metamaterial absorber for sensing applications *Opto-Electron. Rev.* **25** 318–25

[6] Tao H, Strikwerda A C, Fan K, Padilla W J, Zhang X and Averitt R D 2009 Reconfigurable terahertz metamaterials *Phys. Rev. Lett.* **103** 147401

[7] He J, He X, Dong T, Wang S, Fu M and Zhang Y 2021 Recent progress and applications of terahertz metamaterials *J. Phys. D: Appl. Phys.* **55** 123002

[8] Jeong H, Nguyen T T and Lim S 2018 Subwavelength metamaterial unit cell for low-frequency electromagnetic absorber applications *Sci. Rep.* **8** 16774

[9] Alù A, Bilotti F, Engheta N and Vegni L 2007 Subwavelength, compact, resonant patch antennas loaded with metamaterials *IEEE Trans. Antennas Propag.* **55** 13–25

[10] El Sayed M, Ibrahim A, Mirjalili S, Zhang Y D, Elnazer S and Zaki R M 2022 Optimized ensemble algorithm for predicting metamaterial antenna parameters *Comput., Mater. Contin.* **71** 4989–5003

[11] Milias C, Andersen R B, Lazaridis P I, Zaharis Z D, Muhammad B, Kristensen J T and Hermansen D D 2021 Metamaterial-inspired antennas: a review of the state of the art and future design challenges *IEEE Access* **9** 89846–65

[12] Borazjani O, Naser-Moghadasi M, Rashed-Mohassel J and Sadeghzadeh R 2020 Design and fabrication of a new high gain multilayer negative refractive index metamaterial antenna for X-band applications *Int. J. RF Microw. Comput.-Aided Eng.* **30** e22284

[13] Vivek A, Shambavi K and Alex Z C 2019 A review: metamaterial sensors for material characterization *Sensor Rev.* **39** 417–32

[14] Yang J and Lin Y S 2021 Design of tunable terahertz metamaterial sensor with single-and dual-resonance characteristic *Nanomaterials* **11** 1–14

[15] Chen T, Zhang D, Huang F, Li Z and Hu F 2020 Design of a terahertz metamaterial sensor based on split ring resonator nested square ring resonator *Mater. Res. Express* **7** 095802

[16] Feng L, Huo P, Liang Y and Xu T 2020 Photonic metamaterial absorbers: morphology engineering and interdisciplinary applications *Adv. Mater.* **32** 1903787

[17] Mei H, Yang W, Yang D, Yao L, Yao Y, Chen C and Cheng L 2022 Metamaterial absorbers towards broadband, polarization insensitivity and tunability *Opt. Laser Technol.* **147** 107627

[18] Alkurt F O, Altintas O, Ozakturk M, Karaaslan M, Akgol O, Unal E and Sabah C 2020 Enhancement of image quality by using metamaterial inspired energy harvester *Phys. Lett.* A **384** 126041

[19] Karakaya E, Bagci F, Yilmaz A E and Akaoglu B 2019 Metamaterial-based four-band electromagnetic energy harvesting at commonly used GSM and Wi-Fi frequencies *J. Electron. Mater.* **48** 2307–16

[20] Sabban A 2020 New compact wearable metamaterials circular patch antennas for IoT, medical and 5G applications *Appl. Syst. Innov.* **3** 42

[21] Appasani B 2022 A hybrid terahertz metamaterial sensor using a hexagonal ring resonator with bio-medical applications *Plasmonics* **17** 519–24

[22] Sabban A 2019 Compact wearable meta materials antennas for energy harvesting systems, medical and IOT systems *Electronics* **8** 1340

[23] Abdeddaim R, Lecoq P and Enoch S 2019 Metamaterials: opportunities in medical imaging *Metamaterials XII* (SPIE) Vol 11025 pp 29–35

[24] Padilla W J and Averitt R D 2022 Imaging with metamaterials *Nat. Rev. Phys.* **4** 85–100

[25] Roh Y, Lee S H, Kwak J, Song H S, Shin S, Kim Y K and Seo M 2022 Terahertz imaging with metamaterials for biological applications *Sensors Actuators* B **352** 130993
[26] Gui B, Wang J, Zhang L, Zhu Y, Jia Y, Xu C and Qu S 2022 Design of scene-adaptive infrared camouflage emitter based on Au-VO2-Al2O3-Au metamaterials *Opt. Commun.* **512** 128016
[27] Hu R, Xi W, Liu Y, Tang K, Song J, Luo X and Qiu C W 2021 Thermal camouflaging metamaterials *Mater. Today* **45** 120–41
[28] He Y and Eleftheriades G V 2019 A thin double-mesh metamaterial radome for wide-angle and broadband applications at millimeter-wave frequencies *IEEE Trans. Antennas Propag.* **68** 2176–85
[29] Patel S K, Shah K H and Sonagara J S 2020 Broadband liquid metamaterial radome design *Waves Random Complex Medium* **30** 328–39
[30] Otsuji T, Boubanga-Tombet S A, Satou A, Yadav D, Fukidome H, Watanabe T and Ryzhii V 2022 Graphene-based plasmonic metamaterial for terahertz laser transistors *Nanophotonics* **11** 1677–96
[31] Wang Q, Raglione M, Li B, Jin X, Toor F, Arnold M and Ding H 2019 High throughput laser process of transparent conducting surfaces for terahertz bandpass ultrathin metamaterials *Sci. Rep.* **9** 3083
[32] Bashir I and Carley M 2020 Development of 3D boundary element method for the simulation of acoustic metamaterials/metasurfaces in mean flow for aerospace applications *Int. J. Aeroacoust.* **19** 324–46
[33] Li Z, Zhai W, Li X, Yu X, Guo Z and Wang Z 2022 Additively manufactured dual-functional metamaterials with customisable mechanical and sound-absorbing properties *Virtual Phys. Prototyp.* **17** 864–80
[34] AbuHussain M and Hasar U C 2020 Design of X-bandpass waveguide Chebyshev filter based on CSRR metamaterial for telecommunication systems *Electronics* **9** 101
[35] Marcelli R, Capoccia G, Sardi G M, Bartolucci G, Margesin B, Iannacci J and Proietti E 2022 Metamaterials based RF microsystems for telecommunication applications *Ceram. Int.* **49** 24379–89
[36] Kalusniak S, Orphal L and Sadofev S 2015 Demonstration of hyperbolic metamaterials at telecommunication wavelength using Ga-doped ZnO *Opt. Express* **23** 32555–60
[37] Butcher G 2016 *Tour of the Electromagnetic Spectrum* (Government Printing Office)
[38] Bilal R M H, Baqir M A, Iftikhar A, Ali M M, Rahim A A, Akhtar M N and Naqvi S A 2021 A novel omega shaped microwave absorber with wideband negative refractive index for C-band applications *Optik* **242** 167278
[39] Hannan S, Islam M T, Soliman M S, Sahar N B M, Singh M S J, Faruque M R I and Alzamil A 2022 A filling-factor engineered, perfect metamaterial absorber for multiple applications at frequencies set by IEEE in C and X bands *J. Mater. Res. Technol.* **19** 934–46
[40] Zhang Z, Zhang L, Chen X, Wu Z, He Y, Lv Y and Zou Y 2020 Broadband metamaterial absorber for low-frequency microwave absorption in the S-band and C-band *J. Magn. Magn. Mater.* **497** 166075
[41] Chen K, Luo X, Ding G, Zhao J, Feng Y and Jiang T 2019 Broadband microwave metamaterial absorber with lumped resistor loading *EPJ Appl. Metamater.* **6** 1
[42] Wang Q and Cheng Y 2020 Compact and low-frequency broadband microwave metamaterial absorber based on meander wire structure loaded resistors *AEU-Int. J. Electron. Commun.* **120** 153198

[43] Luo Z, Ji S, Zhao J, Wu H and Dai H 2021 A multiband metamaterial absorber for GHz and THz simultaneously *Res. Phys.* **30** 104893

[44] El Assal A, Breiss H, Benzerga R, Sharaiha A, Jrad A and Harmouch A 2020 Toward an ultra-wideband hybrid metamaterial based microwave absorber *Micromachines* **11** 930

[45] Thummaluru S R, Mishra N and Chaudhary R K 2017 Design and analysis of an ultrathin triple-band polarization independent metamaterial absorber *AEU-Int. J. Electron. Commun.* **82** 508–15

[46] Chikhi N, Passarelli A, Andreone A and Masullo M R 2020 Pyramidal metamaterial absorber for mode damping in microwave resonant structures *Sci. Rep.* **10** 19352

[47] Hossain M B, Faruque M R I and Islam M T 2023 Double elliptical resonator based quadruple band metamaterial absorber for EMI shielding applications in microwave regime *Alex. Eng. J.* **69** 193–206

[48] Du L, Shi T, Dong S, Wang X, Zhou M, Zhao J and Huang X 2022 Ultra broadband microwave metamaterial absorber with multiple strong absorption peaks induced by sandwiched water resonators *Appl. Phys.* A **128** 864

[49] Stephen L, Yogesh N and Subramanian V 2019 Realization of bidirectional, bandwidth-enhanced metamaterial absorber for microwave applications *Sci. Rep.* **9** 1–9

[50] Bilal R M H, Baqir M A, Choudhury P K, Karaaslan M, Ali M M, Altıntas O and Sabah C 2021 Wideband microwave absorber comprising metallic split-ring resonators surrounded with e-shaped fractal metamaterial *IEEE Access* **9** 5670–7

[51] Bilal R M H, Baqir M A, Iftikhar A, Naqvi S A, Mughal M J and Ali M M 2022 Polarization-controllable and angle-insensitive multiband Yagi-Uda-shaped metamaterial absorber in the microwave regime *Opt. Mater. Express* **12** 798–810

[52] Moniruzzaman M, Islam M T, Muhammad G, Singh M S J and Samsuzzaman M 2020 Quad band metamaterial absorber based on asymmetric circular split ring resonator for multiband microwave applications *Res. Phys.* **19** 103467

[53] Hossain M B, Faruque M R I, Islam M T, Singh M and Jusoh M 2022 Triple band microwave metamaterial absorber based on double E-shaped symmetric split ring resonators for EMI shielding and stealth applications *J. Mater. Res. Technol.* **18** 1653–68

[54] Kim Y J, Hwang J S, Khuyen B X, Tung B S, Kim K W, Rhee J Y and Lee Y 2018 Flexible ultrathin metamaterial absorber for wide frequency band, based on conductive fibers *Sci. Technol. Adv. Mater.* **19** 711–7

[55] Gao Z, Fan Q, Tian X, Xu C, Meng Z, Huang S and Tian C 2021 An optically transparent broadband metamaterial absorber for radar-infrared bi-stealth *Opt. Mater.* **112** 110793

[56] Zhang Y, Duan J, Zhang B, Zhang W and Wang W 2017 A flexible metamaterial absorber with four bands and two resonators *J. Alloys Compd.* **705** 262–8

IOP Publishing

Advanced Metamaterials for Engineers

Lulu Wang and Muharrem Karaaslan

Chapter 4

Dual-band terahertz metamaterial absorber with high sensitivity for sensing applications

Musa N Hamza, Yadgar I Abdulkarim and Rashid Mirzavand

Symbols

R	Reflection coefficient
T	Transmission coefficient
S_{11}	S-parameters for reflection
S_{21}	S-parameters for transmission

Abbreviations

Al	Aluminum
PET	polyethylene terephthalate
MTMAs	Metamaterial absorbers
E-field	electric field
H-field	magnetic field
MTMs	Metamaterials
MTMPAs	Metamaterial-Based Perfect Absorbers
CAs	circuit-analog absorbers
EM	electromagnetic
IR	infrared
TSRRs	three-split-ring resonators
FIT	finite integration technique
CST	computer simulation technology
PEC	perfect electric conductor
PMC	perfect magnetic conductor
TLT	transmission-line theory
FR4	Flame Retardant type 4
SNR	signal-to-noise ratio

doi:10.1088/978-0-7503-5754-8ch4

A novel terahertz metamaterial absorber design detects changes in refractive index, enabling narrowband absorption for potential sensor applications. The suggested structure is completely capable of absorbing energy in two distinct bands, which are around 0.437 and 0.805 THz, respectively. The design showed two intense absorption peaks of more than 99.8%. Aluminum (Al) metal is employed as the top and bottom layers of the model's three-layered construction, with a polyethylene terephthalate (PET) dielectric spacer serving as the intermediate layer. Al had a conductivity of 3.56×10^7 S m^{-1}, and the dielectric spacer was 10 μm thick. The Al layers were 0.2 μm thick on top and bottom. The dual-band absorption peaks move when the refractive index changes and the suggested sensor is a sensitive metamaterial absorber (MTMA) for studying matter–wave interaction. Further investigation of the E-field, H-field, polarization angle, surface current distribution, electric and magnetic energy density, power flow, and surface power loss density was conducted to elucidate the absorption properties at various resonance frequencies.

4.1 Introduction

Metamaterials (MTMs) are periodic sub-wavelength structures that can be used to alter the laws of optics and electromagnetics [1], allowing for a wide range of applications in biosensing, imaging, antenna engineering [2, 3], and optoelectronics [4]. Metamaterial absorbers (MTMAs) convert input waves into thermal energy, allowing them to be sensed by identifying the absorption [5, 6]. MTM research is important for solving material issues in the THz frequency regime due to the lack of classical electronic and quantum photonic responses. A designed and controllable MTMs resonant response would improve interactions between THz radiation and materials, leading to novel functionalities and unprecedented performance [7, 8]. THz offers wide bandwidth, low scattering, and non-ionizing properties. Metamaterial-based perfect absorbers (MTMPAs) are similar to circuit-analog absorbers (CAs) in that they use a layer of periodically arranged metallic patterns, a dielectric layer, and a continuous metallic layer. High-ε material is preferred to reduce the thickness of the dielectric layer while maintaining the optical path of the incident electromagnetic (EM) wave. The imaginary part of MTMPAs is important for absorption of incident EM waves, while the dielectric layer plays an additional role depending on the design. MTMPAs have two absorption mechanisms: dielectric loss and ohmic loss. Dielectric loss is the dissipation of energy from EM waves in the interior of a dielectric medium, while ohmic loss is the energy dissipation in an electrical circuit. Electric circuits typically heat up as a result of the two losses. Metals are very lossy in the optical range due to inter-band transitions, but in the lower frequency range, especially in the GHz range, metals are usually good conductors with little loss. This is due to the energy difference between the occupied and empty levels [9]. According to the assumption that the permittivity of the majority of metals follows the Drude model and that the loss fraction is of the order of the skin depth to the wavelength, the estimated fraction of ohmic loss in a THz perfect transmitter is 0.1% or less [10]. Particularly in the infrared (IR) and higher frequency ranges, the ohmic loss begins to take precedence over the dielectric loss as frequency rises. THz sensors study nanoscale biological or chemical samples for precise structural behavior measurement using MTMs' extraordinary features, enabling discrimination and precise measurement in this frequency regime. The sensor detects refractive index variations in the medium, monitoring the shift in the dual-band absorption peak.

A novel sensor is presented in this chapter, for which a complex geometric design has been created. Also, how the proposed design is presented with several different models suggests that the sensor was employed to find the shift in refractive index. The proposed sensor is the result of extensive research on all of the critical factors needed to choose a very good sensor, with the findings being provided, such as polarization angle, electric and magnetic field distributions, surface current distribution, electric and magnetic energy density, power flow, and surface power loss density.

4.2 The unit cell model's design

Five models based on metamaterials are presented in this chapter, as seen in figures 4.2 and 4.3. The first model, as shown in figure 4.1(a), and the second model, as shown in figure 4.1(b), are combined to create the suggested metamaterial, as seen

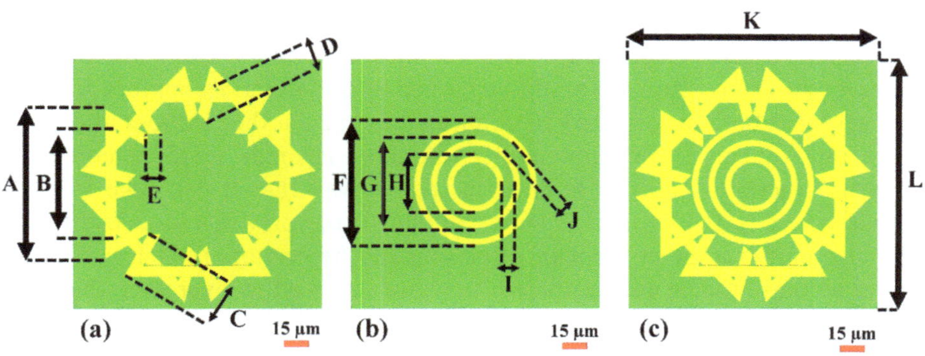

Figure 4.1. A perfect absorber's suggested structural design: (a) model 1, (b) model 2, and (c) model 3 (proposed design).

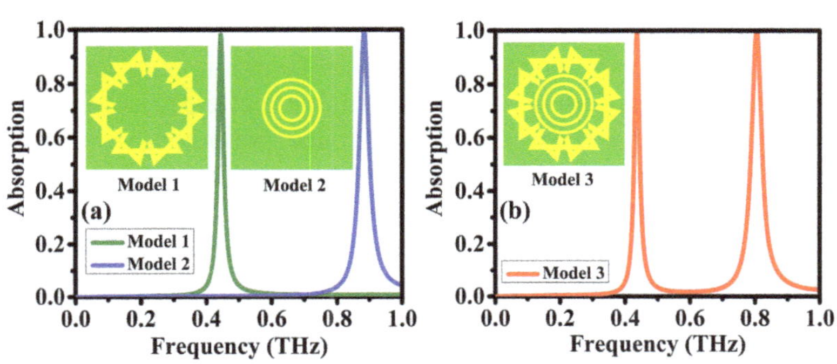

Figure 4.2. The three designs' absorption traits are as follows: (a) models 1 and 2; and (b) model 3.

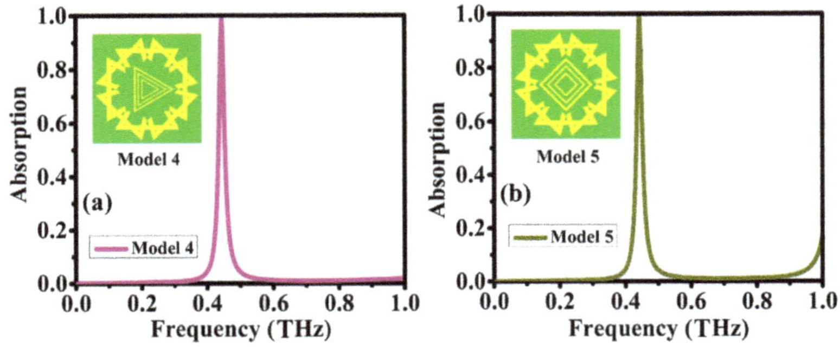

Figure 4.3. The absorption characteristics of the other two designs are as follows: (a) model 4; and (b) model 5.

Table 4.1. The suggested prototype's whole set of optimized parameters.

A	B	C	D	E	F	G	H	I	J	K	L
90 μm	65.8 μm	30 μm	21.9 μm	8 μm	74 μm	54 μm	34 μm	6 μm	5 μm	150 μm	150 μm

in figure 4.1(c). This combination was chosen in order to produce two bands in the terahertz oscillator range. The suggested model consists of three layers, each of which has an excellent conductor on both sides and a dielectric substance in the middle. As shown in figure 4.2(b), the suggested metamaterial resonators are built in such a way that both models may be connected to one another without affecting the fundamental outcome. The first model resonator is made up of a complicated geometric structure that is positioned in such a way that it can take in electromagnetic waves in the terahertz region. Similar to the previous model, the second model also employs the concept of three-split-ring resonators (TSRRs). When the first and second models are combined, they are designed to overcome the overlap and interaction of both electric and magnetic fields. Table 4.1 provides a summary of the models' directional information.

A commercial full-wave finite integration technique (FIT) based on the high-frequency electromagnetic solver computer simulation technology (CST) microwave studio was used to create the suggested absorber. Calculations based on numbers may be used to identify the MTMs' properties and illustrate how MTMs' structures behave at various frequency ranges and boundary conditions. Scientists have used numerical analysis to analyze structures such as perfect electric conductor (PEC)/ perfect magnetic conductor (PMC), PEC, free space, periodic arrays, and unit cells. To simplify the simulation processes, a unit cell was assigned in the *x*- and *y*-directions and an open add space in the *z*-direction. Aluminum (Al) metal is employed as the top and bottom layers of the model's three-layered construction, with a polyethylene terephthalate (PET) dielectric spacer serving as the intermediate layer. Al had a conductivity of 3.56×10^7 S m^{-1}, and the dielectric spacer was 10 μm thick. The Al layers were 0.2 μm thick on the top and bottom. The bottom Al layer blocks all incident electromagnetic (EM) waves, providing zero impedance for transmission-line theory (TLT) interpretation, while the top metal layer was designed to match its impedance with the incident medium, allowing maximum power to penetrate and propagate in the PET. High electrical and/or magnetic losses are required for absorption systems to capture moving waves, and the metallic layer at the bottom blocks transmission. By using an EM wave impinging on the top plane, absorption characteristics were obtained.

4.3 Results and analysis

CST studio was used to design and simulate the sensor, with the ground plane preventing transmission through the model and absorption given by the expression.

$$\boldsymbol{A} = \mathbf{1} - \boldsymbol{R} - \boldsymbol{T} = \mathbf{1} - |S_{11}|^2 - |S_{21}|^2 \tag{4.1}$$

where the S-parameters for reflection and transmission, respectively, are denoted by S_{11} and S_{21}. As a result, to achieve maximal absorption, the reflection coefficient R should be minimized as much as possible.

Five models are shown in figures 4.2 and 4.3 to examine the effect of resonator shape on the absorption capacity in the terahertz band. According to figure 4.2(a), models one and two each feature a single full absorption zone that spans around 0.437 and 0.894 THz, respectively. According to figure 4.2(b), the third model, which is the one that has been suggested, is completely capable of absorbing energy in two distinct bands, which are around 0.437 and 0.805 THz, respectively. The design showed two intense absorption peaks of more than 99.8%.

Three hollow triangles with the same center make up the fourth model. According to figure 4.3(a), it has full absorption capability in a range of around 0.45 THz. Three hollow rhombuses with the same center make up the final model. Similar to the previous model, figure 4.3(b) shows that it can only absorb in a single range, around 0.45 THz.

Another study that has been presented deals with altering the metamaterial's substrate material using various dielectric materials, including PET, Arlon AD 410, Arlon AD 430, FR4, and Rogers RT5780. Figure 4.4(a) displays the outcomes. Al is used as the resonator material in this scenario. Interestingly, FR4 performed poorly compared to all the other materials. The resonators also utilized the same concept of material modification by stabilizing PET, the material used as the substrate. Aluminum, copper, silver, iron, and gold were the materials utilized. According to the findings, there is not much of a difference, as shown in figure 4.4(b).

Polarization can lead to a low signal-to-noise ratio (SNR) in sensing and imaging applications, distorting the results. The suggested structure's symmetry renders the sensor polarization independent [11, 12]. This is seen in figure 4.5, where the absorption properties barely alter with a change in the angle of polarization (I), preventing any obstruction.

For the suggested model, incidence angles ranging from 0° to 90° were used. According to figure 4.5(a), both extreme absorption areas are stated to have remained unchanged for various angles. For polarization, the same approach as before was used. The conclusions demonstrate a change between the zero angle and

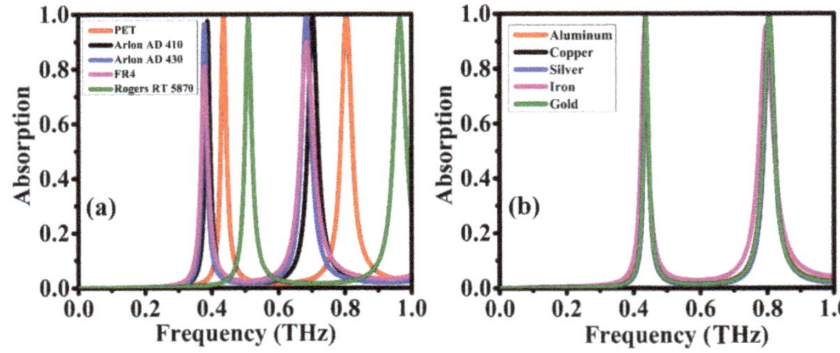

Figure 4.4. Absorption spectra of the suggested design under various (a) substrate material and (b) resonator material conditions.

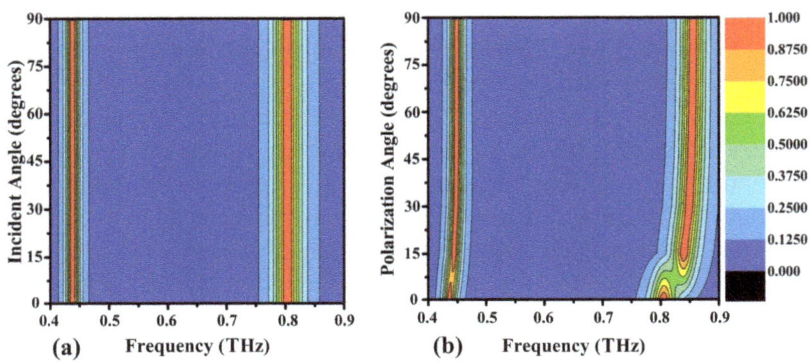

Figure 4.5. Investigation of the impact of angle variation on the absorption rate: (a) incident angle, and (b) polarization angle.

other angles. However, after an angle of 15° up to 90°, there is minimal variation in both entire absorber areas, as shown in figure 4.5(b). This shows that the recommended model performs consistently over a variety of incidence angles and polarizations. The stability of the absorber regions across a variety of angular conditions shows that the model can efficiently absorb and convert incident energy regardless of the orientation or direction of the incoming waves.

The first resonator (the geometric form) and the radius of the TSRRs have both been altered for further study if we obtain better results, as shown in figure 4.6(a). The first case of the suggested model has the best outcomes when compared to the other situations, as demonstrated by the seven cases offered for this purpose. The case summaries are explained in table 4.2. Investigations were also done on how the index of refraction affected absorption (table 4.3). In the numerical program, we entered 25 μm as the thickness of the analyte layer. An increased refractive index caused the resonance frequency value to move to the left. The absorption values of the suggested structure were only slightly changed. As seen in figure 4.6(b), the absorption for the first peak decreased as the refractive index increased. As a result, the design can aid in perceiving the surrounding environment's refractive index.

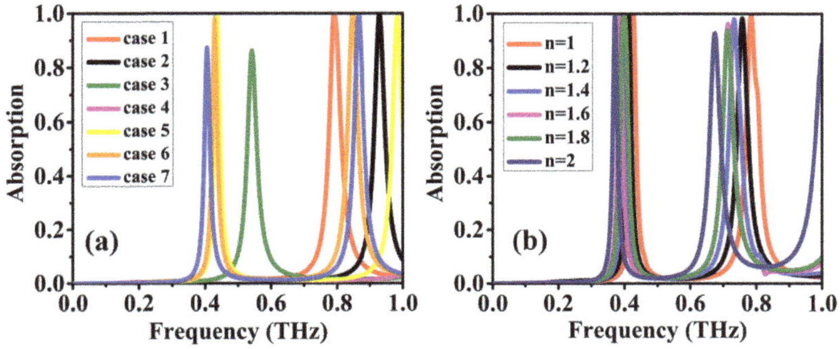

Figure 4.6. Examining the design's absorption spectra under different conditions: (a) resonator lengths, (b) absorption spectrum depends on the refractive index variation in the environment.

Table 4.2. The suggested prototype's optimized parameters.

Case number		Case 1	Case 2	Case 3	Case 4	Case 5	Case 6	Case 7
First shape radius (μm)	Outer	52	50	50	50	50	50	50
	Inner	38	40	40	40	40	40	45
First ring radius (μm)	Outer	37	35	40	30	33	37	37
	Inner	32	30	35	25	30	32	33
Second ring radius (μm)	Outer	27	25	30	20	23	27	27
	Inner	22	20	25	15	20	22	23
Third ring radius (μm)	Outer	17	15	20	10	13	17	17
	Inner	12	10	15	5	10	12	13

Table 4.3. A comparison of the proposed sensor to research perfect metamaterials in the terahertz range.

References	Techniques used	Frequency operating THz	Material substrate	Absorptivity
[13]	Bulk Dirac semimetal/ photonic crystal/Au	1–3	Photonic crystal plate	0.97, 0.98, 0.99
[14]	Graphene/topas//Au	0.5–4.5	Topas spacer	0.99, 0.98, 0.99
[15]	Au/dielectric Teflon/Au	1–2.2	Dielectric Teflon	0.99
[16]	Au/dielectric layer/Au	1–3	Dielectric layer	0.99, 0.99
[17]	Graphene/Au/SiO_2/Au	2–6	SiO_2	0.99
This work	Al/PET/Al	0–1	PET	0.9996, 0.9986

Using the suggested sensor as a refractive index sensor for the first resonance unstably affects the sensor's absorption capacity, as illustrated in figure 4.7(a). However, increasing the refractive index of the material generates a considerable variation in the first resonance, as shown in figure 4.7(b), which is a decreasing curve in the sense that the shift of the curve is to the left, i.e., decreasing.

The second resonance was also studied, and the results demonstrated that the sensor's absorbency varies gradually, as shown in figure 4.8(a). With an increase in refractive index, the absorption dropped. Additionally, altering the material's refractive index results in a significant variation in the first resonance, as seen in figure 4.8(b). That is a diminishing curve, so the second resonance moves to the left and its resonance point diminishes.

To further clarify the resonance mechanism, electric and magnetic field distributions are provided in figures 4.9 and 4.10. The component that causes the initial signal resonance at 0.437 THz is depicted in figure 4.9(a). Figure 4.9(b), which depicts how the electric field in TSRRs is much stronger than other places, reveals which component is responsible for the initial signal resonance at 0.805 THz. These field distributions aid in understanding design performance and offer insights into the resonance process. Analyzing the electric and magnetic field strengths at various frequencies makes it clear how particular elements contribute to signal resonances at various frequencies. This characterization and performance analysis help optimize

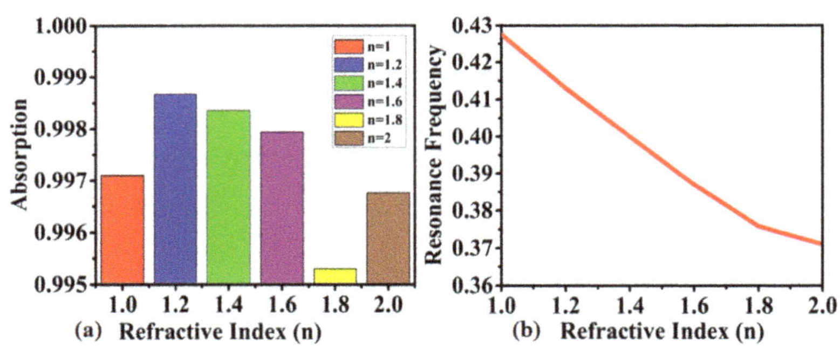

Figure 4.7. First resonance at 0.437 THz: (a) absorption ratio, (b) frequency shift of the resonance.

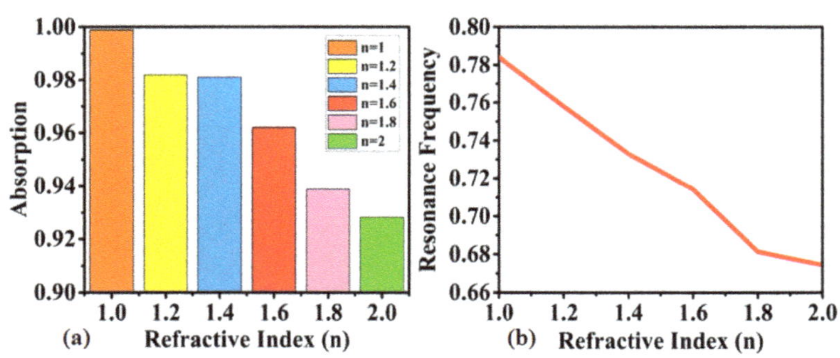

Figure 4.8. Second resonance at 0.805 THz: (a) absorption ratio, (b) frequency shift during resonance.

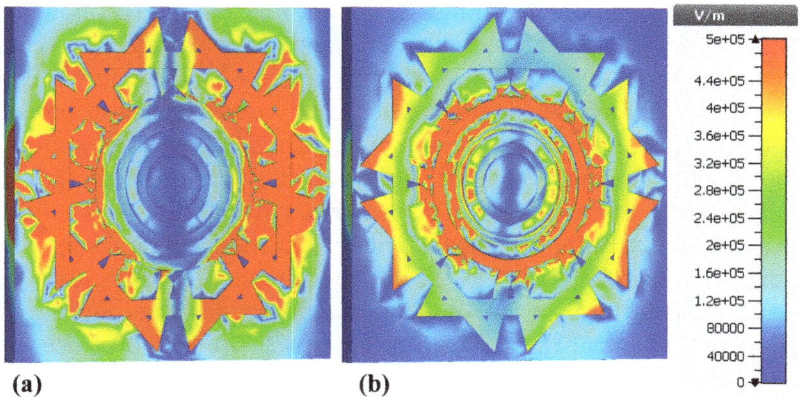

Figure 4.9. Field distributions of the suggested metamaterial structure are depicted in a color map: (a) electric field at 0.437 THz, (b) electric field at 0.805 THz.

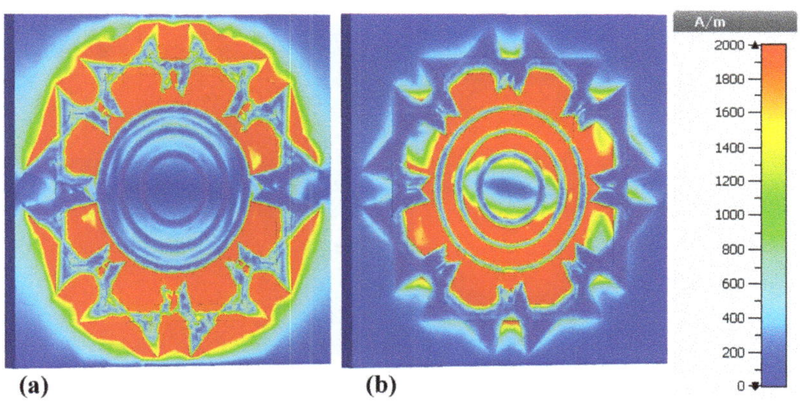

Figure 4.10. A color map shows the hypothesized metamaterial structure's field distributions: (a) magnetic field at 0.437 THz, and (b) magnetic field 0.805 THz.

the design for targeted resonance frequencies and enhance device performance as a whole.

Figure 4.10(a) shows the element responsible for the 0.437 THz initial signal resonance. The component that causes the initial signal resonance at 0.805 THz is shown in figure 4.10(b), which shows how the magnetic field is significantly greater in TSRRs than it is elsewhere. To fully comprehend the behavior and efficiency of THz metamaterials as ideal absorbers, characterization and performance analysis are essential. These graphs offer important new understandings of the precise factors that affect signal resonance at various THz frequencies. In order to achieve the necessary absorption qualities in this frequency range, researchers can further refine the design and manufacturing procedures by examining these components.

The distribution of the current flows, whether they are parallel or antiparallel, is what is responsible for the distribution of the electric and magnetic fields in the aforementioned plots. The internal magnetic field that the parallel flow can create will be in opposition to the external magnetic field (H-field), and vice versa.

The surface current distribution of the top layers of the suggested design was investigated in order to better comprehend the physical absorption process of the proposed metamaterial absorbers. Antiparallel current flows in a circular pattern, suggesting a potent magnetic reaction. Additionally, figure 4.11(a) shows the top layer's parallel and antiparallel surface current patterns for 0.437 THz, which demonstrate a magnetic dipole resonance mode. On the metallic layers, the surface currents might occasionally flow in the reverse direction. The incident H-field is coupled with a magnetic flux as a result. The current distribution for the second resonance mode at 0.805 THz is shown in figure 4.11(b). The presence of circular antiparallel current flows suggests a powerful magnetic reaction.

Another aspect under study is the distribution of electrical energy density over the proposed design's size. The electrical energy density at 0.437 THz demonstrates that the distribution of electrical energy is parallel. This suggests that the red area in figure 4.12(a) has excellent wave suppression and annihilation capabilities.

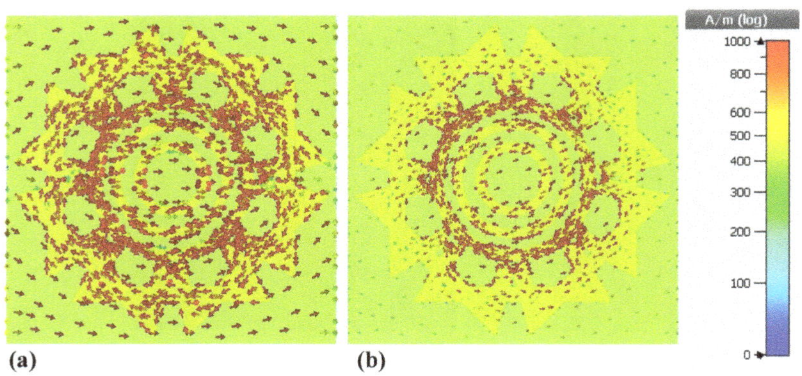

Figure 4.11. The suggested metamaterial design's surface current distribution (a) at 0.437 THz, (b) at 0.805 THz.

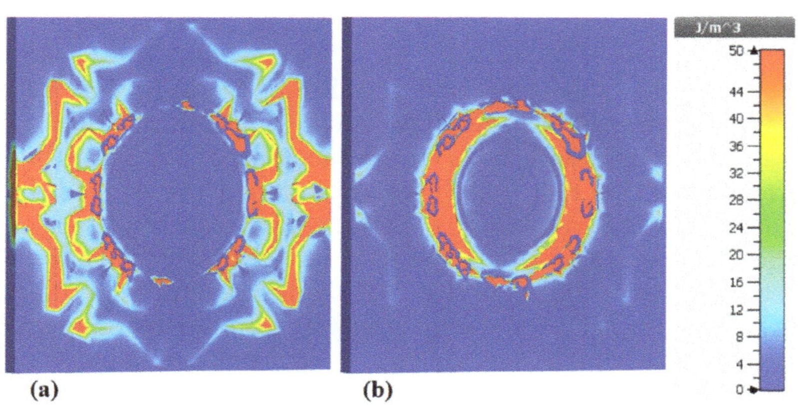

Figure 4.12. Electric energy density at: (a) 0.437 THz, (b) 0.805 THz.

Additionally, figure 4.12(b) demonstrates that the electrical energy density is split between the right and left sides of the TSRRs, which is what causes the electromagnetic wave at 0.805 THz to be entirely destroyed. The TSRRs have completely concentrated the electrical energy at this frequency.

How the magnetic energy density is dispersed across the proposed design volume is another factor that is being looked into. According to the magnetic energy density at 0.437 THz, the magnetic energy distribution over the top and lower components of the proposed design is distributed parallel to one another. Accordingly, it can be concluded that the reddish region in figure 4.13(a) possesses outstanding wave suppression and annihilation qualities. The top and bottom portions of the TSRRs, which are responsible for completely annihilating the electromagnetic wave at 0.805 THz, are shown in figure 4.13(b) to be divided in terms of magnetic energy density. The magnetic energy has been entirely focused at this frequency by the TSRRs. The electric and magnetic energy density distributions have a 90-degree gradient difference, as seen in figures 4.12 and 4.13.

At the suggested sensor's surface area, power flow was investigated. At the resonance point of 0.437 THz, power flow is virtually circularly distributed throughout the surface area of the proposed sensor, as shown in figure 4.14(a). Additionally, a highly dense zone with a resonance point of 0.805 THz can be seen in figure 4.14(b), showing substantial power flow that was produced by TSRRs in complete circular form. This circular distribution of power shows that the suggested sensor design efficiently catches and uses the incoming electromagnetic waves. The high density of power flow at the resonance frequency of 0.805 THz indicates that the sensor is sensitive to signals in this frequency range.

Surface power loss density is another crucial factor needed to recognize and assess a metamaterial as a perfect absorber. The efficiency and efficacy of a metamaterial in absorbing THz radiation may be assessed by researchers by examining the surface power loss density. It also enables a comparison of various designs and aids in the discovery of prospective absorber performance enhancements. The area of electromagnetic wave suppression and annihilation in the THz range is located near the

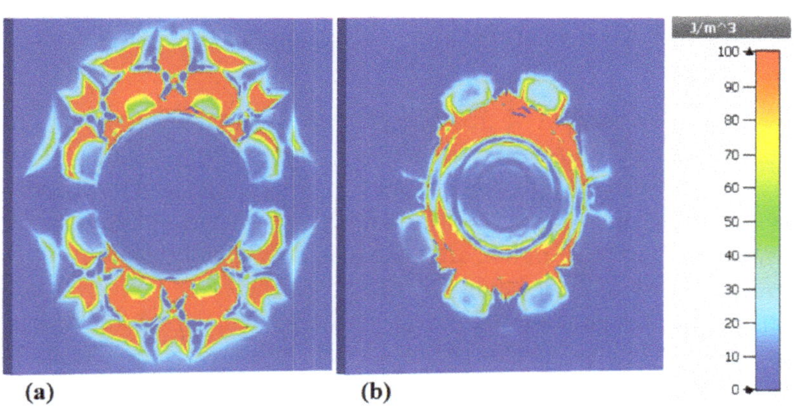

Figure 4.13. Magnetic energy density at: (a) 0.437 THz, (b) 0.805 THz.

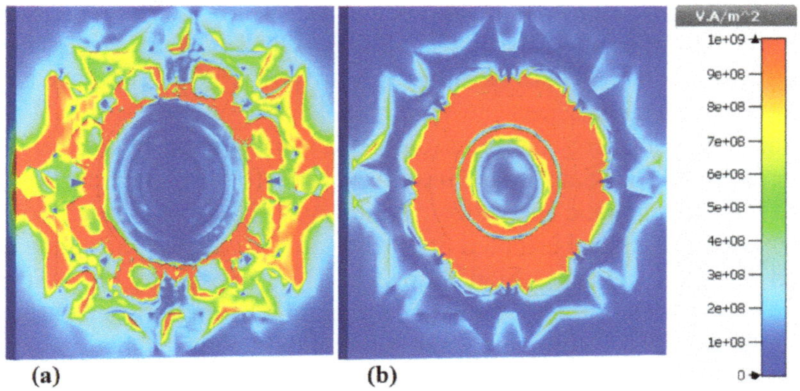

Figure 4.14. Power flow at (a) 0.437 THz, (b) 0.805 THz.

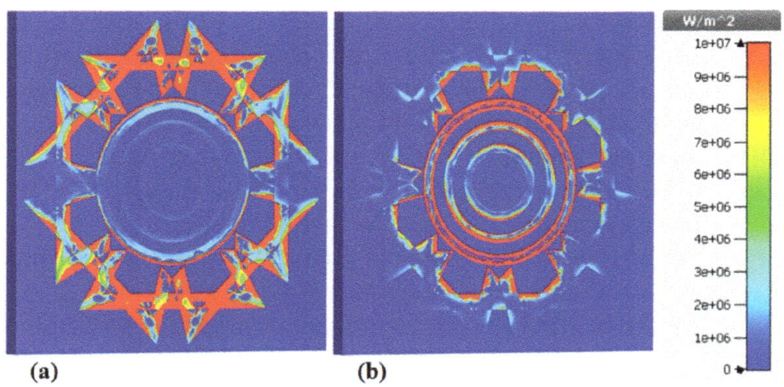

Figure 4.15. Surface power loss density at (a) 0.437 THz, (b) 0.805 THz.

resonance point of 0.437 THz, as shown in figure 4.15(a). Additionally, figure 4.15(b) shows how the TSRRs area may annihilate electromagnetic waves extremely densely, especially at the resonance point of 0.805 THz.

4.4 Conclusions

The innovative terahertz MTM absorber put forth in this chapter is intended to detect changes in the refractive indices of surrounding media. Narrowband absorbers are possible sensors for changes in the surrounding medium because they show a considerable shift in their absorption peak when external factors, such as temperature or refractive index, are changed. This chapter presents five different models, with the third model being deemed to be the best. Following a thorough analysis to enhance the suggested sensor's operational performance, which involved altering the dielectric material, PEC material, and radius resonators in seven distinct scenarios, the suggested sensor is a sensitive MTMA for researching matter–wave interaction, and the dual-band absorption peaks shift as the refractive index changes. The distributions of the E-field, H-field, polarization angle, surface current distribution, electric and magnetic energy density, power flow, and surface power loss density were further examined to clarify the absorption characteristics at different resonance frequencies.

References

[1] Veselago V G 1968 The electrodynamics of substances with simultaneously negative values of img align= absmiddle Alt= ε Eps/Img and μ *Phys. Usp.* **10** 509–14

[2] Hamza M N, Abdulkarim Y I, Saeed S R, Altıntaş O, Mahmud R H, Appasani B and Ravariu C 2022 Low-cost antenna-array-based metamaterials for non-invasive early-stage breast tumor detection in the human body *Biosensors* **12** 828

[3] Zerrad F E, Taouzari M, Makroum E M, El Aoufi J, Islam M T, Ozkaner V, Abdulkarim Y I and Karaaslan M 2022 Multilayered metamaterials array antenna based on artificial magnetic conductor's structure for the application diagnostic breast cancer detection with microwave imaging *Med. Eng. Phys.* **99** 103737

[4] Rashed A N Z, Mohammed A E-N A, Zaky W F, Amiri I and Yupapin P 2019 The switching of optoelectronics to full optical computing operations based on nonlinear metamaterials *Res. Phys.* **13** 102152

[5] Yan Z, Tang C, Wu G, Tang Y, Gu P, Chen J, Liu Z and Huang Z 2020 Perfect absorption and refractive-index sensing by metasurfaces composed of cross-shaped hole arrays in metal substrate *Nanomaterials* **11** 63

[6] Stanley R 2012 Plasmonics in the mid-infrared *Nat. Photonics* **6** 409–11

[7] CHEN H-T, O'HARA J F, AZAD A K and TAYLOR A J 2011 Manipulation of terahertz radiation using metamaterials *Laser Photonics Rev. (Print)* **5** 513–33

[8] Huang L and Chen H-T 2013 A brief review on terahertz metamaterial perfect absorbers *Terahertz Sci. Technol.* **6** 26–39

[9] Lee Y P, Rhee J Y, Yoo Y J and Kim K W 2016 *Metamaterials for Perfect Absorption* (Berlin: Springer)

[10] Lee J, Seo M, Sohn J, Ahn Y, Kim D, Jeoung S, Lienau C and Park Q-H 2005 Invisible plasmonic meta-materials through impedance matching to vacuum *Opt. Express* **13** 10681–7

[11] Li S, Gao L, Zou C, Xie W, Wei Y, Tian C, Wang Z, Liang F, Xiang Y and Yang Q 2018 A polarization-independent fiber-optic SPR sensor *Sensors* **18** 3204

[12] Abdelsalam M, Mahmoud A M and Swillam M A 2019 Polarization independent dielectric metasurface for infrared beam steering applications *Sci. Rep.* **9** 10824

[13] Wang Y, Yi Y, Xu D, Yi Z, Li Z, Chen X, Jile H, Zhang J, Zeng L and Li G 2021 Terahertz tunable three band narrowband perfect absorber based on Dirac semimetal *Phys.* E **131** 114750

[14] Liu L, Liu W and Song Z 2020 Ultra-broadband terahertz absorber based on a multilayer graphene metamaterial *J. Appl. Phys.* **128** 1–6

[15] Saadeldin A S, Hameed M F O, Elkaramany E M and Obayya S S 2019 Highly sensitive terahertz metamaterial sensor *IEEE Sens. J.* **19** 7993–9

[16] Wang B-X, He Y, Lou P and Xing W 2020 Design of a dual-band terahertz metamaterial absorber using two identical square patches for sensing application *Nanoscale Adv.* **2** 763–9

[17] Nickpay M-R, Danaie M and Shahzadi A 2021 Highly sensitive THz refractive index sensor based on folded split-ring metamaterial graphene resonators *Plasmonics* **17** 237–48

Chapter 5

Metamaterial energy harvesters

O Altıntaş

5.1 Introduction

Nowadays, energy requirements and shortfall are at the forefront in terms of the major concerns of modern human life and technological developments. It is well known that fossil energy sources are running out and renewable energy sources are becoming a more valuable concept day by day. Therefore, researchers and engineers consider utilizing entire energy sources that can be converted into electricity.

This chapter discusses the energy harvester concept that benefits from electromagnetic and acoustic wave energy freely traveling in the environment or mechanical systems. At this point, metamaterial-based absorbers help convert these energies to electrical power via some integrated pieces of equipment such as piezoelectric materials, varactor diodes, and load resistance. The studies presented in the literature focus on the absorption and energy conversion performances of the metamaterial structure proposed. One study achieved confinement of the energy in a piezoelectric region and suppression of noise, another group of researchers investigated the thermal behavior of the energy harvester they designed in a wide frequency band. The energy harvester approach has been examined in three subchapters: acoustic, RF, and infrared energy harvesting. These topics cover a major part of the electromagnetic spectrum, which varies from 1 kHz to 750 THz.

In the first section, acoustic and acoustoelastic waves from 1 to 80 kHz have been examined with piezoelectric-based materials. The localized and confined acoustic waves on planar or 3D metamaterial structures are converted into electrical power via piezoelectric materials. From some points of view, the vibrations occurring in the mechanical parts of a system have been transformed into an energy form by filtering them.

In the second section, the electromagnetic waves' energy in the RF band varied from 1 MHz to 300 GHz and was harvested with planar metamaterial absorbers. In this section, many techniques such as rectifying surfaces, rectennas, multiband/broadband absorbers, and wearable and tunable harvesters have been used in the

doi:10.1088/978-0-7503-5754-8ch5

literature to obtain electrical power. Moreover, self-powered Internet-of-Things (IoT) and healthcare monitoring systems have been discussed among researchers with metamaterial energy harvesters at this frequency regime.

In the last section, the conversion of infrared waves and visible light energies has been reported to obtain solar power at a wavelength or a frequency band between 1 mm and 700 nm or 300 GHz to 750 THz, respectively. The essential point of this concept is to improve the metamaterial absorber with temperature- and corrosion-insensitive and high solar-to-thermal conversion efficiency.

5.2 Piezoelectric-based acoustic and acoustoelastic wave energy harvesting

Piezoelectric-based acoustic and acoustoelastic wave energy harvesting is a method among vibration and acoustic wave energy conversion technologies. An instrument made of piezoelectric material or metamaterial can convert the vibrations or acoustic waves from the environment into electrical power. This method makes it possible to generate energy by capturing sound waves from noisy environments. On the other hand, the concept of acoustoelastic wave energy harvesting is exploited from the acoustic properties of a material under mechanical stress. A piezoelectric material can create a certain amount of electrical energy from acoustoelastic waves originating from the material under mechanical stress with this technique.

These methods are investigated by many researchers to be applied in industrial facilities to obtain electrical energy from vibrating equipment or devices. Examples of the sources of vibrations and noise can be bridges, buildings, land vehicles, industrial types of equipment, air movements, and so on. Moreover, this technology can be practiced in many versatile applications when integrated into the environment or equipment that generate acoustic waves, such as acoustic sensors and medical imaging.

Figure 5.1(a) represents the pressure levels of various acoustic sources in ascending order in Pa. A schematic view of the energy harvesting process originating from these types of sources is illustrated in figure 5.1(b). The electrical energy occurs inside the structure described as a metamaterial wall, which is generally composed of piezoelectric materials while the mechanical waves reach it. The pressure resulting from the vibration is converted into electrical charges into piezoelectric materials. A certain amount of charge creates an electrical potential. A bridge rectifier can convert it into DC power to feed electrical devices such as sensors.

Ma *et al* presented a theoretically and numerically studied two-dimensional local resonant metamaterial energy harvester [1]. They verified the results by studying experimentally in a hemi anechoic room and in a natural environment. The results are observed for three different acoustic frequency points at 2441, 3086, and 3262 Hz as illustrated in figure 5.2. The average power is linearly increased while the applied sound pressure is increased at all frequency points. The most major alteration is observed at 3262 Hz. The study results corresponding to 10 Pa of sound pressure were verified with the Fast Fourier transform of output voltage.

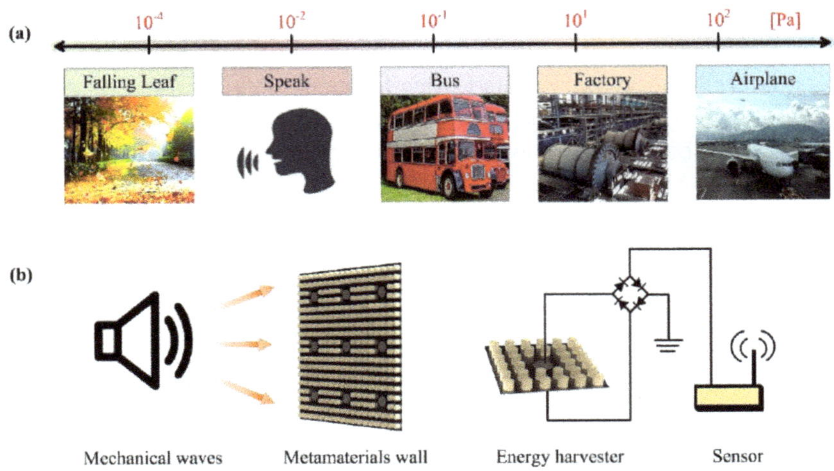

Figure 5.1. (a) Sound pressure levels of different sources, (b) energy harvesting process of the acoustic energy harvester. Reproduced from [1].

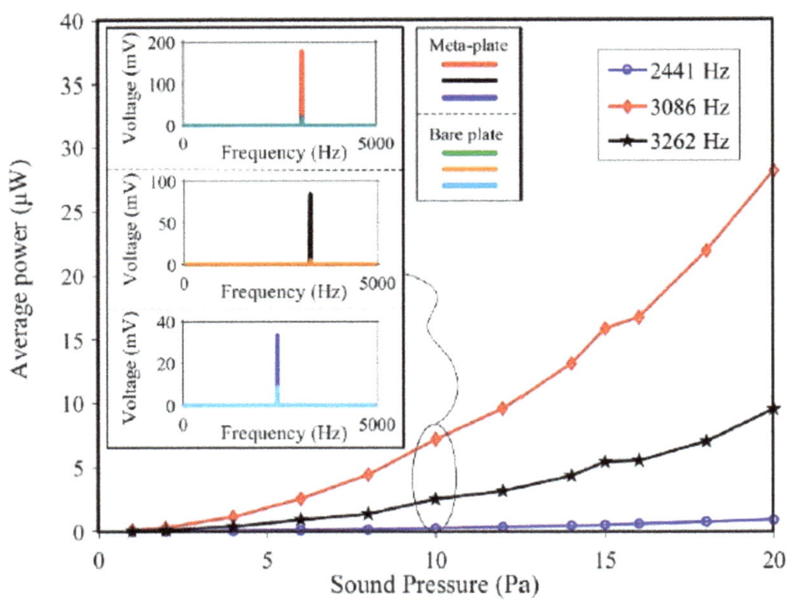

Figure 5.2. Average power results of the meta-plate structure under various sound pressures at optimal resistance values and Fast Fourier transform of output voltage at sound pressure of 10 Pa. Reproduced from [1].

Many researchers have focused on a periodically arranged planar acoustic metamaterial in order to confine incident acoustic waves [2–13]. In their research, the main idea is to utilize acoustic pressure applied to a piezoelectric material. This pressure can be transformed and observed as an output voltage or power density at the resonance frequency of the planar metamaterial composed of piezoelectric

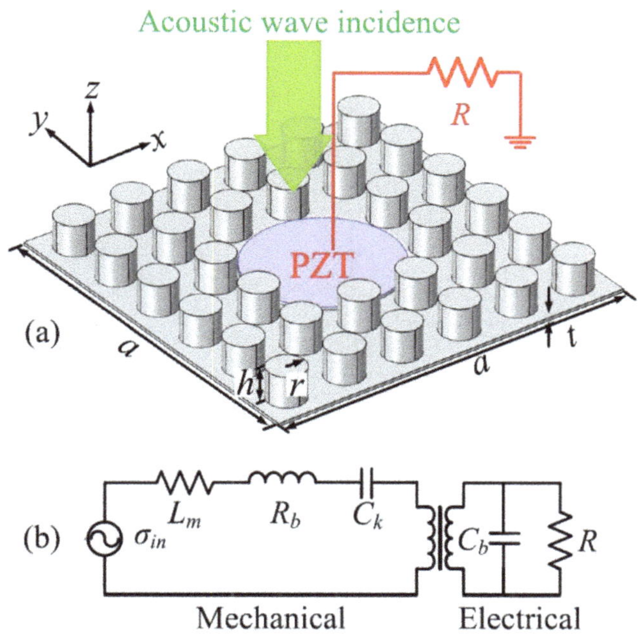

Figure 5.3. (a) An acoustic energy harvester sketch with piezoelectric material and a load resistance and (b) equivalent circuit diagram of the mechanical and electrical models. Reprinted from [2], with the permission of AIP Publishing.

materials. Qi *et al* present their findings on periodically deposited silicon rubber based materials placed in an aluminum plate along the x–y axis as illustrated in figure 5.3(a) [2]. A PZT-5H piezoelectric material patch is inserted into the center of the metamaterial plate. These types of designs consist of two models, which are the acoustic or mechanical model, and the electrical model, as shown in figure 5.3 (b). In the acoustic model, the acoustic pressure incident from the z-axis is confined to energy harvester metamaterial plates. Then, the energy conversion has been realized in the electrical model by piezoelectric materials.

Qi *et al* investigate their proposed structure by using Bloch's theorem and Floquet periodic boundary conditions [2]. As can be seen in figure 5.4, the maximum output voltages occurred at the frequencies of 2257.5 and 2847.5 Hz with a magnitude of 1.3 and 0.78 V, respectively. They have also observed a power density around 0.54 $\mu W\ cm^{-3}$ with 2 Pa of acoustic pressure incident at 2257.5 Hz by optimizing the load resistance as 37.8 kΩ.

Zhang *et al* also studied a silicon rubber based acoustic energy harvester with a PZT patch deposited on an aluminum plate [6]. Their idea was to create a locally resonant region by drilling a certain number of holes around the patch. In this way, the proposed structure can intensely strain energy stemming from acoustic waves. The incident acoustic wave has a frequency of less than 1150 Hz and 2 Pa of acoustic pressure. The observations of output voltage and power are made by experimentally implemented metamaterial structures with various numbers of holes and sizes.

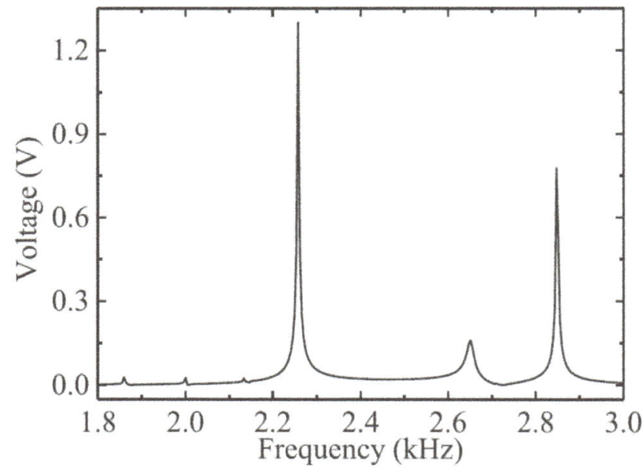

Figure 5.4. Output voltage magnitude versus frequency in kHz. Reprinted from [2], with the permission of AIP Publishing.

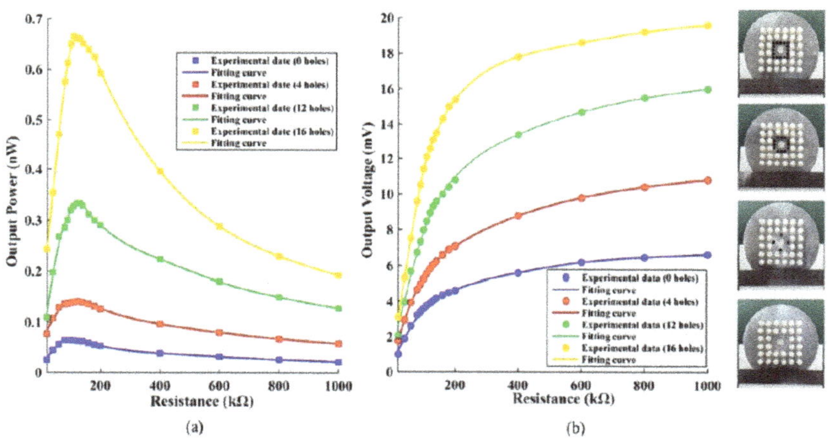

Figure 5.5. (a) Output power and (b) voltage versus load resistance at the defect mode frequency for various holes in the radius of 2 mm. Reproduced from [6]. © The Author(s). Published by IOP Publishing Ltd. CC BY 4.0.

Figures 5.5 and 5.6 are plotted under the frequency of defect mode, which is obtained by removing the central silicon rubber pillars. It is observed as 1090 Hz of the central frequency. Figure 5.5 represents the output power and voltage at different load resistance for various numbers of holes with a radius of 2 mm. The maximum output power and voltage have been observed for 16 holes at the defect mode frequency. The optimum load resistance is around 135 kΩ and the output power is 0.68 nW.

In figure 5.6, Zhang *et al* studied different radii of hole sizes between 1 and 2 mm for the structure with 16 holes. The maximum output power and voltage occurred

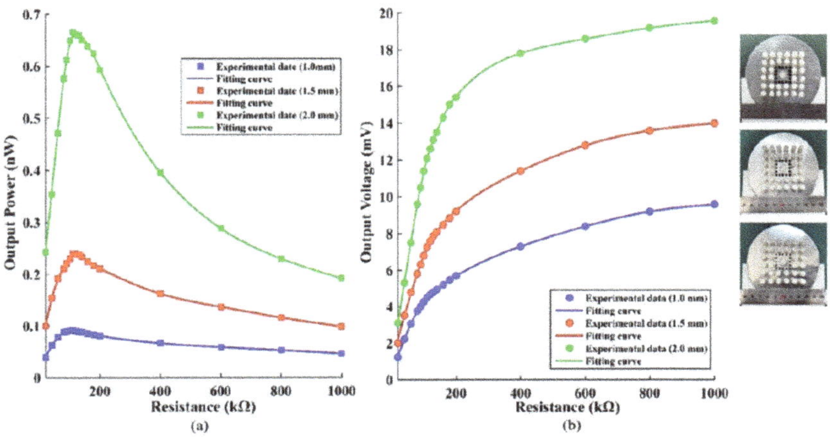

Figure 5.6. (a) Output power and (b) voltage versus load resistance at the defect mode frequency for various radius for the structure having 16 holes. Reproduced from [6].

for 2 mm of radius at the frequency of defect mode. Both the number and size of holes are affected by the energy harvesting performance of the structure.

Li *et al* proposed a Helmholtz acoustic metamaterial (HAM) energy harvester composed of periodically adjusted 5 × 5 Helmholtz resonator cavity (HRC) units on an aluminum plate as illustrated in figure 5.7(a) [7]. The study investigated both structural and local resonance modes. A P-52 piezoelectric sheet is placed at the center of the structure to examine local resonance and transform acoustic wave energy into electric power. Helmholtz's acoustic metamaterial structure was designed based on the Bloch theorem and plane wave expansion method.

The structural and local resonance frequency has been obtained at 381 and 1526 Hz with a 45 and 290 Hz bandwidth, respectively. At the structural resonance frequency, the output power and voltage of HAM occurred at 0.13 mW and 3.2 V with an input sound pressure of 31 Pa. While the sound pressure is 28.83 Pa, the output power and voltage observed at the local resonance frequency of 1526 Hz were at the level of 0.11 mW and 1.5 V. The output power of HAM was found to be 12.7 times and 4.4 times higher than traditional acoustic metamaterial energy harvesters in the structural and local resonance modes, respectively, when the same amplitude of sound pressure was applied.

In figure 5.7 (b), the band characteristics of HAM were presented. When a region at the center of the acoustic metamaterial is arranged as a defect, a defect band has occurred in the local resonance band. Hence, the vibration can be restrained around this region and the defect band inside the local resonance band can be utilized for harvesting energy. The vibration mode, sound pressure level, and strain energy density nephograms have been demonstrated in figure 5.7(c)–(h) for both structural and local resonance bands at 387 and 1493 Hz, respectively. In the vibration mode nephogram, the displacement can be seen along the whole plate at 387 Hz. The displacement in local resonance mode is relatively concentrated at the center of the

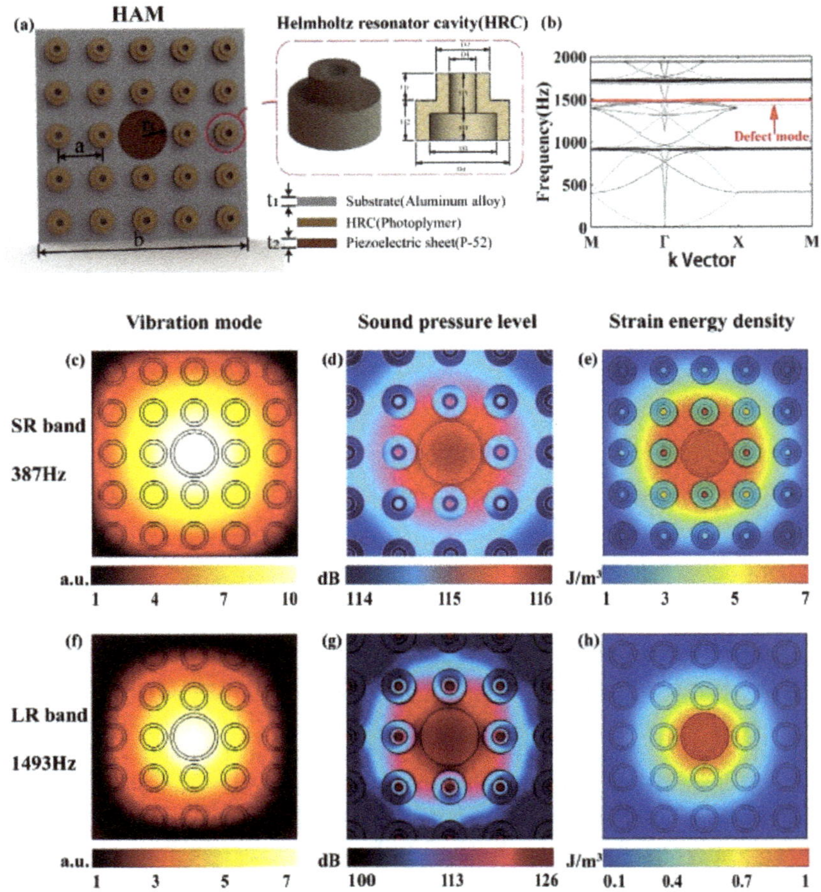

Figure 5.7. (a) The view of the designed Helmholtz acoustic metamaterial and (b) band structure characteristics. (c) Vibration mode, (d) sound pressure level, and (e) strain energy density for the structural resonance band. (f) Vibration mode, (g) sound pressure level, and (h) strain energy density for the local resonance band. Reprinted from [7], Copyright (2021), with permission from Elsevier.

structure, which is the place where the largest deformation is located. This causes the center of the structure to create a maximum applied sound pressure level as can be seen in figure 5.7(d) and (g). Strain energy density, which is demonstrated in figure 5.7(e) and (h), is also focused on the center due to the increasing vibration level at this region. Hence, the most efficacious energy conversion can be achieved at this spot due to the piezoelectric material located at the center.

Alongside the periodic planar metamaterial structures, numerous designs such as mechanical-based, phononic crystal based, elastic metamaterial cavity, graded metamaterial, and embedded piezoelectric material have been studied by researchers as energy harvesters for acoustic and acoustoelectric wave energy conversion [14–47]. Moreover, vibration suppression, attenuation, and isolation are the other purposes for some of these studies.

Ponti *et al* studied a rainbow-based metamaterial that can slow down elastic waves by interacting at resonance [16]. The structure is composed of an array of resonant rods integrated into the elastic beam and it acts as a mechanical delay line. The array design is constructed by the different height sizes of rods that have a 25 mm^2 cross-sectional area as shown in figure 5.8(a). It is considered as a graded metamaterial that can confine energy at the subwavelength of local resonance.

The proposed system slows down wave propagation by means of gradually varied resonant rods that provide an operational frequency range. The moderation of the propagation is achieved by increasing the interaction time between the wave and metasurface as can be seen in figure 5.8(b). Due to this structure characteristic, electromechanical energy conversion is provided between the input and piezoelectric material placed on top of the rod resonators indicated as the red and blue arrows in figure 5.8(a).

The energy harvesting efficiencies for this study have been realized and compared both numerically and experimentally in figure 5.9. At the top of figure 5.9(a), the

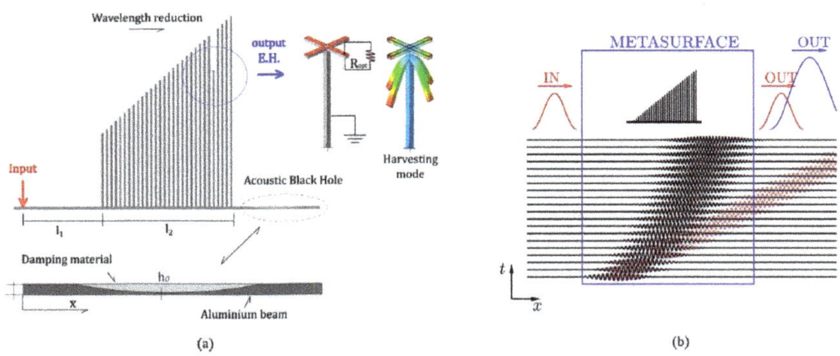

Figure 5.8. (a) Design of the harvester system and (b) graphical presentation of the metasurface effect on the wave propagation. Reprinted from [16], with the permission of AIP Publishing.

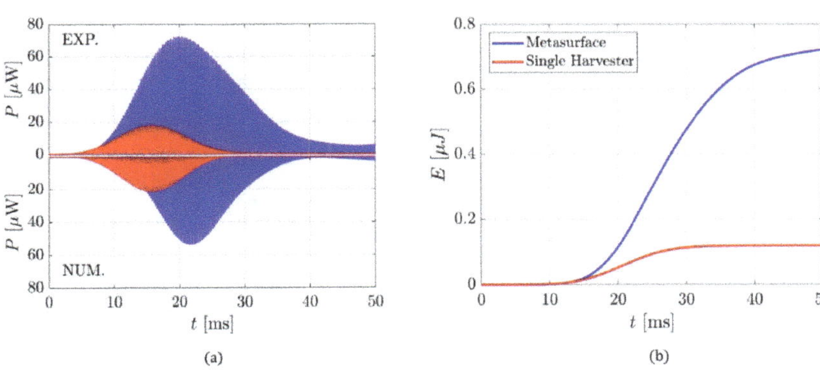

Figure 5.9. (a) Numerical and experimental results for output energy gained from harvester with (blue) and without (red) metamaterial and (b) experimental total electric energy in time. Reprinted from [16], with the permission of AIP Publishing.

measured output power is presented as experimental results of the host beam with one resonator (red area) and full metamaterial (blue area), respectively. At the bottom of figure 5.9(a), the numerical results of the calculated output power have been given in the same color representations. The wave localization provides amplification in time between the wave and the harvester. It is verified from both experimental and numerical results that the system has a rainbow effect as predicted. Moreover, the energy harvesting capabilities have been compared for two scenarios, one of them is a single harvester structure and the other one is harvesters in the metamaterial. The results are observed between the ranges of 0–50 ms and they are obtained as 0.12 and 0.72 μJ, respectively in figure 5.9(b). It illustrates that the harvesting capability is enormously increased.

The challenge of energy harvesting systems is to provide sustainable power generation in practical applications such as health monitoring and wearable biomedical implementations. Park *et al* proposed a structure that is based on two-dimensional octagonal phononic crystals (PNCs) to solve this issue by improving the harvesting capacity up to 22.8 times compared to specimens without the metamaterial [29].

The study compared the performance of two types of structures. One of them is a bare aluminum plate with a thickness of 2 mm and the other is called a meta-plate composed of a 5 × 7 of phononic crystal supercell placed on the center of the aluminum plate with the same size as the bare plate as shown in figure 5.10. The unit

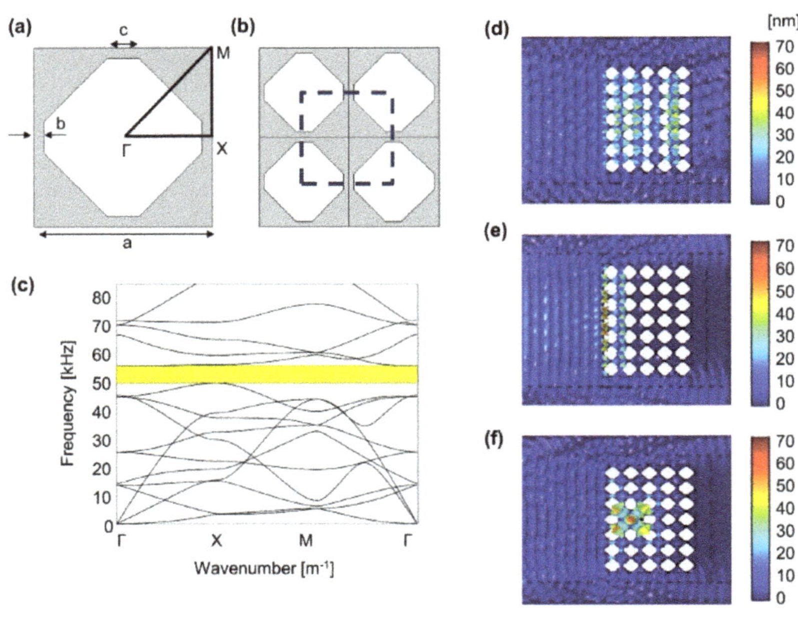

Figure 5.10. (a) Unit cell of the PNC supercell structure, (b) 2 × 2 unit cells designed for defected region, (c) band structure with dispersion relation. Displacement field distribution (d) in the passband at 40 kHz and (e) in the stopband at 50 kHz. (f) Energy localization at the defected region of PNC supercells. Reprinted from [29], Copyright (2019), with permission from Elsevier.

cell of PNC in figure 5.10(a) is parametrically designed for maximizing the band gap relation with dispersion subjected to a target frequency of 50 kHz as illustrated in figure 5.10(c). The numerical studies have been realized to understand the physical behavior of the meta-plate in both the passband at 40 kHz and stopband at 50 kHz as shown in figure 5.10(d) and (e). In figure 5.10(e), the elastic wave in the phononic band gap (at stopband) does not propagate alongside the PnC supercells and is reflected back. On the other hand, in figure 5.10(f), a defective region has been created at a certain region of the periodic PNC supercell structure. The numerical results show that the localized energy is confined to the designed region.

The bare and meta-plate harvesting performances have been compared in measured voltage in figure 5.11(a) and (b). In this comparison, the PZT material has been placed in exactly the same location inside these samples. The wave is reached after about 0.35 msec to the defect location, which is the same position for the bare plate sample. The maximum voltage levels have been observed as 4.55 V at 1.13 msec and 0.97 V at 2.82 msec for meta- and bare plate samples, respectively. The researchers also investigated these two samples' output voltage and power at various load resistance values as illustrated in figure 5.11(c) and (d). In both conditions, the output voltage increases while the output power decreases with the increment of load resistance. They achieved the maximum harvested energies of 1.59 mW and 84.7 μW at about 5 kΩ for meta- and bar plate samples, respectively.

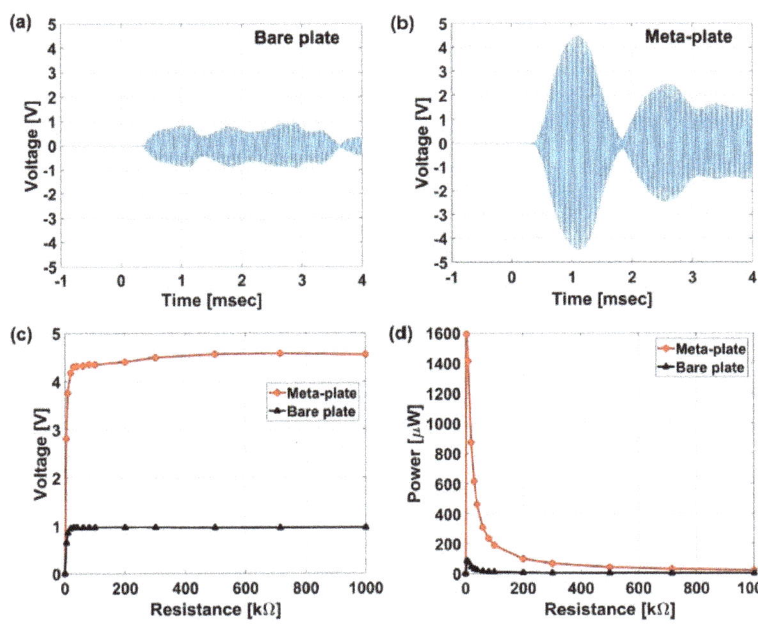

Figure 5.11. Experimental results for output voltages in time (a) bare plate and (b) meta-plate, output voltages at various load resistance for both samples, and (d) output powers at various load resistance for both samples. Reprinted from [29], Copyright (2019), with permission from Elsevier.

5.3 RF regime energy harvesting

The studies about energy harvesting systems have been mainly focused on the RF regime due to the variety of microwave sources that emit electromagnetic waves to the ambient with a certain amount of power [48–88]. In the RF regime, an impedance matching surface as metamaterials or antennas and an energy harvester unit such as resistive load or varactor diodes are required to convert electromagnetic wave energy to DC power. This approach has been discussed in a variety of implementation models such as rectifying circuits, rectennas, broadband/multiband/tunable metamaterial designs, and wearable designs.

One of the energy harvesting approaches in the RF regime, in order to convert the wave power to DC power, is to utilize rectifying circuits embedded into metamaterial structures [48–54]. Duan *et al* presented a study about metamaterial rectifying surfaces designed on a three-layered PCB board composed of split-ring resonators, a metal layer, and rectifying circuit with Schottky diodes [48]. A schematic view is demonstrated in figure 5.12. In this type of design, the most essential part of the study is to provide impedance matching between free space (Z_0 equal to 377 Ω) and metamaterial structure (Z_{RS}). Besides, the complex conjugate part of Z_{RC} and Z_{MM} must be equal to monitor maximum power conversion as illustrated in the block diagram in figure 5.12.

After researchers meet the impedance matching criteria, the harvesting performances of the system can be investigated with respect to various load resistance and incident power density as plotted in figure 5.13. The maximum harvesting efficiencies are observed at 400 Ω of load resistance value for 1, 3, and 5 mW cm^{-2} of incident power densities as 58%, 65%, and 66% in figure 5.13(a). The numerical and

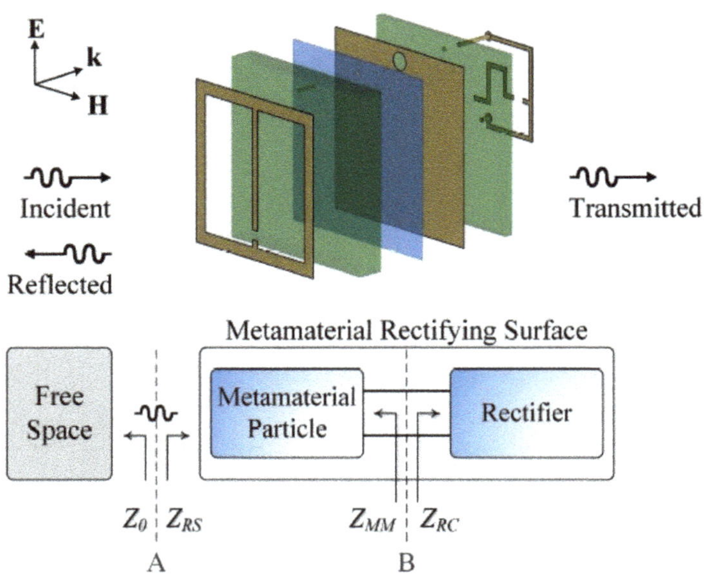

Figure 5.12. Schematic view of the metamaterial rectifying surface and block diagram of its working principle. Reproduced from [48], CC BY 4.0.

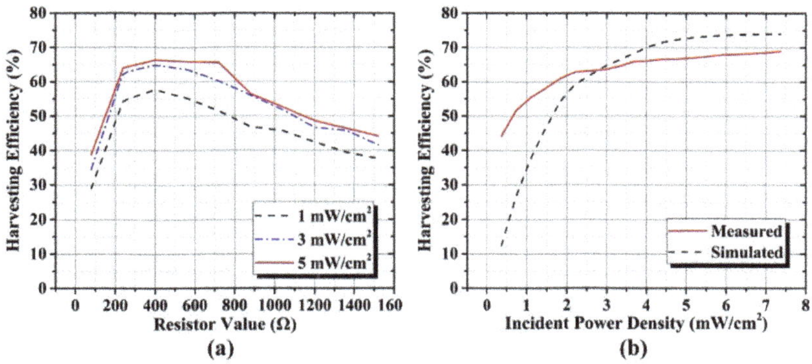

Figure 5.13. The harvesting efficiencies (a) under various load resistance values and (b) comparison of the numerical and experimental results under various incident power density values. Reproduced from [48], CC BY 4.0.

experimental studies are realized at 2.45 GHz and compared with respect to different incident power densities in figure 5.13(b). The results show that the harvesting efficiency suddenly rises up to 5 mW cm^{-2} of power density. It remains stable after this power density level. The harvesting efficiencies are observed at about 73% and 67% for numerical and experimental studies, respectively. The mismatch between the results originated from the nonlinear characteristic of the diode.

Lee *et al* proposed an interesting RF energy harvesting study which aimed to work as a chemical sensor [53]. The suggested network is powered by harvested energy from the ambient at 2.4 GHz. Hence, it does not require any power feed equipment to actuate. This self-powered sensor system comprises three stages, which are the metamaterial unit cell, rectifying circuit, and chemical channel as illustrated in figure 5.14(a).

When the electric field distribution is examined in figure 5.14(b), the distribution is mainly located on the gap of the metamaterial. Therefore, this capacitive region is chosen as the sensor layer because the samples (with a different permittivity characteristic) placed in this layer can dramatically change the resonance frequency and harvesting performance of the system. In this energy harvester design, a Greinacher voltage doubler circuit has been adopted to rectify the current flows into the metamaterial resonator [53].

When various kinds of chemical liquids such as toluene, chloroform, ethanol, acetone, methanol, and water are placed into the sensor layer (or chemical channel), it is shown that the resonance peaks are obtained at different frequency points of 2.32, 2.28, 2.20, 2.08, 1.97, and 1.78 GHz, respectively. The reason for this frequency shift is the different dielectric constant values of these samples. Traditionally, the resonance frequency is a function of the capacitance of the resonator and the capacitance value can alternate by dielectric constant.

The DC output voltage levels are plotted for different concentration levels of toluene, chloroform, ethanol, acetone, and methanol samples with an increment of 20% in figure 5.15(a). While the mixture concentration is increasing with the

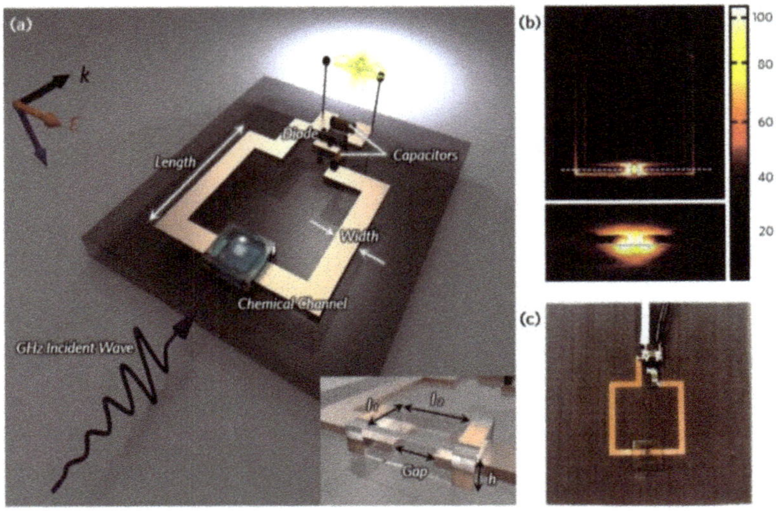

Figure 5.14. (a) Schematic view of the proposed energy harvester based sensor design, (b) electric field distribution of the metamaterial resonator, and (c) manufactured prototype. Reproduced from [53]. CC BY 4.0.

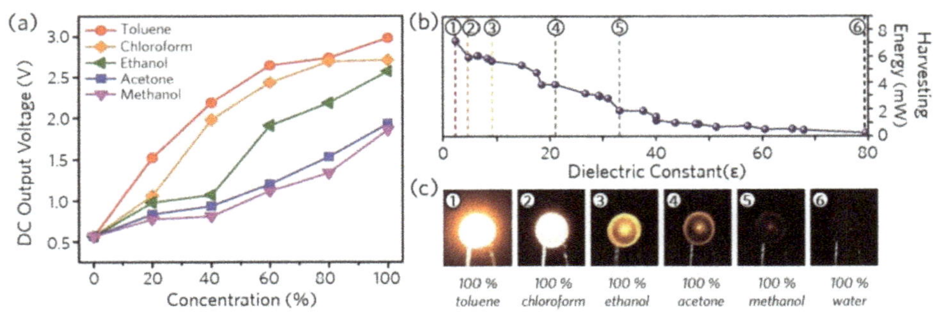

Figure 5.15. (a) DC output voltages of the sensor for different concentration levels of chemical mixtures, (b) energy harvesting performance with various dielectric constant values, and (c) brightness of the LED indicator for different chemical mixture samples. Reproduced from [53]. CC BY 4.0.

decrement in the water percentage, the output voltage is monitored at its maximum levels. The highest value is obtained at 2.98 V for 100% toluene concentration. Moreover, figure 5.15(b) shows that the variations in the dielectric constant of the chemical samples directly affect the energy harvesting performance. Harvested energy can change from 0 to 7 mW with respect to sample substances. Finally, an LED is integrated into the output as an indicator whose brightness depends on the chemical types and substances as shown in figure 5.15(c). While the water level goes to 100% in any substance, the brightness of the LED tends to decrease and eventually dim.

Another method for RF to DC energy harvesting applications that can be considered is rectenna (rectifying antenna) implementations, which has been studied by many scientists [55–59]. Lee *et al* designed a metamaterial-embedded rectifying antenna with a high gain and beam-focusing property at 2.45 GHz [59]. The structure has been created with three layers, which are the metamaterial superstrate, patch antenna, and rectifier as demonstrated in figure 5.16(a) and (b). The metamaterial, which is composed of 4 × 4 unit cells, strongly improves the patch antenna gain and beam-focusing capability. Hence, this causes the increased RF power which is seen at the rectifier input.

The study has been investigated both numerically and experimentally by using an RO5880 type of substrate that has low-loss characteristics. The role of the metamaterial structure on the gain property of the rectenna can be easily seen in figure 5.17. The measured peak gains are observed as 5.75 and 11.87 dBi, without and with the metamaterial, respectively, at 2.45 GHz. This means that 6.12 dBi of improvement is obtained when the metamaterial structure is integrated on the top of the rectenna.

The numerical and experimental RF to DC energy harvesting efficiencies and output DC power is plotted as a function of input power illustrated in figure 5.18.

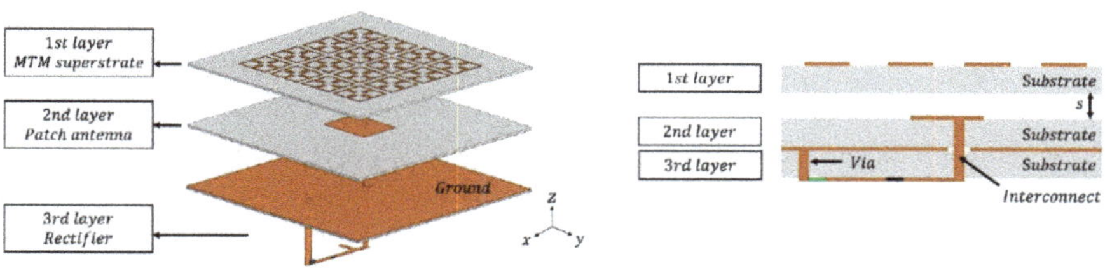

Figure 5.16. (a) Perspective view and (b) side view of the metamaterial integrated rectenna structure. Reproduced from [59]. CC BY 4.0.

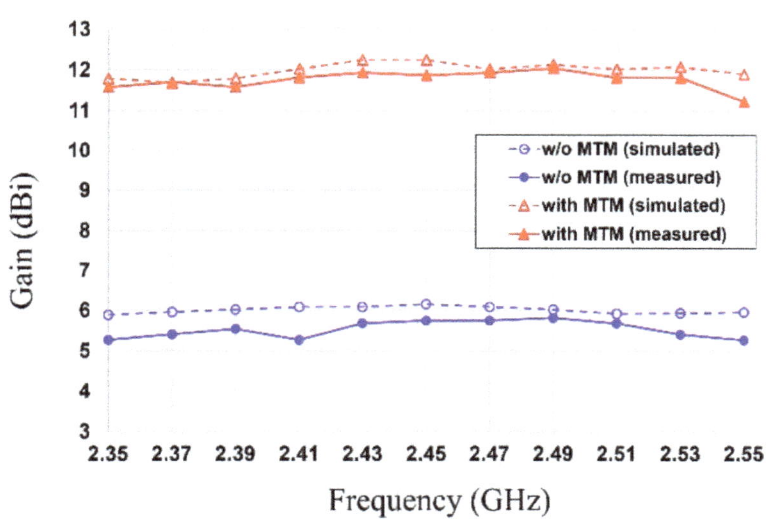

Figure 5.17. Numerical and experimental peak gains of the rectenna with and without the metamaterial superstrate. Reproduced from [59]. CC BY 4.0.

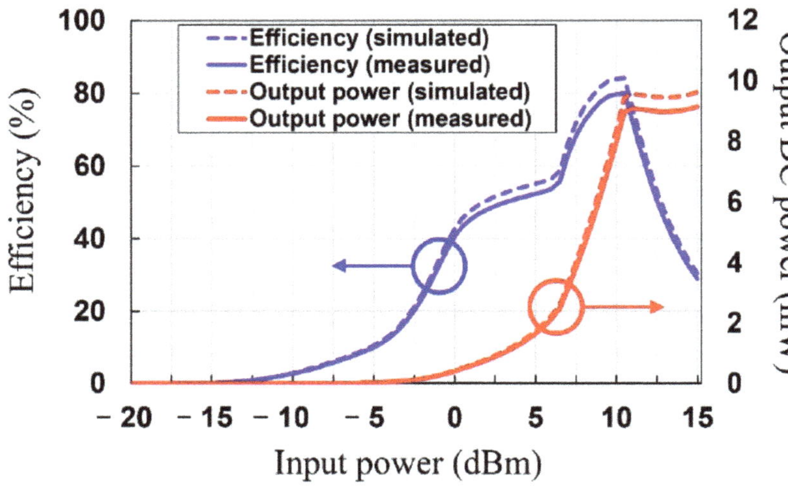

Figure 5.18. Numerical and experimental efficiencies and output powers for the metamaterial integrated rectenna. Reproduced from [59].

The experimental results of the energy harvesting efficiencies are obtained at 2.69%, 40.71%, and 79.7% at −10, 0, and 10 dBm, respectively. The maximum energy harvesting efficiency of 79.7% is observed at the input power of 10.5 dBm [59].

From some points of view, the metamaterial-based absorber design is discussed as an energy harvester by adding resistive or capacitive lumped elements in terms of broadband and multiband features [60–75]. In these types of designs, the metamaterial unit cells are periodically adjusted on top of a planar surface. The back of the surface is covered by a metal plate. Hence, the reflected and transmitted signal is prevented and confined in unit cells along the operating frequency range. The RF signals can be converted into DC power due to the lumped elements located in the capacitive gaps of the unit cells. Moreover, the polarization and incident angle insensitive characteristics are another discussed issue in these kinds of designs.

Karaaslan *et al* studied a microwave energy harvester based on wheel resonators between the frequencies of 2 and 8 GHz as illustrated in figure 5.19 [60]. The structure is firstly discussed as an absorber by covering the backside with a metal ground in figure 5.19(a). Absorbed power can be derived as $A(w)$; $A(\omega) = 1-R(\omega)-T(\omega)$, where $R(\omega) = (|S_{11}|^2)$ and $T(\omega) = (|S_{21}|^2)$ present the reflection and transmission coefficient, respectively, as a function of frequency (w). While the back of the structure is covered by the metal ground, the transmitted signal is prevented and it goes to zero. Therefore, the equation evolves as $A(\omega) = 1-R(\omega)$. If the reflected signal is kept at its minimum level as far as possible, the absorption reaches its maximum level.

Figure 5.20(a) shows the absorption spectrum of the proposed structure having two types of resonators with and without resistive loads. By using these resistive loads, the absorption peak points are reached at maximum level. Besides, the perfect absorption bandwidth has occurred between the frequencies of 3.35 and 5.86 GHz. In figure 5.20(b), the frequency response on the absorption scale of single and two

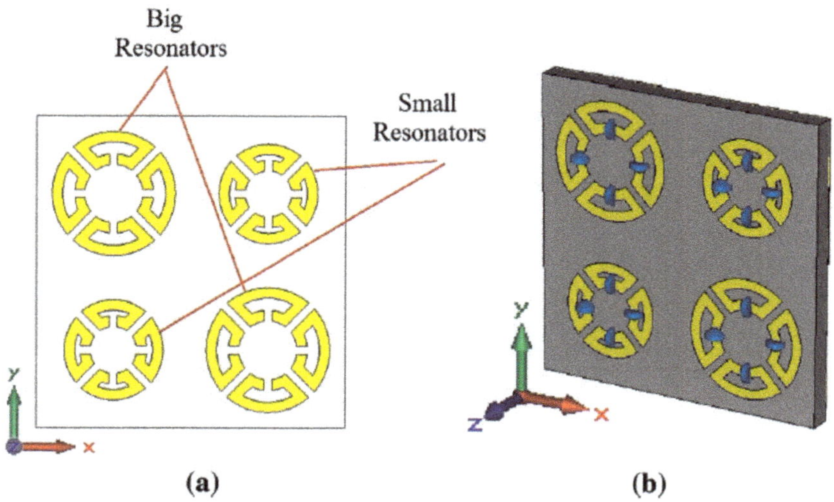

Figure 5.19. Unit cell implementation of the structure having two types of resonators (a) without resistive lumped elements and (b) with resistive lumped elements. Reprinted by permission from Springer Nature [60], Copyright (2018).

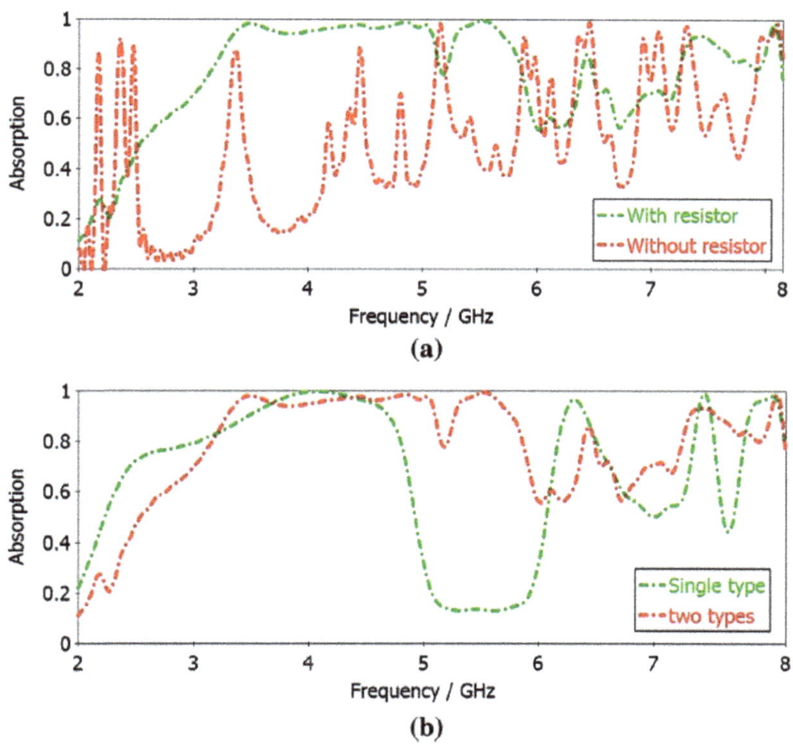

Figure 5.20. (a) Absorption results of the structure having two types of resonators with and without resistive load and (b) absorption results of the structure having a single resonator and two types of resonators with resistive load. Reprinted by permission from Springer Nature [60], Copyright (2018).

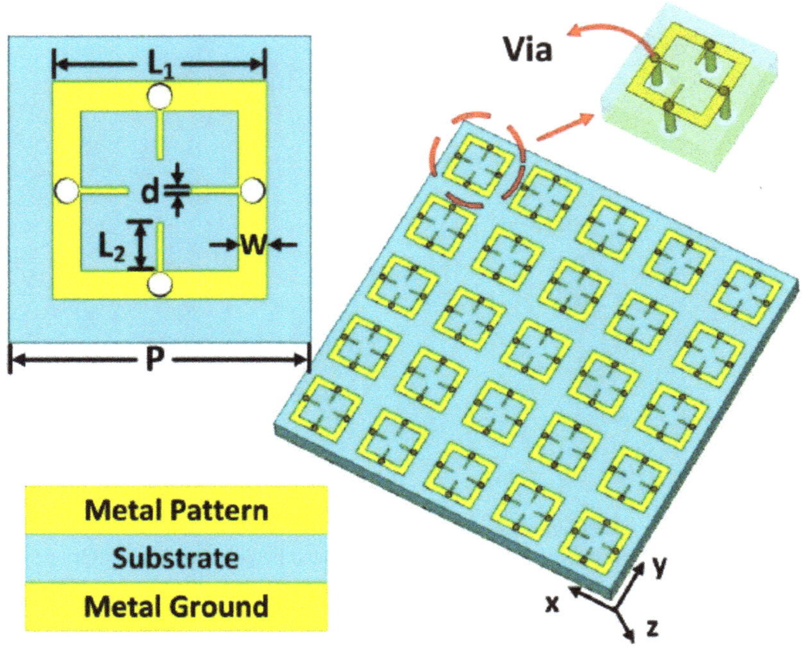

Figure 5.21. Schematic view of the metamaterial array and unit cell. Reprinted from [61], with the permission of AIP Publishing.

types of resonator designs has been discussed. It is observed that the two types of resonator design provide much higher perfect absorption bandwidth than a single one because the frequency response for each individual resonator overlaps and adds up.

In another wideband harvesting application, Zhong *et al* studied polarization insensitive and wide incident angle metamaterial arrays having 110% of half power beam width (HPBW) at frequencies between 6.2 and 21.4 GHz [61]. The metamaterial design consisted of the periodic unit cells that are created with one square ring and four metallic bars as shown in figure 5.21. Each unit cell is connected to the ground by four via connectors from the center of the edges to accumulate the current on the unit cells.

To understand the physical mechanism of this energy harvester structure, the surface current distribution has been investigated as demonstrated in figure 5.22. The electric field component of the wave is assumed to be linearly polarized and at the y-axis direction. The surface currents indicated as black arrows on the top that are passing through the vias show the absorption peak points from figure 5.22(a), (b), and (c). The absorption efficiency can be determined from the strength of the current distribution. It can be observed from figure 5.22 that the surface currents become more intense by increasing the frequency, which provides the third resonance and wide operating frequency range. Moreover, figure 5.22(d), (e), and (f) shows that the current on the bottom is mainly concentrated on the two via loads

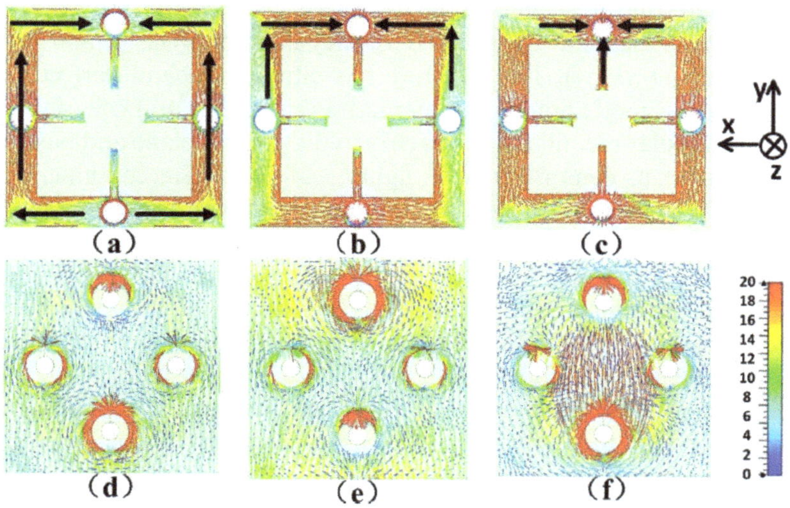

Figure 5.22. The surface current distribution on (a), (b), and (c) top and (d), (e), and (f) bottom at the frequencies of 8.4, 16, and 20.7 GHz, respectively. Reprinted from [61], with the permission of AIP Publishing.

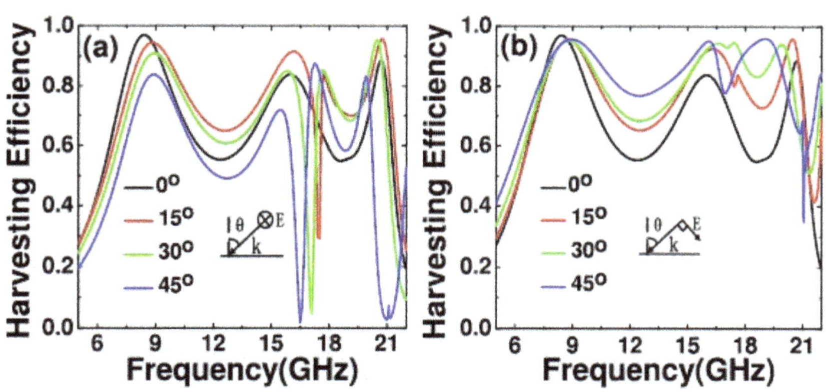

Figure 5.23. Harvesting efficiency of the energy harvester at various incident angles for (a) TE and (b) TM polarizations. Reprinted from [61], with the permission of AIP Publishing.

placed along the y-axis. This means that the current distribution and energy on the loads can be affected by polarization.

The metamaterial energy harvester is also examined with respect to different incident angles for both TE and TM modes in figure 5.23. For TE polarization, the HPBWs remain greater than 80% although there are some resonance frequency shifts among the angles varied from 0° and 45° in figure 5.23(a). For TM polarization, HPBWs are observed as above 100% while incident angles range from 0° to 45° by 15° of increment in figure 5.23(b). It can be explained by the increase of the magnetic resonance while the incident angle increases, which contributes to improving the HPBW.

Another interesting absorber study for energy harvesting applications that is water metamaterial based at the entire frequency band between 12 and 29.6 GHz is proposed by Xie *et al* [67]. The structure has ultra-broadband perfect absorption features and shows good angular and thermal stability. The unit cell of the structure designed with circular metallic holes reaches to the ground plane and passes through the resin container as demonstrated in figure 5.24. The resin container is filled up with water and the bottom of the structure is covered by a copper-type metal.

The absorptivity of the proposed water metamaterial absorber has been investigated in four cases from 10 to 30 GHz in figure 5.25. In the first case, the metamaterial is examined as a full water layer having the same thickness of the metamaterial. As the water has absorptivity characteristics in the microwave regime, it is found that the absorption level is around 35–40%. In the second and third cases, the proposed metamaterial is investigated under conditions without water and a metal background. In the 'without water' cases, the absorption is only observed at around 20% to 40%. In the conditions without a metal background, the wide band absorption and perfect absorption level is not reached, although the absorption level remains above 77% of the entire operating frequency range. In the final case, the metamaterial has been filled up with water as it is designed. It exhibits over 90% absorptivity within the operated frequency range from 12 to 29.6 GHz. In addition, it is shown that the perfect absorptions have been obtained at the frequencies of 13, 19.6, and 27.3 GHz.

Figure 5.26 shows the angular stability of the metamaterial structure under oblique incidents varied from 0° to 75° for transverse electric (TE) and transverse magnetic (TM) polarizations. For TM mode, the absorption characteristic has been observed to remain almost stable up to 60° and it realizes up to 85% for the entire frequency range as illustrated in figure 5.26(a). For TE mode in figure 5.26(b), the absorptivity plot has fluctuated with the increasing angle, especially after 45°.

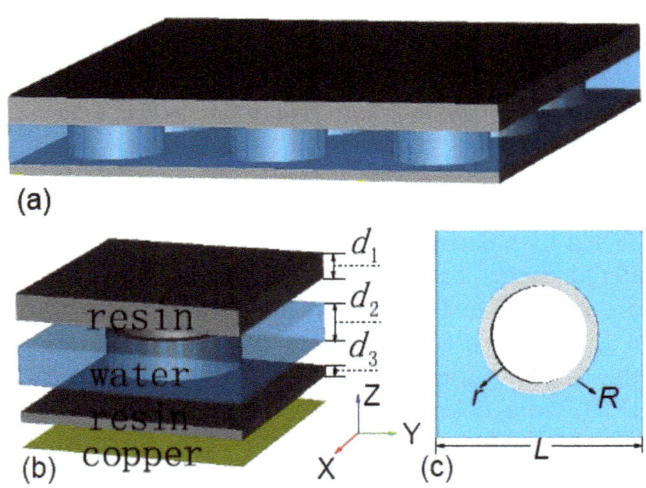

Figure 5.24. (a) Schematic, (b) layer, and (c) cut plane views of the absorber. Reproduced from [67]. CC BY 4.0.

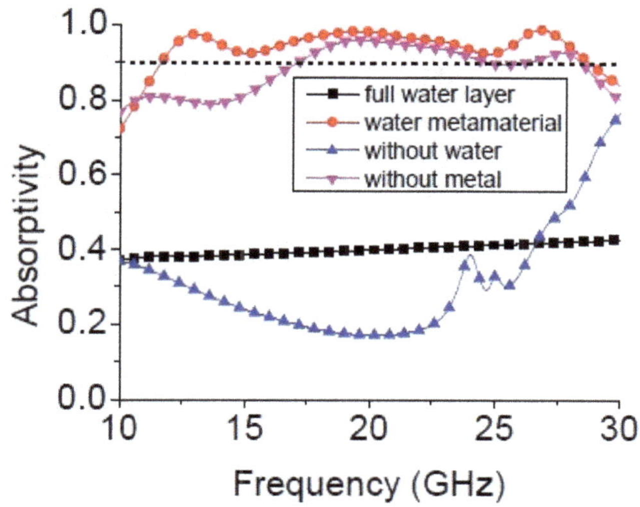

Figure 5.25. The absorptivity plots for the cases of water metamaterial, full water layer, without water, and metal background. Reproduced from [67]. CC BY 4.0.

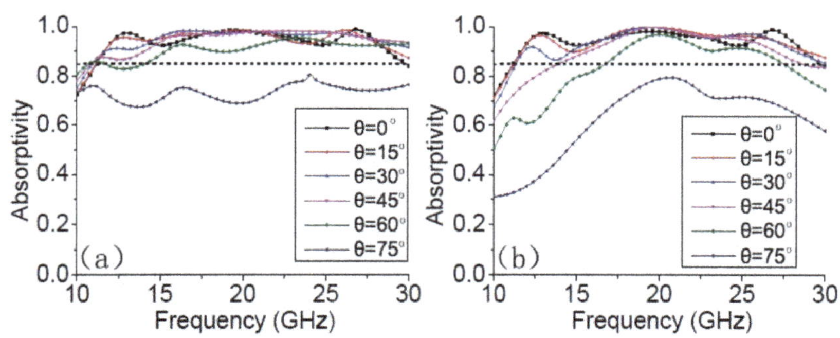

Figure 5.26. Absorptivity of the water metamaterial under oblique incident angle varied from 0° to 75° for (a) TM and (b) TE polarizations. Reproduced from [67]. CC BY 4.0.

However, it still exhibits good performance considering the worst level of 67.5% at 75°. In addition to angular stability examination, the thermal stability of the structure has also been numerically investigated because of the water ingredient. The temperature of the water has been increased from 0° to 100° and the temperature of the resin has been neglected. It is found that the absorption behavior has remained at almost the same level above 90% in the entire frequency range.

Many highly used frequency bands are concrete targets for energy harvesting applications because it is easier and more efficient to collect RF power other than frequency bands, which are not preferred for communication applications. Hence, Zhong *et al* have investigated a triple band, polarization-independent, and wide-angle metamaterial energy harvester for GSM 1800, WiMAX, and WLAN bands [71]. The unit cell of the structure consisted of four split-ring resonators which are

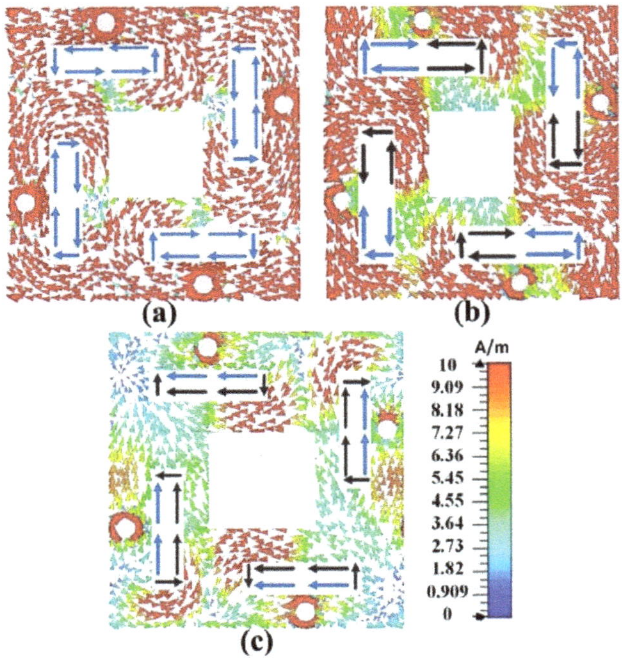

Figure 5.27. Surface current distributions under normal incidence at (a) 1.75 GHz, (b) 3.8 GHz, and (c) 5.4 GHz. Reprinted from [71], with the permission of AIP Publishing.

located to the edges at a 90° of angle. To collect the induced current on the unit cell, four via connectors have been added to the longer edge of each split-ring resonator. Four resistive loads have been connected from via connectors to the ground metal. It is important to determine the value of the load in terms of absorption and harvesting efficiencies. For the proposed structure, 240 Ω of the resistive load has been chosen to obtain absorption efficiencies of 97%, 97%, and 98% at 1.75, 3.8, and 5.4 GHz, respectively.

To observe the resonant mechanism of the structure, the surface current distributions have been analyzed at these frequency points as shown in figure 5.27. The polarization angle of the incident wave for the surface current distribution study is chosen as 45° to obtain uniform distribution with respect to the x- and y-axis. The surface currents have followed a circular path in one direction at 1.75 GHz as illustrated in figure 5.27(a). They have occurred in two reverse directions at 3.8 and 5.4 GHz. It should be noted that these resonance frequencies are calculated from the current path passing through via connectors.

The harvesting efficiencies have been monitored under various incident angles for both TE and TM polarizations in figure 5.28. The results can be adequately accepted for both polarizations although there are slight frequency shifts among the incident angles. For the best performance, 45° of incident angle case has efficiencies of 24%, 73%, and 66% for the TE mode in figure (a) and 36%, 87%, and 70% for the TM mode in figure (b) can be seen.

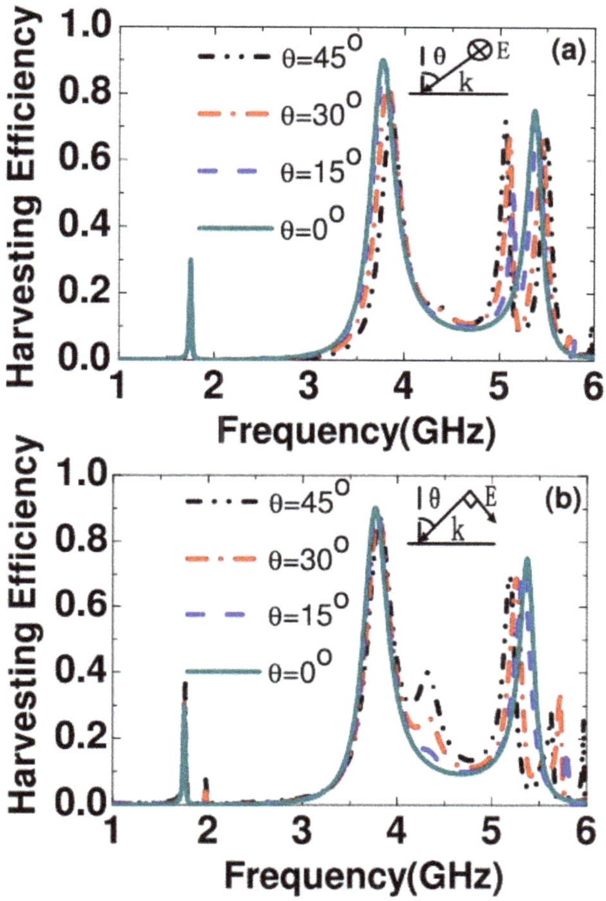

Figure 5.28. Harvesting efficiencies of the proposed metamaterial under oblique incident angles for (a) TE and (b) TM polarizations. Reprinted from [71], with the permission of AIP Publishing.

In another multiband energy harvester case, Gunduz and Sabah investigated a three-nested split-ring resonator as a polarization angle insensitive perfect metamaterial absorber [74]. They implemented two resistive loads for each split ring to convert RF energy to DC power. It is found that the optimum resistive load value is 2 kΩ, as shown in figure 5.29(a). The structure has given acceptable absorption peaks above 45% at six resonance frequencies from 4 to 24 GHz. The highest peak occurred at 5.88 GHz, which has realized a maximum power transfer of 0.25 W with a 2 kΩ resistor as shown in figure 5.29(b).

The researchers have utilized some definitions to determine conversion efficiency (CE), harvesting efficiency (HE), and absorption efficiency (AE). Accordingly, the conversion efficiency represents how much accepted power (AP) is successfully used as consumed power (CP).

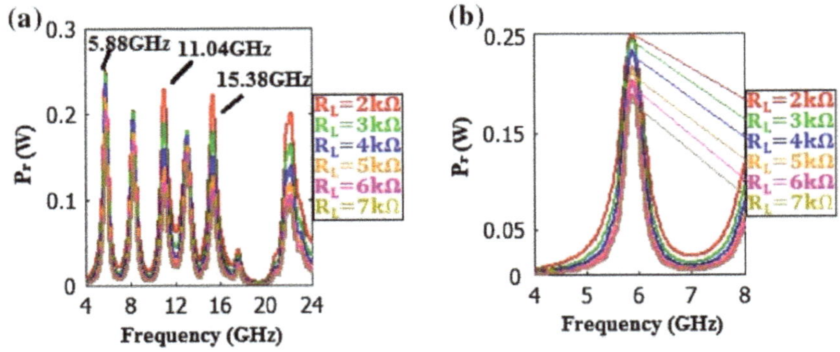

Figure 5.29. (a) Power dissipation under various resistive load conditions at the spectra between 4 and 24 GHz and (b) the highest peak which realizes the maximum power transfer. Reprinted by permission from Springer Nature [74], Copyright (2016).

$$\mathrm{CE} = \left(\frac{\mathrm{PR}}{\mathrm{PA}}\right) \times 100\% \tag{5.1}$$

Harvesting efficiency is the measure of stimulated power (PS) that is successfully delivered to the resistor.

$$\mathrm{HE} = \left(\frac{\mathrm{PR}}{\mathrm{PS}}\right) \times 100\% \tag{5.2}$$

Absorption efficiency indicates the percentage of the incident wave energy that is successfully absorbed by the structure.

$$\mathrm{AE} = \left(\frac{\mathrm{PA}}{\mathrm{PS}}\right) \times 100\% \tag{5.3}$$

From these three definitions, some deductions can be extracted that can show correlations among them. The metamaterial can be an almost perfect or fair harvester if the AE is greater than HE and CE and is smaller than 100%. The ideal scenario that can be defined as a perfect absorber and harvester is only realized at the conditions of AE = HE and CE = 100%. This scenario cannot be observed practically. If AE is smaller than HE and CE is greater than 100%, it can be called an impossible scenario and this case cannot be observed both numerically nor practically.

Figure 5.30 demonstrates the PS, PA, and PR for the entire frequency range. When the calculation for AE, HE, and CE have been made from this figure, it can be found that the first maximum peak has the highest harvesting efficiency of 50.7%. Although the fifth peak has the highest absorption efficiency of 96%, its harvesting efficiency remains at the level of 44.72%. This is caused by the inverse relation between PA and PR. If the resistive load value is raised, it may cause an increase in PA and a decrease in PR due to the current limitation of the resistor. Even though

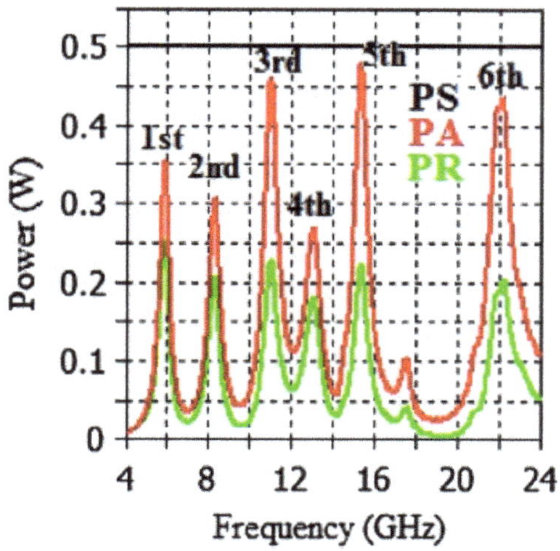

Figure 5.30. PS, PA, and PR versus frequency. Reprinted by permission from Springer Nature [74], Copyright (2016).

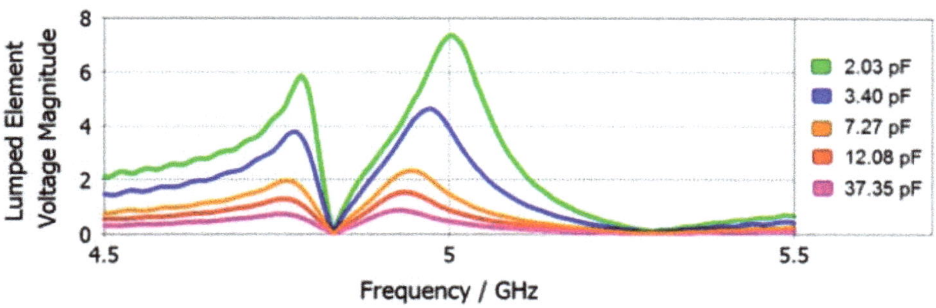

Figure 5.31. Variations of voltage magnitudes on lumped elements via capacitance changes in the varactor diodes [77], (2015), reprinted by permission of the publisher (Taylor & Francis Ltd, http://www.tandfonline.com.).

the absorbed power is sufficiently high, the harvesting efficiency is mainly dependent on the power dissipated in a resistive load.

Some scientists have explored metamaterial energy harvesters with tunable features [76, 77]. In this type of design, a varactor diode has been implemented on the capacitive region of the unit cells, enabling the change of resonance frequency. Dincer utilized the varactor diode not only to easily convert RF energy to DC power but also to adjust the maximum peak point of the energy harvester [77]. Therefore, the proposed structure can be used in applications that require a flexible operating frequency range.

In figure 5.31, it can be seen that the magnitude of the voltage on the lumped element gradually changes when the capacitance value of the varactor diode has been varied from 2.03 to 37.35 pF. This variation provides the means to determine the operating frequency by means of monitoring the maximum voltage magnitude.

The capacitance changes in the varactor diode placed on the capacitive region of the resonator may also provide an enhancement of the energy harvesters' performance as illustrated in figure 5.32. The resonance frequency is decreasing while the capacitance of the varactor is increasing. The reflection coefficient for the minimum peaks has been monitored as 0.11, 0.10, 0.08, and 0.02 for the capacitance value of 37.35, 7.27, 3.4, and 2.03 pF, respectively. It is known that the reflection coefficient is directly affected by AE and HE. Hence, while the operating frequency is adjusting to the desired point, the absorber and harvesting efficiencies can also be tuned by using varactor diode features.

Some of the applications of wearable antennas and wireless body sensors have been discussed in energy harvesting phenomena in order to operate as a self-powered device [78–83]. Within this scope, it can be possible to improve some systems such as remote healthcare monitoring by the IoT. Elwi and Al-Saegh have designed a flexible Hilbert-shaped metamaterial on an antenna to harvest RF energy at 8 GHz [79]. The reason for the flexible design is to integrate the structure into the body efficiently. At this point, one of the essential factors is the investigation of energy harvesting efficiency of the structure with respect to various bending scenarios. Kaur *et al* have also designed a flexible ultrathin structure based on a metamaterial absorber for energy harvesting applications. Besides the design and harvesting performance of the structure, they have also discussed the various incident angle conditions in cases of stretching and bending.

Tian *et al* have realized a comprehensive study that achieves a wireless body sensor network system [83]. The proposed design can be implemented on metamaterial textile products that are made of conductive fabrics. This exciting wireless sensor design powered by energy harvester units can monitor temperature and humidity. Moreover, it can be connected to electronic devices such as smartphones and tablets and can transform the data from the environment. This means that such IoT applications require an energy harvesting system to work independently. The main advantage of this proposed design is a textile-based situation that enables it to be wearable on the human body. Therefore, it offers a huge opportunity to develop a human–machine interface to observe health remotely.

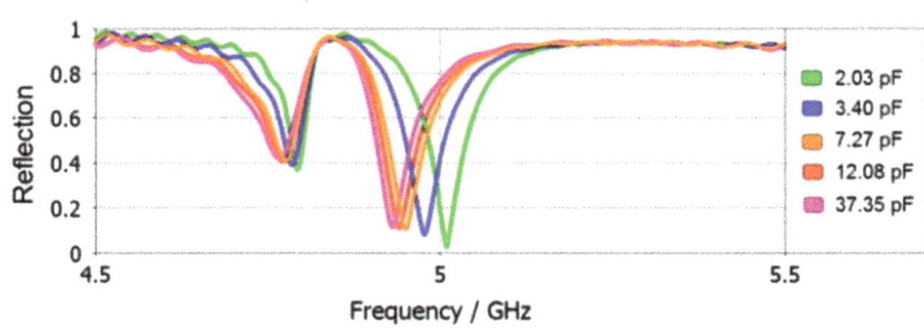

Figure 5.32. The reflection coefficients of the proposed structure under different capacitances of the varactor diode [77], (2015), reprinted by permission of the publisher (Taylor & Francis Ltd, http://www.tandfonline.com.).

5.4 Infrared and visible regime energy harvesting

The energy harvesting phenomenon has been extensively examined in the infrared and visible regimes. which also correspond to the electromagnetic spectrum between 1 mm and 400 nm [89–112]. Within the spectrum, researchers principally have focused on high-temperature solar energy conversion via metamaterial structures whose material types are specifically selected. In energy harvesting, the metamaterials in these regimes can be generally utilized in three layers. The optical metal layer as a resonator should be chosen among the lossless metals such as Au, Ag, and Ti. The dielectric spacer layer (called the stable layer) can be chosen from the material having a higher thermal expansion coefficient (for instance, MgF_2 or SiO_2) to prevent possible cracking, especially for high-temperature implementations. Since thermal efficiency is adversely affected by high temperatures in these applications, it is essential to select the bottom layer as a temperature-insensitive material. Moreover, this layer should be corrosion and oxidation resistive and can be chosen among materials such as tungsten, TiN, titanium, and nickel.

A high-temperature solar energy harvesting implementation designed by a selective metamaterial absorber and solar concentrator unit is studied by Qiu *et al* [89]. The implementation setup with pyramidally placed metamaterial unit cells is given in figure 5.33(a). The above plasmonic mode has been employed to obtain good spectral selectivity. The design has been realized by dielectric metal compositions as illustrated in figure 5.33(b). The bottom of the unit cell consisted of a

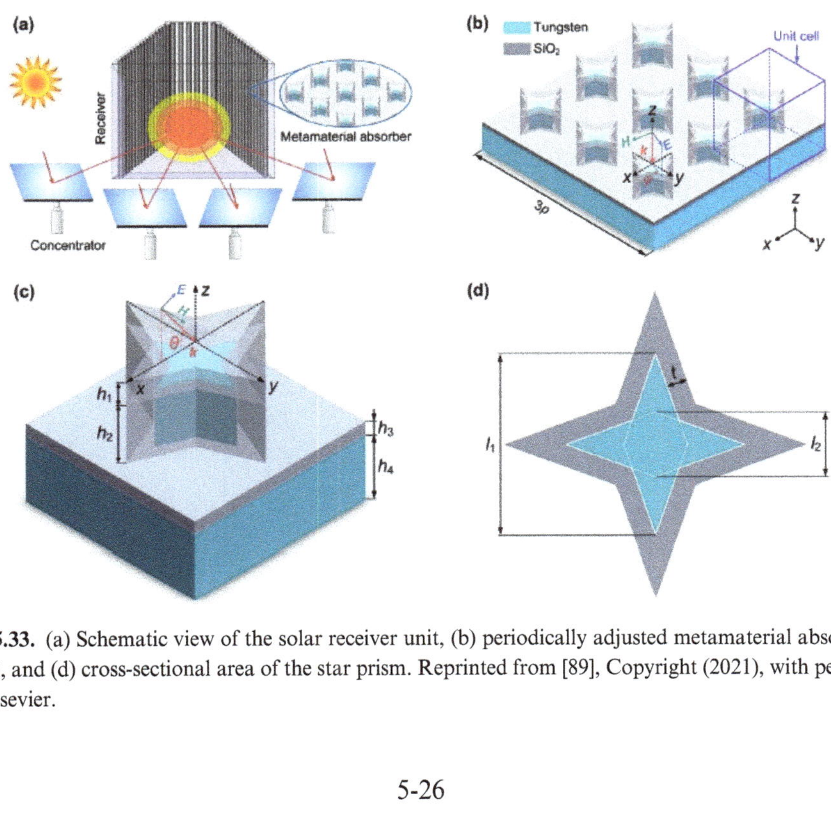

Figure 5.33. (a) Schematic view of the solar receiver unit, (b) periodically adjusted metamaterial absorber, (c) unit cell, and (d) cross-sectional area of the star prism. Reprinted from [89], Copyright (2021), with permission from Elsevier.

tungsten prism and the top layer is designed as a star prism whose inner part is tungsten and the outer part is SiO_2 as illustrated in figure 5.33(c) and (d).

Figure 5.34(a) and (b) shows the spectral absorption rate between 50 and 0.2 μm wavelengths. The spectral absorber performance (blue line) is compared with the solar irradiances (red and black lines) in these figure parts. It can be observed that the absorptivity ratio has been obtained above 0.95 between the wavelength of 1.8 and 0.2 μm, which covers the visible spectrum and near-infrared region in figure 5.34(a). Besides, the perfect absorption band has been monitored from 0.421 to 1.441 μm. While the wavelength is increasing from 1.8 to 4.67 μm, the absorption level is dramatically decreasing by up to 4%. This level has remained almost constant at around 2% along short, medium, and long-wave infrared regions as illustrated in figure 5.34(b). The total emittance and solar-to-heat efficiency as a function of absorber temperature have been also discussed to understand the effect of temperature on the performance of the absorber in figure 5.34(c). The total emittance has been reached from 0.235 to 0.406 between the absorber temperatures 1273 K and 1673 K. The solar-to-heat efficiency has decreased from 93% to 77% between this temperature scale. As can be seen from figure 5.35(c), for the best performance of the proposed absorber, the optimum desired absorber temperature can be selected at 1273 K, which corresponds to 0.235 of total emittance and 93% of solar-to-heat efficiency.

While designing metamaterial solar energy harvesters, one of the significant points is the metal type selection in terms of their melting points. Noble metals have low melting points, although they enable plasmonic resonance at the demanded spectrum. Hence, some researchers tend to prefer higher melting points and efficient metals for solar energy harvesting systems. Ouabida reported a three-layered metamaterial absorber design study that investigates the effect of different high melting point metals (HMPMs) on absorption performance [98]. The bottom layer of the proposed structure consisted of tungsten that can prevent the transmitted signal with temperature-insensitive characteristics in the infrared regime in figure 5.35. MgF_2 has been chosen as the interlayer, which has a high thermal expansion coefficient feature. The HMPMs have been chosen as the top layer, which is a pattern of periodically adjusted trapezoid unit cells as shown in figure 5.35.

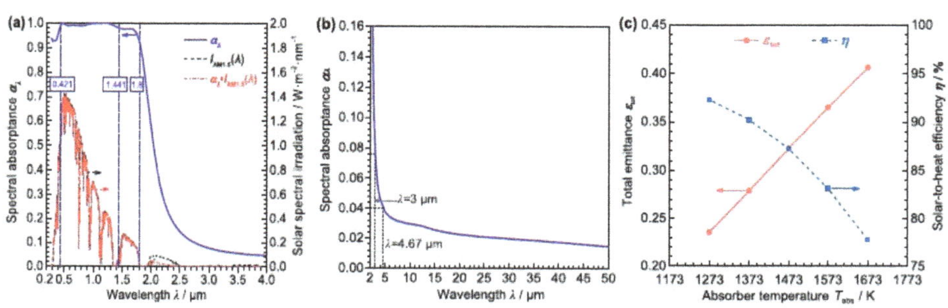

Figure 5.34. Spectral absorbance levels of the perfect selective metamaterial absorber between the wavelengths of (a) 4.0–0.2 μm and (b) 50–2 μm. (c) The total emittance and solar-to-heat efficiency versus absorber temperature. Reprinted from [89], Copyright (2021), with permission from Elsevier.

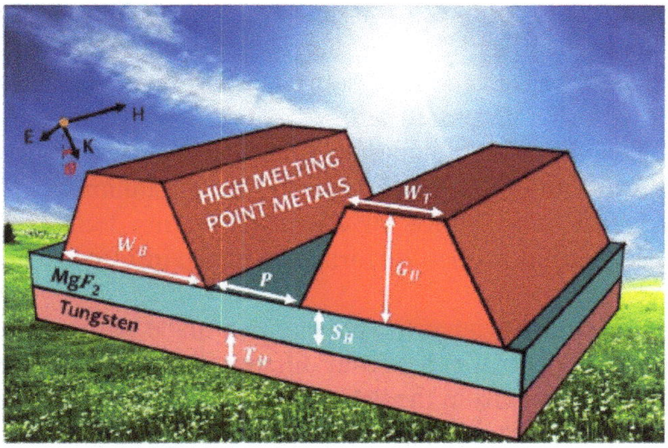

Figure 5.35. Schematic view of the proposed HMPM-based metamaterial solar absorber. Reprinted by permission from Springer Nature [98], Copyright (2022).

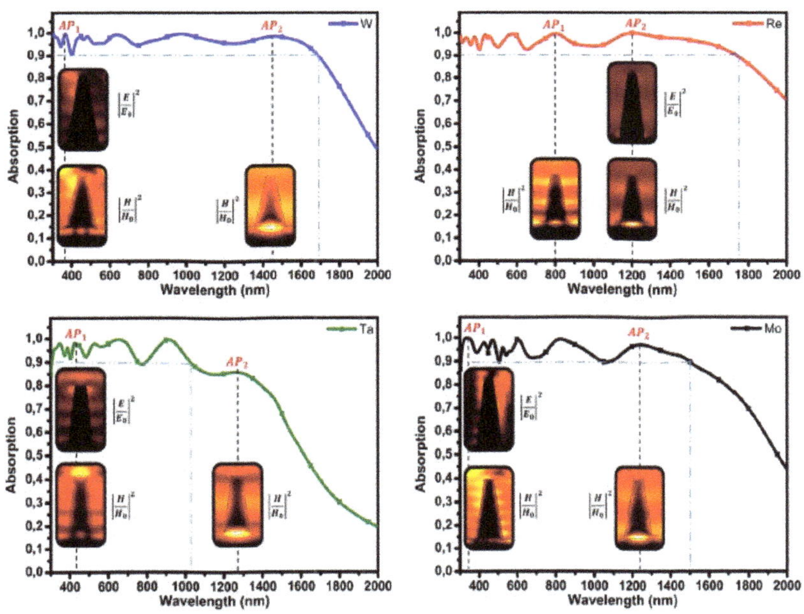

Figure 5.36. The absorption performance versus wavelength for various HMPMs (W, R, Ta, and Mo). Reprinted by permission from Springer Nature [98], Copyright (2022).

In this study, tungsten (W), rhenium (Re), tantalum (Ta), and molybdenum (Mo) types of metals have been chosen as HMPMs due to their high-temperature-insensitive features.

Figure 5.36 shows the absorption performance of the HMPMs (W, R, Ta, and Mo) as grating metals in the visible and near-infrared regimes with 8° of polarization angle. It also indicates the electric and magnetic distribution at cut-off and peak frequency points. All HMPMs exhibit sufficient absorption performance (>90%) at visible regimes varying from 400 to 700 nm. Tungsten (W) has an almost 1100 nm long (400 to 1700 nm) bandwidth higher than 90%. Rhenium also exhibited the same absorption bandwidth characteristics as tungsten. However, the absorption ratio of tungsten is dramatically decreased after greater than 1700 nm of wavelength compared with rhenium. It is a metal that can provide the widest perfect absorption bandwidth (>99%) within the studied spectrum. Tantalum (Ta) has the worst absorption bandwidth ratio above 90% with a 650 nm long (400 to 1050 nm) bandwidth. It also exhibits a poor performance in the infrared regime greater than 1000 nm of wavelength.

The researcher also investigated the polarization angle sensitivity of the HMPM-based metamaterial absorber for these four different types of metals as demonstrated in figure 5.37. It can be seen from the figure that the proposed metamaterial absorber is polarization angle sensitive under 800 nm of wavelength. This can be explained by the fact that, at lower polarization angles, the excitation of the surface plasmon and cavity modes excites the absorption peaks AP1 and AP2. [98] When the studied wavelength is increased from 800 nm to 1400 nm, the polarization angle sensitivity of the structure is dramatically reduced. "This can be explained by the fact that, with the chosen thickness of the dielectric spacer, the cavity modes are maintained at

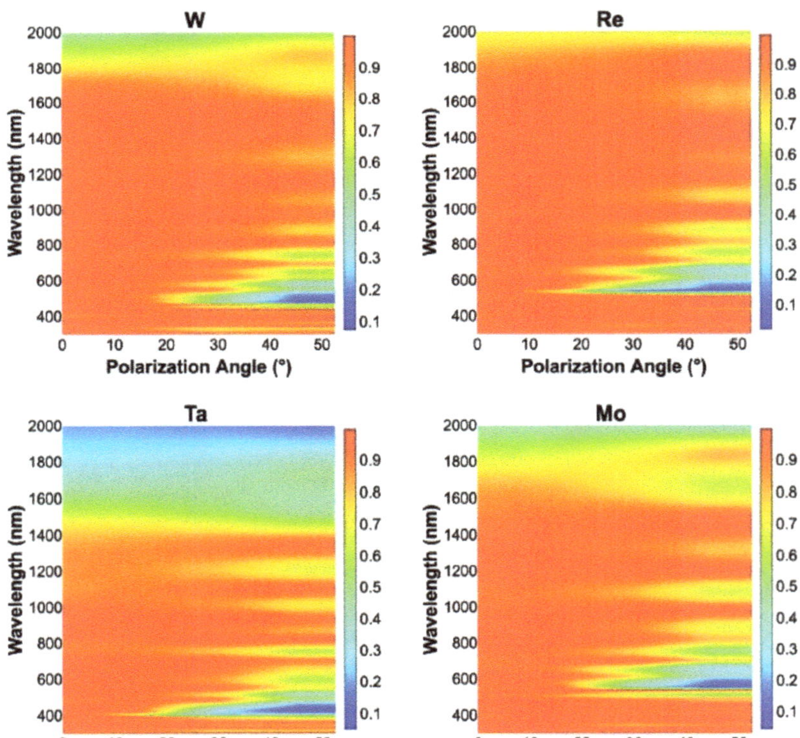

Figure 5.37. Polarization angle sensitivity of the metamaterial absorber with respect to wavelength for W, Re, Ta, and Mo. Reprinted by permission from Springer Nature [98], Copyright (2022).

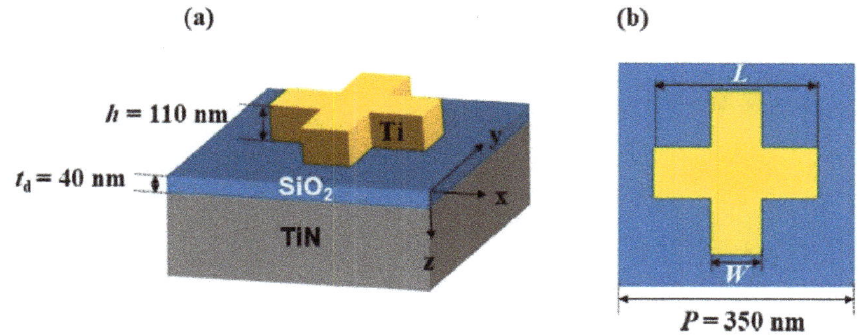

Figure 5.38. (a) Perspective and (b) top view of the proposed metamaterial absorber. Reprinted from [101], Copyright (2021), with permission from Elsevier.

higher wavelengths within spectral region of interest" [98]. The study discovered the versatile usage of HMPMs in infrared regime absorption and energy harvesting applications. The HMPMs can be effectively used instead of noble metals in terms of low cost and high absorption performance considering their temperature-insensitive characteristics.

In solar energy harvesting applications in the infrared spectrum, solar thermal energy conversion efficiency is one of the key factors to determine the performance of the metamaterial structure. The photon energy is first converted into the thermal form of energy before obtaining electricity. Dang *et al* presented a broadband and high solar thermal efficiency metamaterial absorber [101]. It is the combination of three layers which are designed by a cross-shaped grating layer (Ti), substrate layer (TiN), and dielectric layer (SiO_2) to separate the Ti and TiN layers with a 40 nm thickness as demonstrated in figure 5.38.

Figure 5.39 displays the perfect broadband absorption characteristic of the proposed absorber as a function of wavelength. It is well known that a less reflected signal causes more absorption ratio as long as the transmitted signal is prevented. The metamaterial absorber realizes over 95% of absorption from the wavelength of 321 to 1122 nm, which covers the visible and near-infrared regimes. It also enables three perfect absorption peak points or strong resonance peaks at 400, 620, and 1030 nm, respectively.

To understand the solar-to-thermal conversion efficiency, the overall electrical conversion efficiency is firstly written as in equation (5.4) [113].

$$\eta = \eta_{\text{sol-ther}} \times \left(1 - \frac{T_A}{T_B}\right) \tag{5.4}$$

where $\eta_{\text{sol-ther}}$ is solar-to-thermal conversion efficiency, T_A is the air temperature in K, and T_B is the instant temperature of the metamaterial absorber. The solar thermal conversion efficiency can be calculated from the energy balance equation and it is written as in equation (5.5).

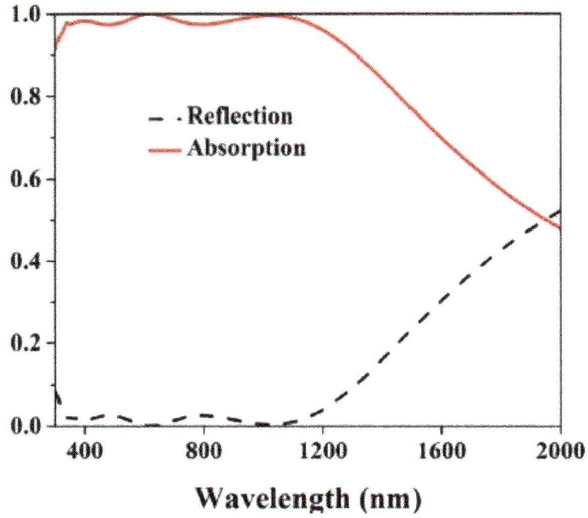

Figure 5.39. The reflection and absorption performance of the proposed metamaterial absorber versus wavelength. Reprinted from [101], Copyright (2021), with permission from Elsevier.

$$\eta_{sol-ther} = \frac{C \times \int d\lambda A(\lambda) E_{solar}(\lambda) - \int d\lambda A(\lambda) E_B(\lambda, T_B)}{C \times \int d\lambda \, E_{solar}(\lambda)} \tag{5.5}$$

where $A(\lambda)$ is the total solar absorption, E_B (λ, T_B) is the black body radiation, $E_{solar}(\lambda)$ is the solar irradiance, and C is the concentration factor. The black body radiation (E_B (λ, T_B)) and the total solar absorption ($A(\lambda)$) can be calculated as follows [114];

$$E_B(\lambda, T_B) = \frac{2\pi hc^2}{\lambda^5} \times \frac{1}{\exp(hc/\lambda k_B T_B) - 1} \tag{5.6}$$

$$A(\lambda) = \frac{\int A(\lambda) E_{solar} \lambda d\lambda}{\int E_{solar} \lambda d\lambda} \tag{5.7}$$

where h is Planck's constant, k_B is Boltzmann's constant, and c is the speed of light.

The solar-to-thermal conversion efficiencies have been investigated with respect to different incident angles and temperatures for both TE and TM modes as shown in figure 5.40. It can be found from figure 5.40(a) and (b) that the solar-to-thermal conversion efficiency is obtained at the level above 93% between the temperatures 373 K and 1073 K. It means that the metamaterial absorber can efficiently work within 700 K of the temperature band. However, the same efficiency cannot be observed for the incident angle results. The conversion level up to 93% can be only obtained in the incident angle varied from 0° to 31° for both TE and TM modes. The maximum conversion efficiency (= 96%) is monitored from 373 K to 770 K for the absorber temperature and from 0° to 11° for the incident angle in TE mode in

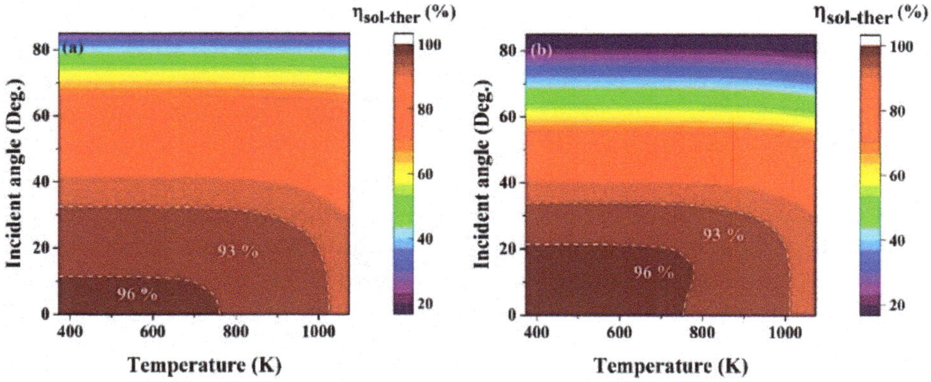

Figure 5.40. The solar-to-thermal conversion efficiencies under various incident angles and temperatures for (a) TE and (b) TM modes. Reprinted from [101], Copyright (2021), with permission from Elsevier.

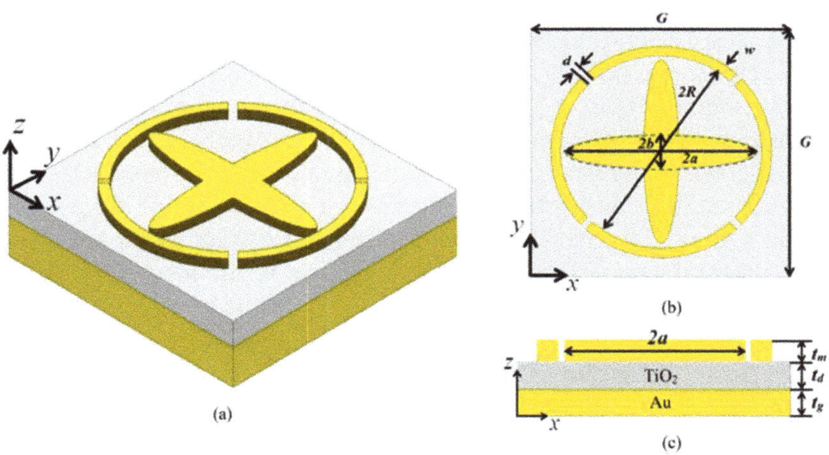

Figure 5.41. (a) Schematic view, (b) top view, and (c) cross-sectional view of the proposed perfect metamaterial absorber. Reproduced from [110]. CC BY 4.0.

figure 5.40(a). It is also observed from 373 K to 750 K of the absorber temperature and 0° to 21° for the incident angle in TM mode in figure 5.40(b).

High absorption efficiency, incident angle, and polarization insensitivity are essential factors while designing metamaterial absorbers in thermal energy harvesting applications due to the polarization, incident direction, and magnitude (in various time interval) of the solar source. From this point of view, Elsharabasy *et al* reported a metamaterial absorber designed by split-ring resonators with crossed ellipses for the near-infrared regime as shown in figure 5.41(a) and (b) [110]. It has a wide incident angle, wide band, and polarization insensitive characteristics. The structure has been composed with the approach of a metal-insulator-metal (MIM) which corresponds to Au-TiO_2-Au as illustrated in figure 5.41(c). They also investigated the energy harvesting performance by integrating the MIM diodes into the multiple gaps of the metamaterial absorber.

The performance of the perfect metamaterial absorber has been analyzed with respect to the angle of incidence in both TE and TM modes in resonance wavelength, two different ground insulators, the resistance and reactance, and diode's resistance sensitivity as shown in figure 5.42.

It is expected that the proposed absorber is polarization insensitive due to the symmetrical design. In TM mode, the perfect absorption performance has been presented in a wide incident angle varied from 0° to 60 in figure 5.42(a). The same feature has been observed in the incident angle from 0° to 30° in TE mode. In this mode, only a maximum of 20% of decrement in absorption ratio has been realized between the angles of 30° and 60°.

To avoid the complexity of the production process, Ti has been selected as a ground plane instead of Au. When their absorptivity performances are compared in figure 5.42(b), a slight shift in resonance wavelength is shown. Besides, the Ti-based ground plane has presented a little bit wider band characteristic compared with the Au-based ground plane. In figure 5.42(c), the impedance of the metamaterial absorber has been calculated. The resistance of the structure has been found to be above 500 Ω over the perfect absorption band, which can enhance the matching efficiency. It is greater than the dipole and bowtie antennas, which are presented in the literature as infrared antennas. In figure 5.42 (d), the effect of the diode's resistance on absorptivity has been investigated in a wide range from 500 Ω to 1 MΩ. It can be observed that the absorption band is not affected by the diode's resistance for the values greater than 1 kΩ. When the resistance value is chosen to be lower than 1 kΩ, almost the same absorption band response is obtained except for the little

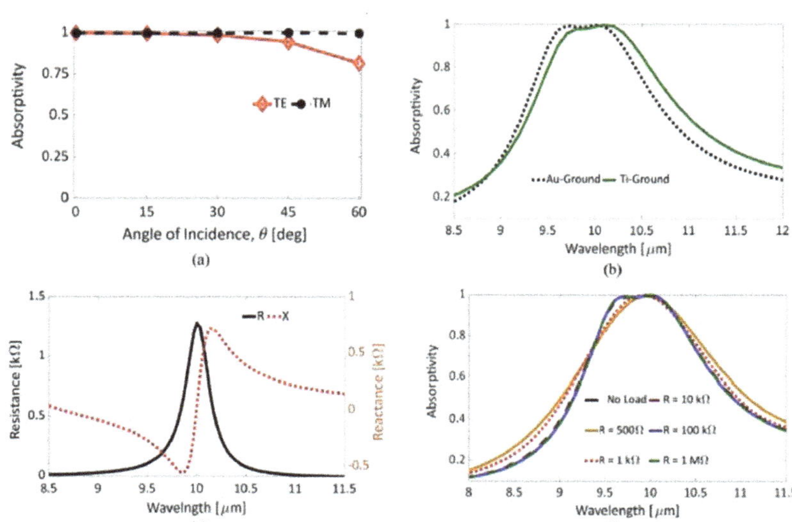

Figure 5.42. (a) Incident angle effect on the absorptivity for both TE and TM modes, (b) Au and Ti ground performance on absorptivity, (c) resistance and reactance versus wavelength, and (d) diode's resistance sensitivity. Reproduced from [110]. CC BY 4.0.

reduction in the upward edge of the plot in figure 5.42(d). This may be caused by the series connection of the diodes, which decreases the capacitance. Hence, it can end up with a shorter time constant and higher cut-off frequency.

5.5 Results and discussions

In this section, the performance of the acoustic, RF, and solar energy harvesters have been discussed by comparing the output voltage and power. In addition, their challenges and future have been reviewed.

Practical research on the acoustic energy harvesting concept is mainly focused on silicon or phononic crystal in terms of the material type selection given in table 5.1. Most of these studies preferred an aluminum plate as a ground plate and a piezoelectric transducer for conversion of the acoustic energy to DC power. A magnitude of difference can be seen in the observed output powers and voltages among the studies given in table 5.1. This especially originates from the magnitude of the source and the surface area of the designed metamaterial. The performances of the studies can be easily observed by comparing their bare plate samples. According to table 5.1, it can be said that the most efficient acoustic energy harvester has been shown by Park *et al*, whose metamaterial can achieve 20 times higher output power comparing the bare plate. On the other hand, Qi *et al* also found an effective metamaterial harvester that provides a 40 dB sound transmission loss to aim at filtering of the noise between 1800 and 3000 Hz. Moreover, Wang *et al* improved a tunable energy harvester that can achieve a multiple band and multiple position.

The optimization of the properties of the piezoelectric material and metamaterial geometry can improve the performance, ensuring maximum energy harness from a wider range of frequencies and amplitudes. In fact, research into the metamaterial and piezoelectric-based acoustic and acoustoelastic wave energy harvesting concept is at an early stage. For future research and innovation, some facets deserve attention. Exploring composite structures, nanostructured materials, and hybrid systems in terms of material advancement can be one of them. System miniaturization and flexibility can be considered for wearable technology, remote sensing, and biomedical applications as one of the future advancements. Furthermore, real-world testing is crucial for future enhancement and assessing mechanical performances such as durability, efficiency, and adaptability.

In table 5.2, some research studies in the field of RF regime energy harvesting have been compared and their research findings have been presented. Among these metamaterial energy harvester studies, one opinion is that the most interesting study is by Xie *et al.* Their wideband structure is temperature insensitive as a resin-type material has been used. On the other hand, Duan *et al* constructed a structure with a PTFE composite dielectric layer at 2.45 GHz. This study suggests an idea to have the energy harvesting capacity in mW per cm^2. Moreover, Lee *et al* studied a structure that collects the electromagnetic wave energy at the GSM band and utilizes an RF sensor. Another interesting study is realized by Dincer. His structure with a

Table 5.1. Comparison of the acoustic energy harvesting systems.

Year	Authors	Operating frequency (Hz)	Resonance frequency (Hz)	Material type	Max. output voltage (V)	Max. output power (uW)	Research findings
2016	Qi *et al* [2]	1800–3000	2257.5	Array of silicone rubber, aluminum plate, PZT-5H	1.3	8.8	40 dB sound transmission loss is achived
2017	Sun *et al* [40]	450–750	600	Coiled up structure, piezoelectric bimorph plate	0.25	0.345	Output voltage is 6.32 times higher than a conventional piezoelectric plate. Cavity dimension is 11 times smaller than subwavelength
2017	Daraby an Leamy [47]	100–500	318	Membrane-type acoustic metamaterial, beam-based PZT transducer	0.0726	0.0731	Output voltage is 4.6 times greater than a bare piezoelectric plate
2019	Park *et al* [29]	0–10 000	5000	Phononic crystal, PZT transducer	4.54	1560	Output power is improved by 20 times comparing the bare plate
2020	Ma *et al* [1]	2000–4000	2713	Silicone rubber, aluminum plate, PZT-5H	0.226	7.2	Voltage improvement is observed up to 950% comparing the bare plate
2020	Zhang *et al* [6]	0–1300	1214	Silicone rubber pillar, aluminum plate, PZT-5H	0.64	1.92	Voltage and power enhancement is achieved by 3 and 10 times comparing the bare plate
2021	Wang *et al* [34]	20 000–80 000	53 336	Aluminum, PZT patch	–	1520 000	Multiple-band and multiple-position piezoelectric energy harvesting is achieved
2021	Li *et al* [7]	200–2000	1581	Phononic crystal, P-52 transducer	–	29	Peak-to-peak output powers of structural and local resonance bands are improved 12.7 and 4.4 times greater than a traditional acoustic metamaterial

Table 5.2. Comparison of the RF regime energy harvesting systems.

Year	Authors	Operating frequency (GHz)	Resonance frequency (GHz)	Material type	Absorption efficiency (%)	Max. output voltage (V)	Max. output power (mW)	Research findings
2015	Dincer [77]	4.5–5.5	5.01	FR4	99.96	7.34	–	Tunable characteristic with a lumped varactor element
2016	Zhong *et al* [71]	1–6	1.75, 3.8, and 5,4	Rogers RO4003	97, 97, and 98	–	–	Multiband characteristic with lumped resistor element
2016	Gunduz and Sabah [74]	5–25	5.98, 8.42, 11.12, 13.66, and 14.76	FR4	99.16, 99.52, 99.95, 99.99, and 99	–	480	Multiband characteristic with lumped resistor element
2016	Duan *et al* [48]	2.2–5.2	2.45	Polytetrafluoroethylene (PTFE) composites	95.3	–	7.4 per cm^2	Rectifying surface with single diode for each unit cell
2018	Xie *et al* [67]	10–30	12–29.6	Resin	>85	–	–	Wideband and temperature-insensitive characteristics
2018	Lee *et al* [53]	1.6–2.6	2.4	Taconic TLC	–	3.4	–	Chemical sample sensing feature
2018	Karaaslan *et al* [60]	2–8	3.3–6.3	FR4	>80	–	900	Broadband characteristic with resistive lumped element
2021	Lee *et al* [59]	2–2.8	2.45	Rogers RO5880	–	–	6.27	High gain and RF sensing features

lumped varactor element can be tunable and perfectly absorb energy at certain frequency points between 4.5 and 5.5 GHz.

In recent years, the metamaterial energy harvesters in the RF regime have made a significant advancement. They offer an opportunity to obtain highly feasible energy harvesting by tuning up the metamaterial geometry and resonance frequency. Some future works in this field can be listed as material optimization, multiband harvesting, system integration, and environmental impact. Future efforts can be made to enhance tunability, efficiency, and durability by developing material properties. Simultaneous energy capturing by multiband harvesters can be also considered as the next phase. Integration of the metamaterial energy harvesters to systems such as rectification circuits, energy storage systems, and power management units will be an essential topic in the future. Future studies can also be focused on decreasing carbon footprints and waste originating from the metamaterials' production and deployment phase.

Table 5.3 compares recent literature research efforts for the infrared and visible light regime metamaterial energy harvesters in operating wavelength, material selection, absorption bandwidth, and solar-to-heat efficiency. As shown in the table, researchers have preferred tungsten, SiO_2, ZnO, Au, Ti, TiN, Re, Ta, and Mo materials for material selection in different layer designs. The operating wavelengths cover between 200 and 11 500 nm. As given in table 5.3, Bagmanci *et al* reached over 90% absorption in the visible light region that has a 310 nm bandwidth. Similarly, Unal *et al* presented a wideband solar absorber that reaches absorption over 98% with 1200 nm bandwidth in high angular stability. Elsharabasy *et al* presented a relatively narrow band absorption but it has nearly perfect absorption, and also has a rectifying circuit to harvest thermal radiations at 228 K by MIM diodes. In addition, Qiu *et al* and Dang *et al* have proposed different types of designs with high solar efficiencies. Finally, Elhoussaine presented a multiband design which has four resonance points with an absorption level over 90%.

In particular, metamaterial-based energy harvesting in the infrared and visible regimes has gained considerable attention due to the high potential in energy sources and the expanding range of applications. The integration of energy harvesting technologies in wearable electronics, IoT devices, and autonomous sensors has been notable. Moreover, harvesting energy from ambient light and heat presents an opportunity to power electronic devices sustainably, eliminating the need for any battery replacements and reducing environmental effect. By the ongoing research efforts, the energy conversion rates will be maximized and reach unity in the future by metamaterial integration with absorber structures. In addition, the absorption bandwidth in these wavelengths will be ultra-wideband with the advancement of material science. Smartphone screens, textile materials, etc, need flexible surfaces to harvest and convert incident infrared and visible light energy. In the future, this phenomenon will be reached and more efficient technological devices will be proposed. Furthermore, energy storage is another important point for future advancements. The energy storage and battery life problem will be exceeded by the help of metamaterial-based energy harvesters.

Table 5.3. Comparison of the infrared and visible regime energy harvesting systems.

Year	Authors	Operating wavelength (nm)	Material type	Absorption bandwidth (nm)	Solar-to-heat efficiency	Research findings
2017	Bagmanci *et al* [102]	390–700	Tungsten and SiO_2	310 (>90%)	–	Absorbing behaviors for derivations of silicon
2017	Unal *et al* [107]	300–3000	Tungsten and ZnO	1200 (>98%)	–	Perfect absorber, high angular stability
2020	Elsharabasy *et al* [110]	8500–11 500	TiO_2 and Au	500 (>99%)	–	MIM diodes designed to rectify the harvested thermal radiations at 288 K
2021	Qiu *et al* [89]	200–4000	Tungsten and SiO_2	1500 (>99%)	92.31–77.78% at 1000 suns and 1273–1673 K	High spectral absorptance, incident angle insensitivity under 60° both TE and TM modes
2021	Dang *et al* [101]	300–2000	Ti, TiN, and SiO_2	792 (>97%)	96% at 1000 suns and 800 K	High solar thermal conversion efficiency
2022	Elhoussaine [98]	200–2000	Tungsten (W), rhenium (Re), tantalum (Ta), and molybdenum (Mo)	1390, 1440, 720, and 1200 (>90%)	–	High-temperature solar absorbing features of some high melting point metals

5.6 Conclusion

In conclusion, this book chapter presents an in-depth investigation into metamaterial energy harvesting technology and its advancing and future perspective. Design processes, numerical analysis, mathematical backgrounds, fabrication techniques, experimental investigations, and application technologies of metamaterial energy harvesters have been demonstrated throughout this exploration. Metamaterials offer remarkable capability and have huge potential for different technologies because of their ability to control structural parameters such as constitutive parameters, resonant frequencies, bandwidths, etc. By their unique behavior, metamaterials pave the way for an improvement for energy harvesting applications with optimized integration. The energy harvesting concept is based on the conversion of electromagnetic waves scattered around into usable energy. This mentioned scattered electromagnetic energy covers the frequency band from acoustic waves to the visible wave region that is varied from 1 kHz to 750 THz. This chapter has examined metamaterial energy harvesters in three subparts: acoustics, RFs, and infrared frequencies.

From urban environments with a low noise ratio to industrial locations with machine generated sounds, metamaterial acoustic energy harvesters offer a novel solution for powering wireless sensors, low-energy electronics, and other energy efficient technologies. Moreover, it is clearly seen that these energy harvesters have the capability to manipulate acoustic waves and convert ambient sound energy or vibrations into usable electrical energy. Other important advantages of these devices are scalability, durability, and cost-effectiveness in an application area. According to their advantages, metamaterial acoustic energy harvesters will play a crucial role in everyday technology; however, material scientists, engineers, and researchers are still working to unlock their full potential.

Furthermore, another metamaterial energy harvesting subpart is for RF frequencies as microwave and mm-waves. By high precision engineering structures at the micro and millimeter scale, metamaterial energy harvesters have a great capability to convert incident electromagnetic energy to usable electrical energy levels, and the converted electrical energy can be used in any electronic circuits. In the process of harvesting and converting to incident electromagnetic waves, the arrayed structures are mostly preferred because each cell can have an energy level and the total energy level can reach high levels. In addition, resistors, varactors, diodes, switches, and any other microwave circuit components can be used in a metamaterial RF energy harvester design to obtain high-level electrical energy. This book chapter presents the background of RF metamaterial energy harvester design processes and their experimental analysis. These harvesters play a role in more energy efficient and green energy technologies.

The last subpart of this book chapter presents metamaterial energy harvesters for infrared and visible light frequency regions that cover frequencies between 300 and 750 GHz. By the development of nanoscale technologies, these harvester devices enable the conversion of infrared and visible light energy into usable electrical power, that helps to improve self-powered electronics, smart surfaces, and photovoltaic enhancements. Finally, these metamaterial energy harvesters have a huge potential to reshape our energy technology by using the scattered energy of light across the electromagnetic spectrum.

References

[1] Ma K J, Tan T, Liu F R, Zhao L C, Liao W H and Zhang W M 2020 Acoustic energy harvesting enhanced by locally resonant metamaterials *Smart Mater. Struct.* **29** 075025

[2] Qi S, Oudich M, Li Y and Assouar B 2016 Acoustic energy harvesting based on a planar acoustic metamaterial *Appl. Phys. Lett.* **108** 263501

[3] Deng T, Zhang S and Gao Y 2019 Tunability of band gaps and energy harvesting based on the point defect in a magneto-elastic acoustic metamaterial plate *Appl. Phys. Express* **13** 015503

[4] Chen Q, Zhu Y, Zhang K and Feng K 2023 Broadband low-frequency acoustic energy harvesting amplified by sonic crystal metamaterial with double defects *J. Vib. Eng. Technol.* 1–12

[5] Mir F, Mandal D and Banerjee S 2023 Metamaterials for acoustic noise filtering and energy harvesting *Sensors* **23** 4227

[6] Zhang Z, Li Q, Oudich M, Pan Y and Li Y 2020 Experimental demonstration of enhanced acoustic energy harvesting with a subwavelength metamaterial plate *New J. Phys.* **22** 123019

[7] Li T, Wang Z, Xiao H, Yan Z, Yang C and Tan T 2021 Dual-band piezoelectric acoustic energy harvesting by structural and local resonances of Helmholtz metamaterial *Nano Energy* **90** 106523

[8] Sun W, Zhong K, Liu Y, Xiao H, Zhao D, Yan Z and Tan T 2022 Enhanced metamaterial vibration for high-performance acoustic piezoelectric energy harvesting *Compos. Commun.* **35** 101342

[9] Li Y, Baker E, Reissman T, Sun C and Liu W K 2017 Design of mechanical metamaterials for simultaneous vibration isolation and energy harvesting *Appl. Phys. Lett.* **111** 251903

[10] Chen Z, He J and Wang G 2019 Vibration bandgaps of piezoelectric metamaterial plate with local resonators for vibration energy harvesting *Shock Vib.* 2019

[11] Xu X, Wu Q, Pang Y, Cao Y, Fang Y, Huang G and Cao C 2022 Multifunctional metamaterials for energy harvesting and vibration control *Adv. Funct. Mater.* **32** 2107896

[12] Oudich M and Li Y 2017 Tunable sub-wavelength acoustic energy harvesting with a metamaterial plate *J. Phys. D: Appl. Phys.* **50** 315104

[13] Ma K, Tan T, Yan Z, Liu F, Liao W H and Zhang W 2021 Metamaterial and Helmholtz coupled resonator for high-density acoustic energy harvesting *Nano Energy* **82** 105693

[14] De Ponti J M, Colombi A, Ardito R, Braghin F, Corigliano A and Craster R V 2020 Graded elastic metasurface for enhanced energy harvesting *New J. Phys.* **22** 013013

[15] Lee T G, Jo S H, Seung H M, Kim S W, Kim E J, Youn B D and Kim M 2020 Enhanced energy transfer and conversion for high performance phononic crystal-assisted elastic wave energy harvesting *Nano Energy* **78** 105226

[16] De Ponti J M, Colombi A, Riva E, Ardito R, Braghin F, Corigliano A and Craster R V 2020 Experimental investigation of amplification, via a mechanical delay-line, in a rainbow-based metamaterial for energy harvesting *Appl. Phys. Lett.* **117** 143902

[17] Li K, Rizzo P and Bagheri A 2015 A parametric study on the optimization of a metamaterial-based energy harvester *Smart Mater. Struct.* **24** 115019

[18] Ma T X, Li Z Y, Zhang C and Wang Y S 2022 Energy harvesting of Rayleigh surface waves by a phononic crystal Luneburg lens *Int. J. Mech. Sci.* **227** 107435
[19] Ma T X, Fan Q S, Li Z Y, Zhang C and Wang Y S 2020 Flexural wave energy harvesting by multi-mode elastic metamaterial cavities *Extreme Mech. Lett.* **41** 101073
[20] Afework M B and Manimala J M 2022 Exploring mechanical assonance for impact energy harvesting using acoustic metamaterials *J. Intell. Mater. Syst. Struct.* **33** 1126–46
[21] Hu G, Tang L, Liang J, Lan C and Das R 2021 Acoustic-elastic metamaterials and phononic crystals for energy harvesting: a review *Smart Mater. Struct.* **30** 085025
[22] Bukhari M and Barry O 2020 Simultaneous energy harvesting and vibration control in a nonlinear metastructure: a spectro-spatial analysis *J. Sound Vib.* **473** 115215
[23] Hyun J, Choi W and Kim M 2019 Gradient-index phononic crystals for highly dense flexural energy harvesting *Appl. Phys. Lett.* **115** 173901
[24] Wang X, Wang L, Shu H and Zhang L 2022 Research on dual-functional properties of an improved piezoelectric metamaterial beam for simultaneous vibration suppression and energy harvesting *Appl. Phys.* A **128** 906
[25] Lee G, Lee D, Park J, Jang Y, Kim M and Rho J 2022 Piezoelectric energy harvesting using mechanical metamaterials and phononic crystals *Commun. Phys.* **5** 94
[26] Lu Z Q, Zhao L, Ding H and Chen L Q 2021 A dual-functional metamaterial for integrated vibration isolation and energy harvesting *J. Sound Vib.* **509** 116251
[27] Zhao B, Thomsen H R, De Ponti J M, Riva E, Van Damme B, Bergamini A, Colombi … and A 2022 A graded metamaterial for broadband and high-capability piezoelectric energy harvesting *Energy Convers. Manage.* **269** 116056
[28] Ichige R, Kuriyama N, Umino Y, Tsukamoto T and Suzuki T 2021 Size optimization of metamaterial structure for elastic layer of a piezoelectric vibration energy harvester *Sensors Actuators* A **318** 112488
[29] Park C S, Shin Y C, Jo S H, Yoon H, Choi W, Youn B D and Kim M 2019 Two-dimensional octagonal phononic crystals for highly dense piezoelectric energy harvesting *Nano Energy* **57** 327–37
[30] Dwivedi A, Banerjee A and Bhattacharya B 2020 Simultaneous energy harvesting and vibration attenuation in piezo-embedded negative stiffness metamaterial *J. Intell. Mater. Syst. Struct.* **31** 1076–90
[31] Jiao P, Yang Y, Egbe K I, He Z and Lin Y 2021 Mechanical metamaterials gyro-structure piezoelectric nanogenerators for energy harvesting under quasi-static excitations in ocean engineering *ACS Omega* **6** 15348–60
[32] Hwang M and Arrieta A F 2018 Input-independent energy harvesting in bistable lattices from transition waves *Sci. Rep.* **8** 3630
[33] Sugino C and Erturk A 2018 Analysis of multifunctional piezoelectric metastructures for low-frequency bandgap formation and energy harvesting *J. Phys. D: Appl. Phys.* **51** 215103
[34] Wang B, Huang Y, Zhou W and Yang Z 2021 Metamaterial beam for flexural wave resonance rainbow trapping and piezoelectric energy harvesting *J. Appl. Phys.* **129** 064505
[35] Hu G, Tang L and Das R 2018 Internally coupled metamaterial beam for simultaneous vibration suppression and low frequency energy harvesting *J. Appl. Phys.* **123** 055107
[36] Lei Y, Wu J H, Huang Z and Yang S 2021 Enhanced broadband monopole emission and acoustic energy harvesting via a dual anisotropic metamaterial *J. Phys. D: Appl. Phys.* **55** 065301

[37] Bansal S, Choi C, Hardwick J, Bagchi B, Tiwari M K and Subramanian S 2023 Transmissive labyrinthine acoustic metamaterial-based holography for extraordinary energy harvesting *Adv. Eng. Mater.* **25** 2201117

[38] Mir F, Saadatzi M, Ahmed R U and Banerjee S 2018 Acoustoelastic MetaWall noise barriers for industrial application with simultaneous energy harvesting capability *Appl. Acoust.* **139** 282–92

[39] Zhang X, Zhang H, Chen Z and Wang G 2018 Simultaneous realization of large sound insulation and efficient energy harvesting with acoustic metamaterial *Smart Mater. Struct.* **27** 105018

[40] Sun K H, Kim J E, Kim J and Song K 2017 Sound energy harvesting using a doubly coiled-up acoustic metamaterial cavity *Smart Mater. Struct.* **26** 075011

[41] Li J, Zhou X, Huang G and Hu G 2016 Acoustic metamaterials capable of both sound insulation and energy harvesting *Smart Mater. Struct.* **25** 045013

[42] Ahmed R U and Banerjee S 2014 Low frequency energy scavenging using sub-wave length scale acousto-elastic metamaterial *AIP Adv.* **4** 117114

[43] Lan C, Hu G, Tang L and Yang Y 2021 Energy localization and topological protection of a locally resonant topological metamaterial for robust vibration energy harvesting *J. Appl. Phys.* **129** 184502

[44] Wang Z, Cheng J, Xie Y, Wang Y, Yu Z, Li S and Wang H 2022 Lead-free piezoelectric composite based on a metamaterial for electromechanical energy conversion *Adv. Mater. Technol.* **7** 2200650

[45] Alshaqaq M and Erturk A 2020 Graded multifunctional piezoelectric metastructures for wideband vibration attenuation and energy harvesting *Smart Mater. Struct.* **30** 015029

[46] Darabi A and Leamy M J 2017 Analysis and experimental verification of multiple scattering of acoustoelastic waves in thin plates for enhanced energy harvesting *Smart Mater. Struct.* **26** 085015

[47] Wang X, Xu J, Ding J, Zhao C and Huang Z 2019 A compact and low-frequency acoustic energy harvester using layered acoustic metamaterials *Smart Mater. Struct.* **28** 025035

[48] Duan X, Chen X and Zhou L 2016 A metamaterial electromagnetic energy rectifying surface with high harvesting efficiency *AIP Adv.* **6** 125020

[49] Oumbé Tékam G T, Ginis V, Danckaert J and Tassin P 2017 Designing an efficient rectifying cut-wire metasurface for electromagnetic energy harvesting *Appl. Phys. Lett.* **110** 083901

[50] Hawkes A M, Katko A R and Cummer S A 2013 A microwave metamaterial with integrated power harvesting functionality *Appl. Phys. Lett.* **103** 163901

[51] Coskuner E and Garcia-Garcia J J 2021 Metamaterial impedance matching network for ambient RF-energy harvesting operating at 2.4 GHz and 5 GHz *Electronics* **10** 1196

[52] Nowak M 2021 Metamaterial-based sub-microwave electromagnetic field energy harvesting system *Energies* **14** 3370

[53] Lee W, Jung Y, Jung H, Seo C, Choo H and Lee H 2018 Wireless-powered chemical sensor by 2.4 GHz Wi-Fi energy-harvesting metamaterial *Micromachines* **10** 12

[54] Fowler C, Silva S, Thapa G and Zhou J 2022 High efficiency ambient RF energy harvesting by a metamaterial perfect absorber *Opt. Mater. Express* **12** 1242–50

[55] Elwi T A 2019 Novel UWB printed metamaterial microstrip antenna based organic substrates for RF-energy harvesting applications *AEU-Int. J. Electron. Commun.* **101** 44–53

[56] El Badawe M and Ramahi O M 2018 Efficient metasurface rectenna for electromagnetic wireless power transfer and energy harvesting *Prog. Electromagn. Res.* **161** 35–40

[57] Elwi T A, Jassim D A and Mohammed H H 2020 Novel miniaturized folded UWB microstrip antenna-based metamaterial for RF energy harvesting *Int. J. Commun. Syst.* **33** e4305

[58] Elwi T A 2020 A further realization of a flexible metamaterial-based antenna on nickel oxide polymerized palm fiber substrates for RF energy harvesting *Wirel. Pers. Commun.* **115** 1623–34

[59] Lee W, Choi S I, Kim H I, Hwang S, Jeon S and Yoon Y K 2021 Metamaterial-integrated high-gain rectenna for RF sensing and energy harvesting applications *Sensors* **21** 6580

[60] Karaaslan M, Bağmancı M, Ünal E, Akgol O, Altıntaş O and Sabah C 2018 Broad band metamaterial absorber based on wheel resonators with lumped elements for microwave energy harvesting *Opt. Quantum Electron.* **50** 1–18

[61] Zhong H T, Yang X X, Song X T, Guo Z Y and Yu F 2017 Wideband metamaterial array with polarization-independent and wide incident angle for harvesting ambient electromagnetic energy and wireless power transfer *Appl. Phys. Lett.* **111** 213902

[62] Cheng Y, Luo H and Chen F 2020 Broadband metamaterial microwave absorber based on asymmetric sectional resonator structures *J. Appl. Phys.* **127** 214902

[63] Alavikia B, Almoneef T S and Ramahi O M 2015 Wideband resonator arrays for electromagnetic energy harvesting and wireless power transfer *Appl. Phys. Lett.* **107** 243902

[64] Bağmancı M, Karaaslan M, Altıntaş O, Karadağ F, Tetik E and Bakır M 2018 Wideband metamaterial absorber based on CRRs with lumped elements for microwave energy harvesting *J. Microw. Power Electromagn. Energy* **52** 45–59

[65] Akgol O, Bağmancı M, Karaaslan M and Ünal E 2017 Broad band MA-based on three-type resonator having resistor for microwave energy harvesting *J. Microwave Power Electromagn. Energy* **51** 134–49

[66] Bağmancı M, Karaaslan M, Unal E, Özaktürk M, Akgol O, Karadağ F and Bakır M 2019 Wide band fractal-based perfect energy absorber and power harvester *Int. J. RF Microw. Comput.-Aided Eng.* **29** e21597

[67] Xie J, Zhu W, Rukhlenko I D, Xiao F, He C, Geng J and Premaratne M 2018 Water metamaterial for ultra-broadband and wide-angle absorption *Opt. Express* **26** 5052–9

[68] Zhao J and Cheng Y 2016 Ultrabroadband microwave metamaterial absorber based on electric SRR loaded with lumped resistors *J. Electron. Mater.* **45** 5033–9

[69] Duan X, Chen X, Zhou Y, Zhou L and Hao S 2018 Wideband metamaterial electromagnetic energy harvester with high capture efficiency and wide incident angle *IEEE Antennas Wirel. Propag. Lett.* **17** 1617–21

[70] Liu C, Wang W, Olivier D N, Wang Z, Ding B and Feng J 2022 Target driven design of electromagnetic metamaterial for dual-band Wi-Fi energy harvester *Sensors Actuators* A **345** 113815

[71] Zhong H T, Yang X X, Tan C and Yu K 2016 Triple-band polarization-insensitive and wide-angle metamaterial array for electromagnetic energy harvesting *Appl. Phys. Lett.* **109** 253904

[72] Wei Y, Duan J, Jing H, Lyu Z, Hao J, Qu Z and Zhang B 2022 A multiband, polarization-controlled metasurface absorber for electromagnetic energy harvesting and wireless power transfer *IEEE Trans. Microw. Theory Tech.* **70** 2861–71

[73] Chandrasekaran K T, Agarwal K, Alphones A, Mittra R and Karim M F 2020 Compact dual-band metamaterial-based high-efficiency rectenna: an application for ambient electromagnetic energy harvesting *IEEE Antennas Propag. Mag.* **62** 18–29

[74] Gunduz O T and Sabah C 2016 Polarization angle independent perfect multiband metamaterial absorber and energy harvesting application *J. Comput. Electron.* **15** 228–38

[75] Karakaya E, Bagci F, Yilmaz A E and Akaoglu B 2019 Metamaterial-based four-band electromagnetic energy harvesting at commonly used GSM and Wi-Fi frequencies *J. Electron. Mater.* **48** 2307–16

[76] Bakır M, Karaaslan M, Altıntaş O, Bagmancı M, Akdogan V and Temurtaş F 2018 Tunable energy harvesting on UHF bands especially for GSM frequencies *Int. J. Microwave Wireless Technolog.* **10** 67–76

[77] Dincer F 2015 Electromagnetic energy harvesting application based on tunable perfect metamaterial absorber *J. Electromagn. Waves Appl.* **29** 2444–53

[78] Sabban A 2022 Wearable circular polarized antennas for health care, 5G, energy harvesting, and IoT systems *Electronics* **11** 427

[79] Elwi T A and Al-Saegh A M 2021 Further realization of a flexible metamaterial-based antenna on indium nickel oxide polymerized palm fiber substrates for RF energy harvesting *Int. J. Microw. Wirel. Technol.* **13** 67–75

[80] Alaukally M N N, Elwi T A and Atilla D C 2022 Miniaturized flexible metamaterial antenna of circularly polarized high gain-bandwidth product for radio frequency energy harvesting *Int. J. Commun. Syst.* **35** e5024

[81] Kaur K P, Upadhyaya T, Palandoken M and Gocen C 2019 Ultrathin dual-layer triple-band flexible microwave metamaterial absorber for energy harvesting applications *Int. J. RF Microw. Compt.-Aided Eng.* **29** e21646

[82] Tetik E 2022 Flexible metamaterial design based on wearable material for microwave energy harvesting *Bull. Mater. Sci.* **45** 178

[83] Tian X, Lee P M, Tan Y J, Wu T L, Yao H, Zhang M and Ho J S 2019 Wireless body sensor networks based on metamaterial textiles *Nat. Electron.* **2** 243–51

[84] Shang S, Yang S, Shan M, Liu J and Cao H 2017 High performance metamaterial device with enhanced electromagnetic energy harvesting efficiency *AIP Adv.* **7** 105204

[85] AlShareef M R and Ramahi O M 2014 Electrically small particles combining even-and odd-mode currents for microwave energy harvesting *Appl. Phys. Lett.* **104** 253906

[86] Alkurt F O, Altintas O, Ozakturk M, Karaaslan M, Akgol O, Unal E and Sabah C 2020 Enhancement of image quality by using metamaterial inspired energy harvester *Phys. Lett. A* **384** 126041

[87] Alkurt F O, Altintas O, Bakir M, Tamer A, Karadag F, Bagmanci M and Akgol O 2018 Octagonal shaped metamaterial absorber based energy harvester *Mater. Sci.* **24** 253–9

[88] Costanzo S and Venneri F 2020 Polarization-insensitive fractal metamaterial surface for energy harvesting in IoT applications *Electronics* **9** 959

[89] Qiu Y, Zhang P, Li Q, Zhang Y and Li W 2021 A perfect selective metamaterial absorber for high-temperature solar energy harvesting *Sol. Energy* **230** 1165–74

[90] Ji Q, Chen X, Liang J, Laude V, Guenneau S, Fang G and Kadic M 2021 Designing thermal energy harvesting devices with natural materials through optimized microstructures *Int. J. Heat Mass Transfer* **169** 120948

[91] Bazgir M, Naser-Moghadasi M, Zarrabi F B, Arezomand A S and Heydari S 2017 Nano particle implementation in nano loop antenna for energy harvesting application and light trapping *Optik* **132** 127–33
[92] Rufangura P and Sabah C 2016 Design and characterization of a dual-band perfect metamaterial absorber for solar cell applications *J. Alloys Compd.* **671** 43–50
[93] Hoque A and Islam M T 2020 Numerical analysis of single negative broadband metamaterial absorber based on tri thin layer material in visible spectrum for solar cell energy harvesting *Plasmonics* **15** 1061–9
[94] Wang H, Sivan V P, Mitchell A, Rosengarten G, Phelan P and Wang L 2015 Highly efficient selective metamaterial absorber for high-temperature solar thermal energy harvesting *Sol. Energy Mater. Sol. Cells* **137** 235–42
[95] Hossain I, Samsuzzaman M, Moniruzzaman M, Bais B B, Singh M S J and Islam M T 2021 Polarization-independent broadband optical regime metamaterial absorber for solar harvesting: a numerical approach *Chin. J. Phys.* **71** 699–715
[96] Wu J, Sun Y, Wu B, Sun C and Wu X 2022 Perfect metamaterial absorber for solar energy utilization *Int. J. Therm. Sci.* **179** 107638
[97] Almoneef T and Ramahi O M 2017 Dual-polarized multi-band infrared energy harvesting using H-shaped metasurface absorber *Prog. Electromagn. Res.* C **76** 1–10
[98] Ouabida E 2022 Broadband metamaterial solar absorbers based on the earth first-rate temperature-insensitivity metals for high temperature energy harvesting *Plasmonics* **17** 1691–8
[99] Yao Y, Liu H and Wu W 2014 Spectrum splitting using multi-layer dielectric meta-surfaces for efficient solar energy harvesting *Appl. Phys.* A **115** 713–9
[100] AlShareef M R and Ramahi O M 2013 Electrically small resonators for energy harvesting in the infrared regime *J. Appl. Phys.* **114** 223101
[101] Dang P T, Kim J, Nguyen T K, Le K Q and Lee J H 2021 Ultra-broadband metamaterial absorber for high solar thermal energy conversion efficiency *Phys.* B **620** 413261
[102] Bağmancı M, Karaaslan M, Ünal E, Akgol O and Sabah C 2017 Extremely-broad band metamaterial absorber for solar energy harvesting based on star shaped resonator *Opt. Quantum Electron.* **49** 1–14
[103] Mulla B and Sabah C 2016 Multiband metamaterial absorber design based on plasmonic resonances for solar energy harvesting *Plasmonics* **11** 1313–21
[104] Lin K T, Lin H, Yang T and Jia B 2020 Structured graphene metamaterial selective absorbers for high efficiency and omnidirectional solar thermal energy conversion *Nat. Commun.* **11** 1389
[105] Li Y, Lin C, Li K, Chi C and Huang B 2022 Nanoparticle-on-mirror metamaterials for full-spectrum selective solar energy harvesting *Nano Lett.* **22** 5659–66
[106] Hossain M J, Faruque M R I and Islam M T 2018 Perfect metamaterial absorber with high fractional bandwidth for solar energy harvesting *PLoS One* **13** e0207314
[107] Ünal E, Bağmancı M, Karaaslan M, Akgol O, Arat H T and Sabah C 2017 Zinc oxide–tungsten-based pyramids in construction of ultra-broadband metamaterial absorber for solar energy harvesting *IET Optelectron.* **11** 114–20
[108] Langlais M, Hugonin J P, Besbes M and Ben-Abdallah P 2014 Cooperative electromagnetic interactions between nanoparticles for solar energy harvesting *Opt. Express* **22** A577–88

[109] Feng H, Li X, Wang M, Xia F, Zhang K, Kong W and Yun M 2021 Ultrabroadband metamaterial absorbers from ultraviolet to near-infrared based on multiple resonances for harvesting solar energy *Opt. Express* **29** 6000–10

[110] Elsharabasy A, Bakr M and Deen M J 2020 Wide-angle, wide-band, polarization-insensitive metamaterial absorber for thermal energy harvesting *Sci. Rep.* **10** 16215

[111] Lin Y, Cui Y, Ding F, Fung K H, Ji T, Li D and Hao Y 2017 Tungsten based anisotropic metamaterial as an ultra-broadband absorber *Opt. Mater. Express* **7** 606–17

[112] Mohamadi T and Yousefi L 2019 Metamaterial-based energy harvesting for detectivity enhanced infrared detectors *Plasmonics* **14** 815–22

[113] Li P, Liu B, Ni Y, Liew K K, Sze J, Chen S and Shen S 2015 Large-scale nanophotonic solar selective absorbers for high-efficiency solar thermal energy conversion *Adv. Mater.* **27** 4585–91

[114] Zhou J, Liu Z, Liu G, Pan P, Liu X, Tang C and Wang J 2020 Ultra-broadband solar absorbers for high-efficiency thermophotovoltaics *Opt. Express* **28** 36476–86

IOP Publishing

Advanced Metamaterials for Engineers

Lulu Wang and Muharrem Karaaslan

Chapter 6

Frequency selective surfaces (FSSs) in metamaterials

Fatima-Ezzahra Zerrad and Mohamed Taouzari

Symbols

ε	Permittivity
λ	Wavelength (m)
f	Frequency (Hz)
θ	Angle of incidence
C	Capacitance
L	Inductance
R	Resistance
ƨ	Reflection coefficient
T	Transmission coefficient
TE	Transverse electrical
UC	Unit Cell

6.1 Introduction

Metasurfaces are widely recognized as planar metamaterials with a subwavelength thickness, offering the advantage of facile fabrication through lithography and nano-printing techniques. The domains of metamaterials and metasurfaces have witnessed rapid growth as research frontiers, enabling precise control over spatially varying electromagnetic or optical responses encompassing scattering phase, amplitude, and polarization. Remarkably, the ultra-thin structure of metasurfaces, when thoughtfully designed and composed of suitable materials, can effectively mitigate deleterious losses along the direction of wave propagation.

Regarding polarization behavior, metasurfaces can be classified according to the fundamental operational principle of their individual array components. This categorization encompasses a variety of functions, including but not limited to

doi:10.1088/978-0-7503-5754-8ch6

frequency selective surfaces (FSSs), high-impedance surfaces, perfect absorbers, reflecting surfaces, and additional capabilities.

As per the definition provided in [1], frequency selective surfaces (FSSs) are categorized as metasurfaces that exclusively display an electric response. This means that, in order to manipulate the frequency selectivity in terms of transmission and reflection characteristics, an electric polarization alone may suffice. Drawing from the principles of antenna and microwave engineering, these surfaces are constructed using planar and periodic arrays of metallic patches or strips with varying shapes.

The thickness of the metallic patch is negligible compared to the wavelength, although it remains significantly larger than the skin depth of the metal. Consequently, this structure can be accurately approximated as a minute, thin array of perfectly conducting resonant elements.

This approximation is also relevant for complementary frequency selective surface structures, particularly those of the aperture type. Nevertheless, these aperture-type FSSs face a constraint when the dimensions of the cavity or aperture approach the size of the unit cell, which is frequently depicted as a wire-mesh arrangement. Wire-mesh unit cells, often referred to as capacitive grids, are commonly used, including square and hexagonal configurations. The existence of resonating components within the array leads to the emergence of side lobes in both transmitted and reflected fields, distinguishing features of FSSs.

In contrast, metasurfaces boast significantly smaller resonating elements and unit cells relative to the wavelength, resulting in the elimination of grating lobes in the frequency response. This feature sets them apart from traditional FSSs, prompting the term "metasurfaces" to be commonly used when discussing FSSs in the terahertz domain.

Frequency selective surfaces are extensively studied in the field of electromagnetic science. They are two-dimensional periodic structures consisting of metallic array elements, such as patches or apertures, on a dielectric substrate. These structures exhibit transmission and reflection characteristics at specific resonant frequencies. Depending on the design of the array elements, an incoming plane wave can be either completely or partially transmitted or reflected. This occurs when the frequency of the plane wave matches the resonance frequency of the FSS elements. Consequently, FSSs act as spatial filters, allowing or blocking electromagnetic waves of specific frequencies in free space.

Conventionally, the size, shape, and periodic arrangement of the FSS elements determine the resonance behavior. Over the past six decades, FSSs have been extensively investigated, leading to the evolution of various microwave and optical FSS structures. They have been commonly used in applications such as reflector antennas, diplexers for quasi-optical microwave devices, resonant beam splitters, and antenna radomes. More recently, FSSs have found application in dichroic sub-reflectors, radio frequency identification (RFID), lens antennas, and electromagnetic interference protection. Presently, the most prominent applications of FSSs are in antenna radomes and controlling radar cross section (RCS).

FSSs present a range of benefits, including extensive design flexibility, small footprint, versatility, simplified fabrication, and integration. These attributes position the FSS as a practical substitute for conventional filters, and its exceptional filtering attributes make it an apt choice for communication filtering functions.

Beyond filtering, the FSS serves a multitude of roles in diverse areas such as antenna radomes, reflectors, electromagnetic shielding, polarizers, absorbers, maintaining angular stability, enabling reconfiguration, and more. Nonetheless, it is important to acknowledge that these structures do have certain limitations:

- The large electrical length of the unit cell necessitates a large array of FSSs, which can result in angular instability and affect the performance.
- Fabricating complex FSS structures can be challenging.
- Implementing multiple layers of FSSs increases both the size and fabrication cost required to attain the desired outcome.

Consequently, scientists are working towards reducing the dimensions and structural intricacy of FSSs to simplify the fabrication and production processes. Significant studies have been conducted to improve FSS performance, and this chapter explores a selection of techniques that have been investigated.

6.2 Operational principles of periodic structures

An infinite arrangement of identical components leads to the formation of a periodic surface in either one or two dimensions [2]. The activation of a periodic array can occur through an incident plane wave (in the passive array configuration) or by connecting generators to each element (in the active array setup). In the preceding instance, the incoming plane wave (Ei) experiences a dual outcome: it is partly transmitted (Et) in the forward direction and partly reflected (Er) with a specular reflection effect. In the absence of grating lobes, the amplitude of the reflected wave Er may equal the amplitude of the incident wave Ei, while the transmitted signal Et is equal to zero. Equations (6.1), (6.2) give the specular reflection coefficient (Γ) and the transmission coefficient (T), respectively.

$$\Gamma = Er/Ei \tag{6.1}$$

$$T = Et/Ei \tag{6.2}$$

For an active array to qualify as a periodic surface, its voltage generators need to share identical amplitudes and exhibit consistent linear phase variation across all elements within the array. Dipole and slot arrays with identical element shapes form complementary arrays, and when these arrays are stacked on top of one another (cascaded), a perfect conducting plane is formed. Babinet's principle asserts that the transmission coefficient of an array is equivalent to the specular reflection coefficient of its complementary array. To achieve a stable resonance with changes in the angle of incidence for a periodic surface, it is crucial to maintain a very small inter-element spacing ($<0.4\ \lambda$) [2].

FSSs are commonly composed of periodic arrangements of metallic elements on a dielectric substrate. These structures exhibit distinctive transmission and reflection characteristics centered around a specific resonant frequency. A specific type of FSS, known as a mushroom-like structure, is engineered to resonate at lower frequencies. This resonance is achieved by capitalizing on the inductive effect produced by the spacing between a metal array element and a metal plane in the microwave domain.

Contrastingly, perfect absorbing surfaces are compact layered configurations that have the ability to absorb nearly all incident electromagnetic wave power at a specific frequency and angle of entry. These properties are particularly pronounced within the microwave to optical frequency ranges.

Importantly, it should be noted that when the periodicity of array elements (spacing between elements) is not significantly smaller in comparison to the wavelength in free space, periodic structures may struggle to maintain a stable resonance when exposed to varying angles of incidence [2].

6.3 Explanation of the functional mechanism of frequency selective surfaces

As per circuit theory principles, FFSs encompassing both capacitive and inductive elements, often referred to as spatial filters, possess an equivalence to microwave filters. The distinguishing properties of FSS filtering can be primarily classified into four distinct categories: low pass, high pass, passband, and stopband. Low pass FSS filters facilitate the passage of frequencies within a lower range through the structure, effectively obstructing the transmission of higher frequencies. Employing the Babinet principle, the operation of high pass FSS filters stands as the converse of low pass filters, allowing higher frequencies to pass while attenuating lower frequencies.

Likewise, a stopband FSS filter eliminates undesired frequencies, whereas a passband FSS filter permits transmission for just a limited frequency range. FSSs are created using periodic arrays of metal patches and/or slots etched on a dielectric material to provide the necessary resonant operation. The most significant aspect of the design process is the suitable selection of FSS array elements, form, dimension, and substrate material.

Figure 6.1 depicts FSS functionality provided by a complementary self-resonating network. In brief, when EM waves strike an FSS structure, they induce electric currents in the array elements. The amplitude of the generated currents is determined by the level of coupling energy. These created currents, however, act as EM sources and generate further scattered fields. The resultant field in the vicinity of the FSS is made up of incident EM fields merged with these distributed fields. As a result, properly constructed elements may obtain the required currents and field characteristics and generate the filter response.

Patch elements, known as dipole FSSs, are designed to function as stopband filters for incoming plane waves and as fully reflecting surfaces within a specific frequency range. Conversely, their counterparts, the aperture elements or slot FSSs, demonstrate passband characteristics, serving as transparent surfaces for electromagnetic waves within the designated frequency band. However, conventional FSSs have restricted effectiveness in fulfilling the functional needs of diverse

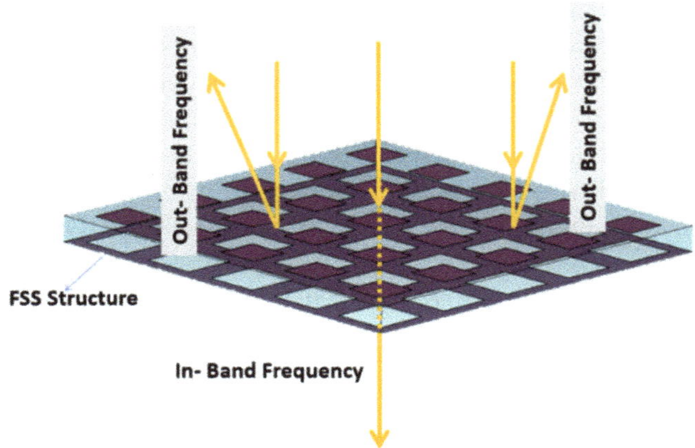

Figure 6.1. FSS functioning.

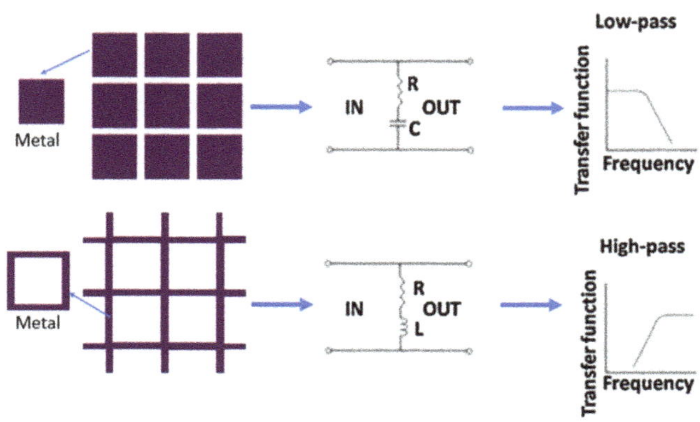

Figure 6.2. FSS periodic structure with complementary array elements, equivalent circuits, and frequency responses.

electromagnetic applications due to their inadequate filter response, limited angular stability, and narrow bandwidth.

Figure 6.2 shows illustrations of two fundamental types of elements, along with their corresponding equivalent circuits and filter responses for low and high pass frequency selective surfaces. The FSS patches generate resistance (R) and inductance (L), while the gaps between the FSS elements generate capacitance (C). To adapt these passive values for various FSS elements, such as using the electrostatic principle to relate L to two parallel wires and C to a parallel plate capacitor, the physical interpretation is adjusted. This approach allows the combination of capacitive and inductive elements to produce the desired filter response. However, any alteration in the dimensions of the FSS leads to corresponding changes in the values of L and C. When an electromagnetic wave interacts with an FSS unit cell, it transforms into an analogous resonance circuit.

Equation (6.3) can be used to calculate the resonance frequency.

$$f = \frac{1}{2\pi\sqrt{LC}} \tag{6.3}$$

Choosing a proper array element is critical in FSS construction. While different unit cell designs have been developed, some are easily adjustable and thus more well-known in the FSS community. The work in [3] summarizes a taxonomy of commonly utilized FSS element types based on their resonant characteristics.

There are two approaches to determining the impedance and scattering features of periodic structures [2]. The primary one is the mutual impedance method, which is tied to the method of moments (MoM), and the alternative is the plane wave expansion approach, commonly known as the spectral method.

Floquet's principle, which stipulates that each element in a planar structure, or each unit cell, should have the same amount of field and current distribution when an infinite plane wave contacts with an infinite periodic structure, forms the foundation of the FSS's basic operation. A genuine periodic arrangement (infinite × infinite) is a planar array that is infinite in the x and z directions with identical inter-element spacing ($Dx = Dz$), as seen in figure 6.3. Assume that the array is struck by an incident EM wave moving in the $\hat{s}$ direction provided in (6.4).

The element currents in column b and row a are defined by Floquet's theorem as (6.5)[4].

Ohm's law for the reference element 0,0 yields (6.6), and (6.7), which gives the array scan impedance.

$$\hat{s} = \hat{x}s_x + \hat{y}s_y + \hat{z}s_z \tag{6.4}$$

$$I_{ba} = I_{0,0}e^{-j\beta b D_x s_x}e^{-j\beta a D_z s_z} \tag{6.5}$$

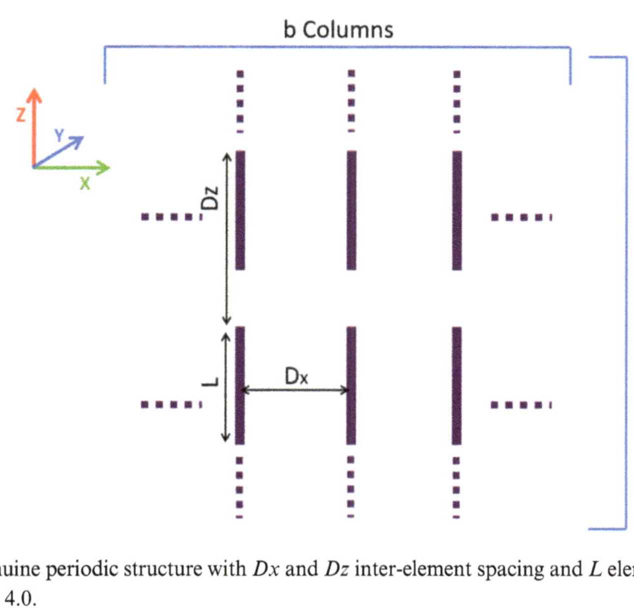

Figure 6.3. A genuine periodic structure with Dx and Dz inter-element spacing and L element length. Adapted from [1]. CC BY 4.0.

$$V^{0,0} = \left[Z_L + \sum_{b=-\infty}^{\infty} \sum_{a=-\infty}^{\infty} Z_{0,\,ba} e^{-j\beta b D_x s_x} e^{-j\beta a D_z s_z} \right] I_{0,\,0} \tag{6.6}$$

$$Z^{0,0} = \sum_{b=-\infty}^{\infty} \sum_{a=-\infty}^{\infty} Z_{0,\,ba} e^{-j\beta b D_x s_x} e^{-j\beta a D_z s_z} \tag{6.7}$$

6.4 Equivalent circuit of FSS

Equivalent circuits are developed to comprehend the physical backdrop of the specified FSS architecture. Using a similar model, the square loop FSS is the easiest to examine. This model is supported by the transmission line analogy, where the FSS is represented by equivalent inductive (*L*) and capacitive (*C*) lumped components. The quasi-static equivalent circuit approximation, developed by Marcuvitz [5], enables the calculation of *L* and *C* values by considering conducting strips. In the case of transverse electrical (TE) wave incidence, the vertical strips function as an *L* impedance, while the horizontal gratings act as a *C* impedance within the equivalent circuit model. To determine the values of *L* and *C* in the equivalent circuit, the dimensions of the square loop, including *d*, *p*, *w*, and *g*, are commonly used in calculations.

Figure 6.4 depicts the square loop unit cell construction and its corresponding circuit. Table 6.1 summarizes the characteristics for various FSS structures [6]. In the case of equivalent inductance, X_L represents the inductive reactance and Z_0 represents the characteristic impedance, but in the case of equivalent capacitance, B_c represents the susceptance, and B_g and B_d represent the susceptance between the horizontal and vertical arms, respectively. The wavelength of the FSS is determined by λ, and *F* is a function that depends on the FSS's design parameters.

The double and triple square loop structures offer several advantages over others. They stand out due to the simple geometry, ease of fabrication, high selectivity, and

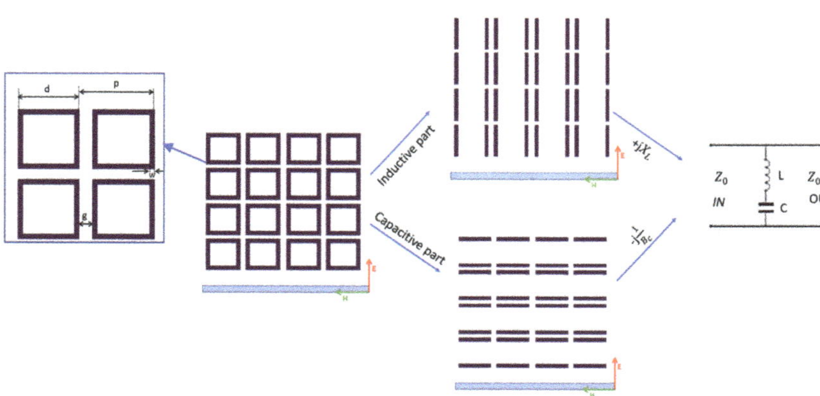

Figure 6.4. FSS with square loop and its equivalent circuit.

Table 6.1. Overview of geometries and equivalent circuits.

FSS	Configuration	Circuit equivalence	Equivalent inductance	Equivalent capacitance
Single square loop	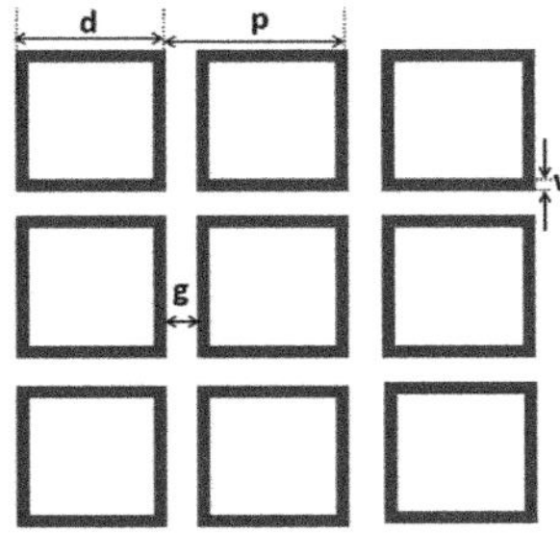	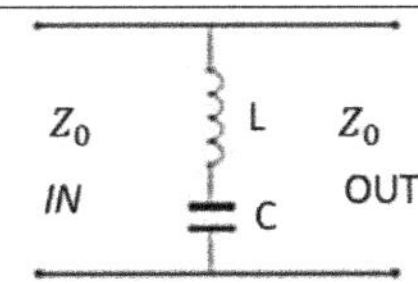	$\frac{X_L}{Z_0} = \frac{d}{p}F(p, 2w, \lambda)$	$\frac{B_c}{Z_0} = 4\frac{d}{p}F(p, g, \lambda)$
Double square loop	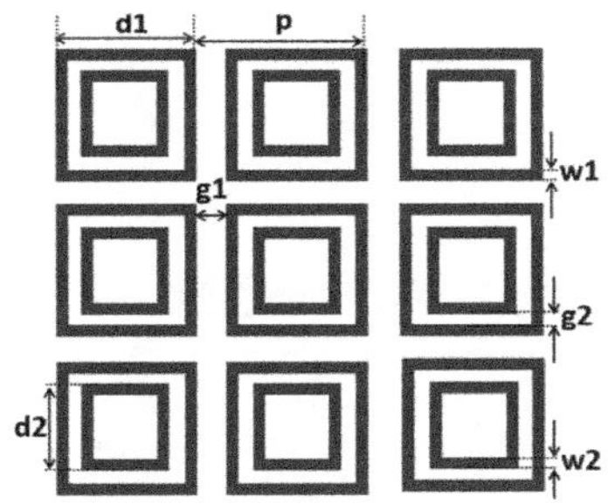	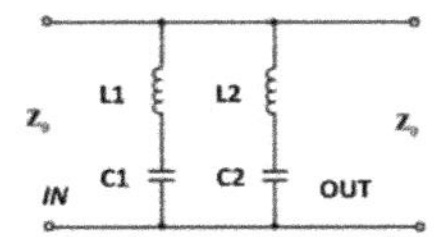	$\left(\frac{X_L}{Z_0}\right)_1 = \frac{d}{p}F(p, w1, \lambda)$ $\left(\frac{X_L}{Z_0}\right)_2 = \frac{d}{p}F(p, w2, \lambda)$	$\left(\frac{B_c}{Z_0}\right)_1 = 4F(p, g1, \lambda)$ $\left(\frac{B_c}{Z_0}\right)_2 = 4F(p, g2, \lambda)$
Triple square loop	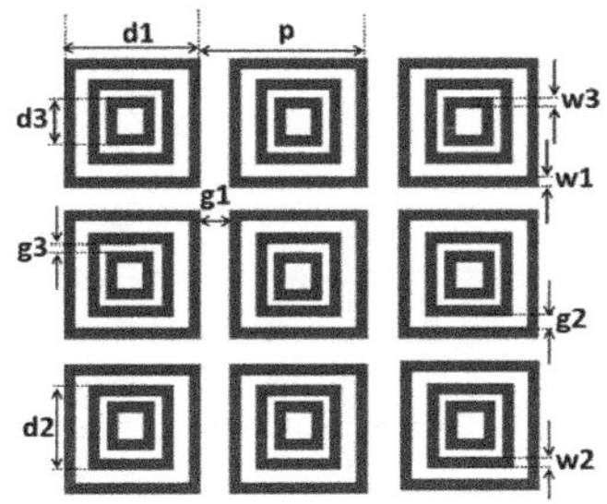	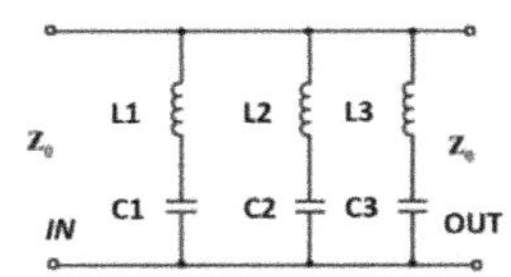	$\left(\frac{X_L}{Z_0}\right)_1 = \frac{d}{p}F(p, w1, \lambda)$ $\left(\frac{X_L}{Z_0}\right)_2 = \frac{d}{p}F(p, w2, \lambda)$ $\left(\frac{X_L}{Z_0}\right)_3 = \frac{d}{p}F(p, w3, \lambda)$	$\left(\frac{B_c}{Z_0}\right)_1 = 4F(p, g1, \lambda)$ $\left(\frac{B_c}{Z_0}\right)_2 = 4F(p, g2, \lambda)$ $\left(\frac{B_c}{Z_0}\right)_3 = 4F(p, g3, \lambda)$

Cross dipole	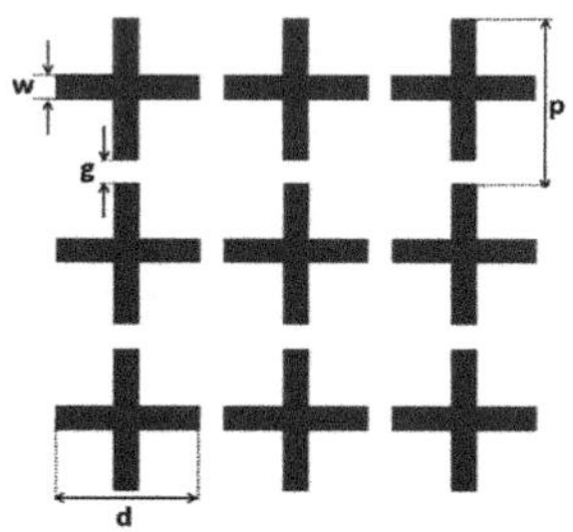	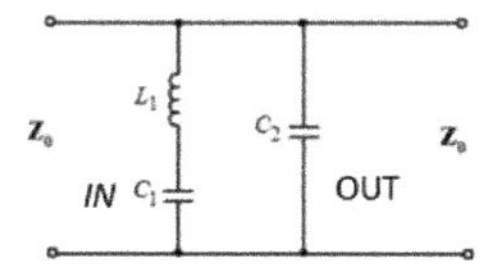	$\frac{X_L}{Z_0} = \frac{d}{p}F(p, w, \lambda)$	$B_g = 4\frac{w}{p}F(p, g, \lambda)$ $B_d = 4\frac{d}{p}F(p, p - w, \lambda)$
Jerusalem cross array	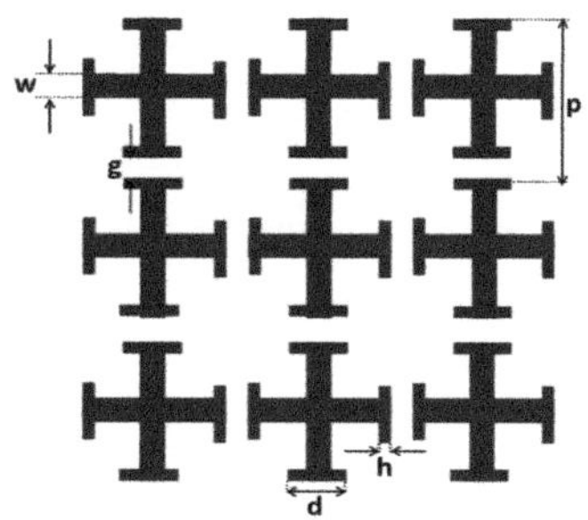	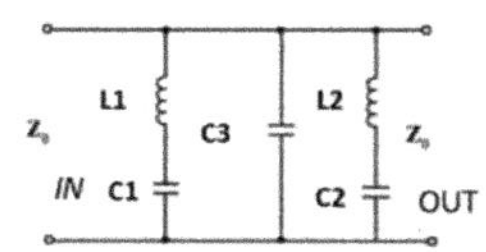	$\frac{X_L}{Z_0} = \frac{d}{p}F(p, w, \lambda)$	$B_g = 4\frac{d}{p}F(p, g, \lambda)$ $B_d = 4\frac{(2h+g)}{p}F(p, p - d, \lambda)$

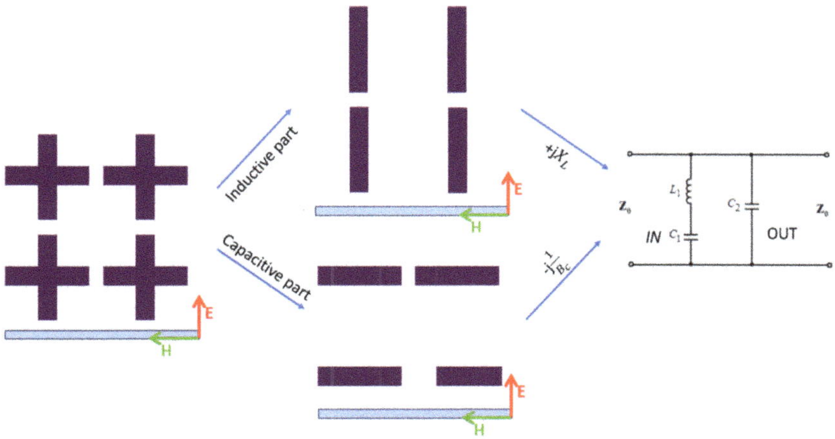

Figure 6.5. FSS with cross dipole and its equivalent circuit.

ability to operate in multiple frequency bands. Additionally, modifying these FSS structures is easily accomplished, and they typically remain unaffected by changes in polarization and angle of incidence.

Figure 6.5 illustrates a periodic array utilizing crossed dipoles. The array consists of crossed dipoles with specific parameters, including the periodicity (p), dipole length (d), strip width (w), and spacing between the two dipole arms (g). The structure of the crossed dipole can be divided into perpendicular and parallel conducting strips based on the orientation of the electric field. The parallel strips, aligned with the electric field, act as inductive elements, while the perpendicular strips represent capacitive elements. By breaking down the structure of the crossed dipole into vertical and horizontal strips, as depicted in figure 6.5, this configuration is obtained.

Due to the similarity in the manner in which the strips can be divided, it was initially utilized to describe the crossed dipole FSS as a straightforward LC serial circuit composed of an array of square loops. But according to the bandwidth and resonance frequency, we found that the curves for transmitted power and reflected power were symmetric and did not match the identical measurement curves. A capacitor acting in parallel with the LC branch is a factor in the suggested equivalent circuit.

A crucial point to note is to avoid changing the main branch LC (L1-C1) since it controls the configuration reject-band and the resonant frequency. A subsequent rule is that no element should be placed in series or parallel with C1 or L1. The capacitor C2 is connected to the branch L1-C1 in parallel. Within this arrangement, capacitor C2 exerts no influence over the resonant frequency; rather, its role lies in enhancing the bandwidth.

The equivalent circuit for the crossed dipole structure considers the inductive reactance of width (w), a capacitive susceptance derived from the spacing (p–w), and another susceptance derived from the spacing (g). The inductive reactance and capacitive susceptance are reduced by a factor of d/p due to the non-continuous nature of the strips. However, it was observed that for a crossed dipole, a finite conducting strip parallel to the electric field does not behave as an inductor, contrary to what was previously assumed in [2].

Instead, based on Munk's demonstration in [2], it was concluded that these finite strips should be associated with capacitors in the equivalent circuit for non-continuous periodic structures. This alternative was chosen because it was observed that increasing the frequency resulted in an increase in the resonance frequency, which contradicted the findings in [2]. The Munk theory aligns with the requirement of using the factor *d*/*p*.

In the design of narrow band, multiband filters, and multi-resonant high-impedance surfaces (HIS), it is often necessary to use more complex FSS elements. Some FSS elements have the ability to exhibit a double resonant or even a multi-resonant behavior within the zero-order Floquet propagating zone. To analyze these structures, additional lumped elements are introduced into the resonant circuit. One commonly used FSS shape is known as the Jerusalem Cross, which is essentially a cross shape with loading ends that enhance the capacitance value. In order to account for the electric field coupling with the two end-loading dipoles, an additional series LC circuit is required in parallel with the original circuit. Figure 6.6 illustrates the layout of the double resonant unit cell's geometry and its corresponding equivalent circuit.

Selecting the appropriate element is critical when designing a frequency selective surface. There are numerous possible geometries that can be chosen as the unit cell for a periodic screen, offering a virtually infinite selection [3]. However, certain geometries are more widely used and easier to manipulate than others. In table 6.2, a classification of the most popular geometries based on their resonant properties and

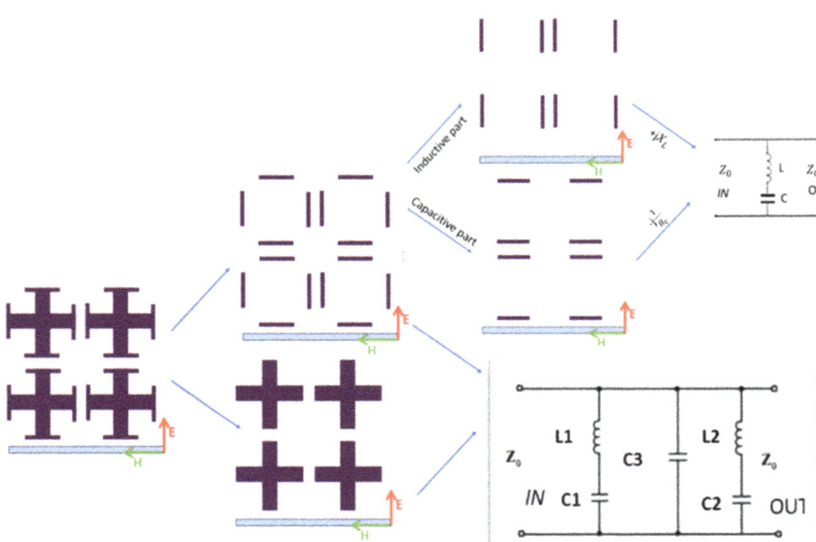

Figure 6.6. FSS with Jerusalem cross array and its equivalent circuit.

Table 6.2. Categorization of FSS elements based on resonant properties.

Resonant property	Geometry	Circuit equivalence
Non-resonant	Strip, patch, wire grid	C
Single resonant	Loop, dipole, cross, tripole, dogbone	LC
Double resonant	Double cross, double loop, Jerusalem cross, etc.	(LC)//(LC) or (LC)(L//C)
Multi-resonant (geometrical)	Concentric loops, fractal elements	(LC)//(LC)//(LC) ....
Multi-resonant (current distribution)	Quadrifilar spiral, meandered dipole or loop, genetically optimized	Not physical

equivalent circuits is presented. The number of lumped elements required is directly related to the number of resonances exhibited by the chosen geometry.

6.5 Applications of FSS

6.5.1 Spatial filter based on FSS

Filtering plays a vital role in communication systems by eliminating undesired frequencies, allowing only the necessary frequencies to pass through. The FSS acts as either a bandstop or a bandpass filter, depending on whether the metallic component is a patch or an aperture [7].

Numerous bandpass and bandstop configurations have been suggested in previous studies [8, 9]. For instance, polarization-independent bandstop filters employ multi-stacked metallic patches, while combinations of spiral elements are utilized as compact bandpass filters [10] through the implementation of FSS. The frequency selective surface is utilized to directly create a dual bandstop filter on glass windows, specifically designed for shielding Wi-Fi and WLAN signals. This design incorporates an annular ring and a modified flower-shaped geometry to achieve the desired filtering capabilities [11].

6.5.2 Integration of the FSS with antennas

Researchers are focused on enhancing the antenna design to improve gain, directivity, broaden bandwidth, and achieve elevated efficiency. In this context, a FSS not only facilitates filtering but also contributes to enhancing the antenna's gain and bandwidth, leading to improved communication capabilities [12].

A dual-polarized antenna that exhibits unidirectional radiation capabilities in the ultra-wideband (UWB) frequency range is presented in [13]. In this design, a FSS serves as the backing reflector, upon which the dual-polarized radiator is mounted. The antenna demonstrates an improved average gain of 9.5 dBi with two distinct and favorable polarizations. It achieves impedance coverage across the entire UWB with a notable isolation of 20 dB.

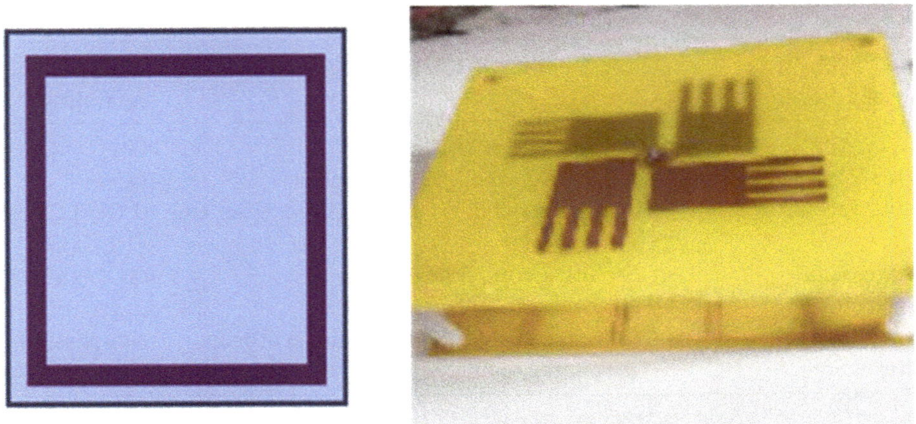

Figure 6.7. FSS unit cell, and the fabricated antenna presented in [15]. Adapted from [15]. CC BY 4.0.

Another study [14] presents an electromagnetic bandgap (EBG) antenna. To enhance both the impedance bandwidth and gain, a layer of FSS is employed as a superstrate. Additionally, a multisource feeding technique is utilized to enhance directivity, enabling operation in the Ku-band. The antenna achieves a maximum gain of 22.5 dB with an impedance bandwidth of 14%.

Additionally, researchers in [15] present a wideband circularly polarized dipole antenna with a low radar cross section (RCS) using a FSS (figure 6.7). To enhance the antenna's radiation performance and reduce RCS, a bandstop FSS is employed as the antenna reflector. The presented FSS consists of a grid of 5×5 unit cells and is positioned beneath the planar slotted dipole antenna. The FSS effectively acts as a reflector for electromagnetic waves within the desired frequency range. The antenna achieves an impedance bandwidth of 0.95 to 3.1 GHz ($S_{11} \leqslant -10$ dB) and a -3dB axial ratio bandwidth of 1.9 to 2.35 GHz. Results from simulations and measurements confirm that incorporating the FSS increases the maximum antenna gain by 5.9 dB.

6.5.3 MIMO system based on FSSs

In addition to microwave applications, FSSs can also be implemented in millimeter-wave applications. In reference [16], a wideband FSS structure is employed in the U and V bands. The structure consists of a Jerusalem cross pattern on the top of the RO4003 substrate and a FAN shape pattern on the bottom. These two different shapes are combined to form a unified unit cell FSS. Individually, the Jerusalem cross resonates at 48 GHz, while the FAN shape resonates at 60 GHz. Nevertheless, when these two structures are integrated, the resonant frequency shifts due to the mutual coupling effect. The bandwidth is also evaluated by varying the loop radius of the FAN shape and the arm length of the Jerusalem cross. It is noted that as the loop radius changes from 0.4 to 0.25 mm, the bandwidth reduces from 56.75% to 44%. Additionally, increasing the arm length causes the resonance frequency to shift to lower values.

The increasing significance of millimeter MIMO systems has been observed in recent years. In reference [17], a multiband MIMO system integrating 3D antennas and a FSS is introduced for diverse applications (figure 6.8). To achieve efficient isolation between antenna ports, a decoupling structure based on a grid of FSS patches is implemented, resulting in a measured isolation of more than 30 dB.

6.5.4 Electromagnetic shielding based on FSS

In order to safeguard wireless receiving devices and circuits from undesirable radiating field effects, electromagnetic shielding is necessary. The utilization of a FSS represents one approach towards achieving this objective. Reference [18] introduced a configuration where a cross dipole is positioned on one side of the FR4 substrate, while a ring is placed on the other side. This structure exhibits a stopband spanning from 6.5 to 14 GHz, with a center frequency of 7.5 GHz. It provides shielding effectiveness within the X-band and Ka-band communication bands, offering a minimum attenuation of 20 to 35 dB.

Nonetheless, an alternative approach involving a reconfigurable FSS has been explored, wherein a pair of square loops are interconnected to form the unit cell. The electromagnetic shielding capabilities of this reconfigurable FSS model have been evaluated [25]. This particular structure offers effective electromagnetic shielding within the WiMAX (3.5 GHz), WLAN (5 GHz), and ISM/WiMAX bands (5.8 GHz). For tuning purposes, a configuration of 1×20 elements of FSS has been implemented. The reconfiguration range spans from 3.5 to 8.2 GHz and is achieved by adjusting the spacing between the unit cells. Notably, by increasing the horizontal spacing (k), the resonant frequency shifts towards lower frequencies, resulting in optimal shielding effectiveness at lower values of k. Consequently, this prototype demonstrates usefulness in wideband frequency tuning applications.

6.5.5 Meta-skin

One potential application involves integration and recently there has been an incorporation of liquid metals as active components in the development of electronic

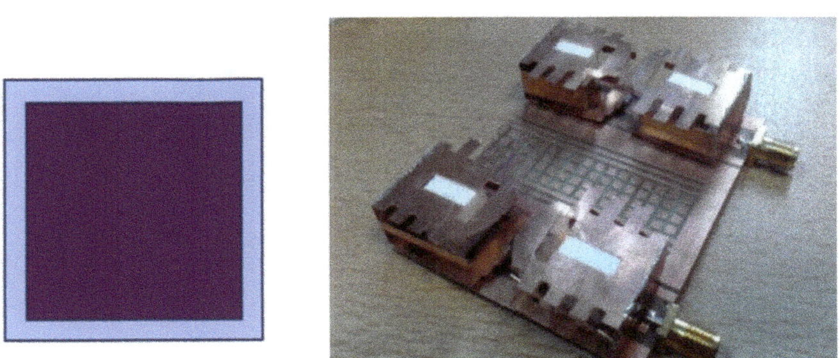

Figure 6.8. FSS unit cell, and the fabricated MIMO antenna presented in [17]. Adapted from [17]. CC BY 4.0.

devices that can be expanded and bent. Through the introduction of liquid metal into a predetermined arrangement, it can be molded into a diverse range of deformable gadgets, including electrical connections, probes, antennas, sensors emulating artificial skin, micro-electrode-driven micro-fluidic systems, and alterable millimeter-wave micro-fluidic structures [19]. Traditional stealth technologies aim to minimize the backscattering phenomenon experienced by probing radar. However, a new meta-skin technology has emerged to bridge the gap between single stretchable meta-atoms and large-scale metamaterials [20]. Researchers are currently exploring the potential of adjusting the resonance of planar metamaterials and investigating cloaking effects in curved metamaterials.

In reference [19], a microwave-frequency meta-skin is introduced, which possesses flexibility and stretchability properties. The meta-skin demonstrates frequency selectivity and cloaking effects. It consists of an array of liquid metal split-ring resonators encased within an elastomer. By stretching multiple layers of meta-skins along their surfaces in a planar direction and adjusting the distance between the meta-skin sheets vertically, a highly adaptable tunable FSS is achieved, as illustrated in figure 6.9. Moreover, the meta-skin can be wrapped around objects with arbitrary shapes, such as curved dielectric materials, resulting in significant scattering suppression (cloaking effect) in a designed frequency band. This has been verified through experimental investigations conducted from various directions.

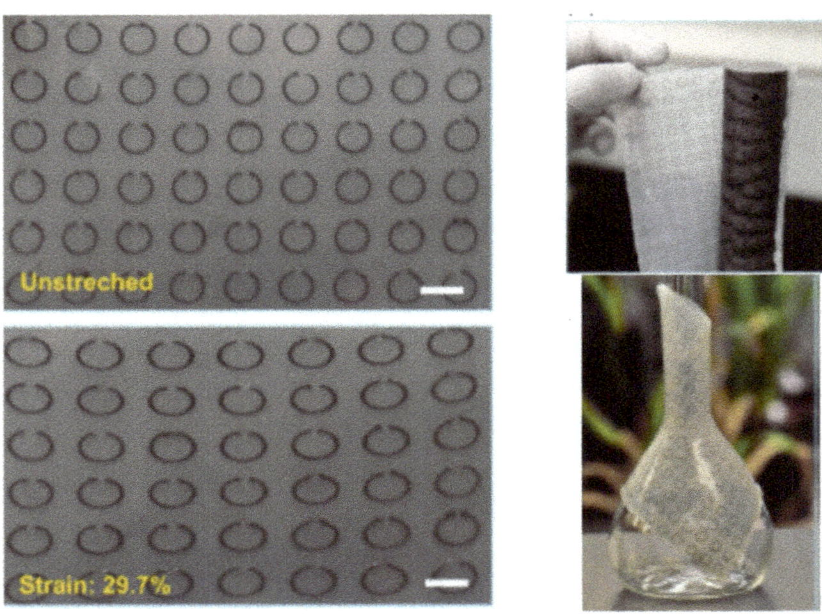

Figure 6.9. Flexibility demonstration: meta-skin characteristics in relaxed and extended states, enveloping a dielectric nylon rod and glass flask. Reproduced from [19]. CC BY 4.0.

6.5.6 3D FSS structures

One of the key objectives in designing a FSS is to accomplish a wide bandwidth and a sharp transition between the passband and the stopband. Traditional FSSs typically employ a two-dimensional (2D) structure composed of patch or aperture elements. While these structures are relatively easy to fabricate, they suffer from limitations such as narrow bandwidth, poor selectivity, susceptibility to polarization changes, and sensitivity to incident angles. To overcome these shortcomings, novel 3D-FSS structures have been presented in [21]. The 3D-FSS consists of a 2D periodic array of multimode cavities that facilitate coupling to control the desired frequency response. The utilization of electromagnetic structures based on 3D-FSS has become increasingly favored in diverse real-world scenarios, owing to their beneficial attributes such as enhanced flexibility, improved filtering capabilities, and consistent angular response.

Reference [22] introduces the concept of employing conformal FSSs in the near-field region with the aim of enhancing bandpass performance compared to conventional finite planar FSS structures. To validate this idea, a Ka-band spherical dome FSS array was 3D-printed on PLA, metalized, and its bandpass performance and fabrication technique were verified (see figure 6.10). The results demonstrated that the use of a conformal FSS offers improved angular stability and is less sensitive to the distance between the FSS and the receiver, as opposed to the finite planar case. The choice of a spherical dome shape was based on its ability to support the quasi-spherical wavefront from the open-ended waveguide. Moreover, this simple shape can be easily fabricated in-house using a low-cost fused filament fabrication 3D printer. Although more complex shapes are feasible, they would require more intricate supporting structures. Addressing the challenge of deploying the FSS on the conformal surface without element distortion, the authors employed the additive manufacturing (AM) technique. This approach overcomes the limitations associated with traditional manufacturing methods. Future advancements could involve

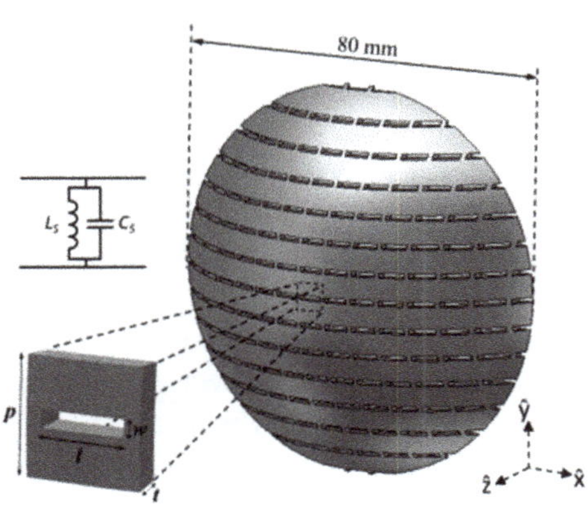

(a)

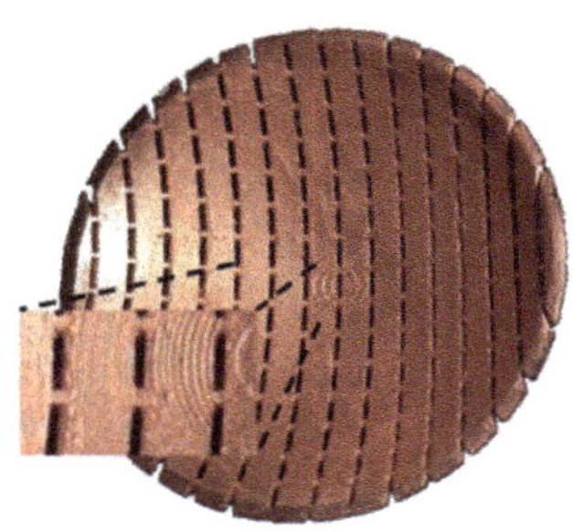

(b)

Figure 6.10. (a) 3D-FSS design, and (b) fabricated 3D-printed FSS. Reproduced from [22]. CC BY 4.0.

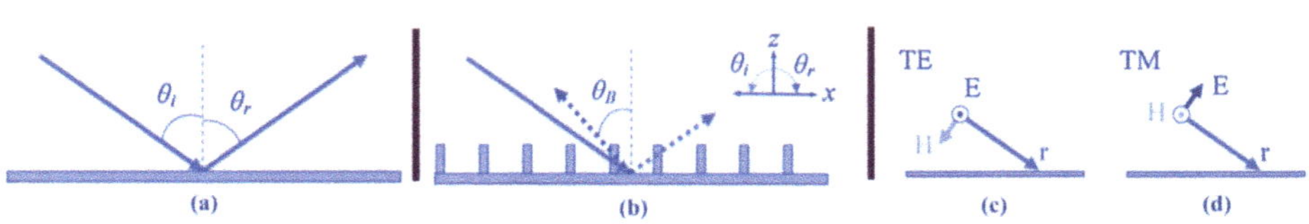

Figure 6.11. Principle of operation for three-dimensional frequency selective surfaces (3D-FSS).

enhancing the resolution of 3D-printed elements and exploring various geometries. AM also provides the flexibility to use a mix of different FSS elements, such as slots, crosses, etc, with varying dimensions along the curved array from the center to the edges. This mitigates the edge effect. This concept of a curved FSS has broad applications, including constructing other types of metasurfaces like transmit arrays, lenses, etc. Additionally, it could serve as a tunable polarizer, allowing adjustment to widen or narrow its passband properties. Overall, this research opens up exciting possibilities for the development of advanced conformal FSS structures and their potential use in diverse electromagnetic applications.

The operational principle of a 3D-FSS can be elucidated by considering normal and oblique incidences of the EM wave, as depicted in figure 6.11. When an EM wave impinges upon a flat perfect electric conductor (PEC), it is reflected as specular scatter. In this context, θi and θr denote the incident and reflected angles of the electromagnetic (EM) wave, respectively. When periodic elements are introduced, as depicted in figure 6.11(b), the presence of these elements reduces the specular scatter, causing the remaining energy to be redirected as backscatter at an angle of θB. Furthermore, figure 6.11(c) defines the TE polarized propagating mode, while figure 6.11(d) defines the TM polarized propagating mode, both in relation to the direction of propagation.

6.5.7 Reconfigurable FSS

In addition to its spatial filtering applications, FSSs can also serve as tunable devices. Various techniques are employed to achieve reconfiguration in a FSS [23]. One such approach involves the creation of a concentric threaded cylinders unit cell in a 3D-FSS [24]. This technique utilizes a sliding mechanism, where the outer cylinder can be adjusted over the inner cylinder to achieve frequency reconfiguration within a tuning range of 1.86–3.10 GHz.

Additionally, reference [25] introduces a versatile and adaptable FSS with multiple functionalities. The concept involves positioning two meandered patterns with orthogonal polarization on opposite sides of a dielectric substrate. To realize this design, a laser-etched micro-fluidic channel of customizable shape is created on a polydimethylsiloxane substrate. The reconfiguration aspect is facilitated by incorporating a liquid metal, eutectic gallium indium, into the structure.

By injecting the liquid metal along the meandering lines, the FSS can operate in four distinct reconfiguration modes: dual-polarized all pass, dual-polarized band-pass, single-polarized bandpass, and single-polarized low pass. This innovative configuration not only ensures a high angular stability of up to 60° but also retains mechanical flexibility. Consequently, this structure proves suitable for a wide range of potential applications.

Reference [26] presents the design of a FSS with a cross structure (see figure 6.12), specifically tailored to achieve the conversion from reflection to transmission by manipulating the state of two PIN diodes. Building upon this concept, the study goes on to develop a multi-functional reconfigurable antenna with both reflection and transmission capabilities. The antenna consists of 9 × 9 units and utilizes a four-layer substrate structure. The key innovation lies in incorporating a reconfigurable FSS layer at the bottom, allowing for seamless switching between reflect array and transmit array configurations. Importantly, this proposed antenna array stands out from previous studies due to its unique ability to reconfigure between transmission and reflection modes at the same frequency.

6.5.8 FSS impacted textiles

Despite being uncommon, there is a limited amount of valuable research available on textile frequency selective surfaces in the literature. Nevertheless, the textile FSS represents another emerging category following wearable FSSs. Various methods such as screen printing, lithographic processes, inkjet printing, and stamping have been employed to demonstrate the printing of FSSs on flexible and smooth surfaces like paper or polyethylene terephthalate [27]. One potential application involves integrating these components into garments to enable both on-body and off-body communication [28].

Textile formations provide benefits such as pliability, reduced weight, sleekness, and an elevated ratio of strength to weight. Manufacturing techniques for fabrics have been utilized to create conductive elements. Similarly, textile surfaces can be used to construct fabrics that control the propagation of electromagnetic waves in the microwave-frequency range, which can be embedded into buildings [29]. These fabric-based FSSs can have short-term use and can be easily controlled, making them suitable for various applications such as conferences, museums, art galleries.

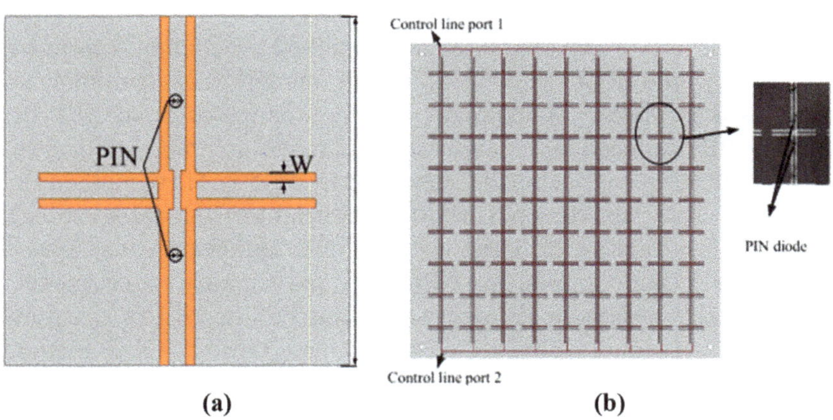

Figure 6.12. (a) FSS structure, and (b) FSS layer with control lines. Reproduced from [26]. CC BY 4.0.

An innovative FSS, composed of an array of square scattering elements, is screen-printed onto a woven textile using conductive inks [27]. Each printed square corresponds to a specific textile weave pattern, presenting a challenge in achieving high printing reliability. The squares collectively exhibit a response close to their resonant wavelength.

6.6 Effective approaches for analyzing, optimizing, and fabricating frequency selective surfaces

Typical methods used for fabricating FSSs include micromachining, wafer coating technology [30], printed circuit technology [31, 32], standard lithography, and substrate bonding techniques [33], the SU-8 based micromachining technique [34], photolithography, thin-film deposition, and lift-off techniques [35]. These techniques are commonly employed to manufacture FSS structures with desired properties.

An alternative approach for FSS fabrication involves the utilization of 3D printing technology [36]. This method enables the direct manufacturing of complex EM structures based on a digital model [37]. By using a 3D printer equipped with plaster-based material, the desired FSS can be created through layering and shaping processes [38].

Several widely used optimization strategies for EM structures based on FSSs include the genetic algorithm (GA) [39], particle swarm optimization (PSO) [40], artificial neural networks (ANNs) [41, 42], finite element method (FEM) [43], method of moments (MoM) [44], finite-difference time-domain (FDTD) [45], and fast Fourier transform (FFT) [46]. Another approach for EM optimization involves combining full wave MoM, multilayer perceptron (MLP), ANN, and population-based search algorithms, which can be applied in the design of a Vicsek fractal motif [47, 48]. These techniques offer the advantage of faster processing times, which is highly desirable for EM applications.

Various techniques such as finite integration technology (FIT) and finite element method (FEM) can be employed to calculate the phases, magnitudes, and transmittances of FSS structures [49, 50]. Additionally, other efficient numerical models have recently been utilized for the analysis of FSSs.

Researchers have also explored alternative methods, including transmission line (TL) [51] theory and RLC equivalent circuit models [52], for analyzing EM structures based on FSSs. To compute the input impedance of EM structures incorporating PIN diodes [53], a multiscale approach combined with generalized equivalent circuit (GEC) modeling has been employed [54]. These methods take into account the intrinsic characteristics, width, and height of the diodes to accurately calculate the surface impedance. These methods offer significant advantages in terms of computational resources, reducing CPU time and memory requirements.

Another valuable technique for analyzing active FSS-based EM structures is the combination of renormalization group theory and the MS-GEC method. This approach provides a comprehensive understanding of the behavior and properties of active FSS structures. By integrating renormalization group theory, which deals with scaling transformations, researchers can effectively analyze and optimize the performance of active FSS-based EM structures with the MS-GEC method [53]. This combined approach yields valuable insights and contributes to the development of advanced EM structures for various applications. However, it should be noted that the MS-GEC method is not applicable to infinite structures.

In addition to the MS-GEC method, researchers have employed various other techniques for analyzing FSS-based EM structures. These include the wave concept iterative procedure, Impedance type boundary condition-based model, physical optics and ray tracing, Monte Carlo method, algebraic multigrid, Hurst's and Prony's methods [55], characteristic basis function, and spectral rotation approaches. Each method provides alternative approaches for analyzing FSS structures, each with its own strengths and limitations. Researchers continue to explore and refine these techniques to gain deeper insights into the behavior and performance of FSS-based EM structures.

6.7 Results and discussion

By examining table 6.3, it becomes evident that altering the shape parameter of the FSS, as depicted in references [9, 11, 16, 56–58], results in distinct resonant frequencies emerging. It is noticeable that augmenting the dimensions of the FSS leads to reduced angular stability and heightened attenuation. Thus, achieving improved performance necessitates the utilization of a compact FSS design. References [24, 59] underscore the FSS's reconfigurable nature, showcasing a broad range of tuning capabilities.

Based on current trends, we can anticipate the future advancements of FSSs in the coming years, given the evident ongoing progress in research. The potential uses of FSSs encompass a broad range of applications, including but not limited to:

- Dynamic reconfiguration of FSS: FSS that can be adjusted on-the-fly.
- Wearable FSS for sensing: FSS integrated into wearable devices for sensing purposes.
- Tunable and software-defined FSS: FSS with adjustable properties through software control.
- Mechanically adaptable FSS: FSS with mechanical tuning capabilities.
- High-performance tunable FSS absorbers: FSS designed for efficient absorption of specific frequencies.
- Multi-functional FSS: FSS serving multiple purposes simultaneously.
- Smart absorbers using FSS: FSS-based absorbers with smart capabilities.
- Frequency selective absorbers: FSS employed for selective absorption and reflection.
- Curved FSSs: Frequency selective surfaces with double curvature.
- FSSs for high-power radar: FSSs used in high-power radar applications.
- Antenna-FSS system integration: FSSs integrated into antenna systems.
- Graphene-based FSS: FSS utilizing graphene materials.
- 3D-printed all-dielectric FSS: FSS created through 3D printing with broad bandwidth.

Table 6.3. Contrasts among various FSS configurations.

Ref	Dimension (mm) of the UC	Resonant frequency (GHz)	No. of operating bands	Frequency bands	Attenuation (dB)	Angular stability	Dimension (mm) of the FSS	Bandwidth	Tunable
[9]	6.8 × 6.8 × 0.127	10	1	X-band	56	0°–60°	–	4 GHz	NO
[11]	20 × 20 × 2	2.45, 5.5	2	Wi-fi and WLAN	More than 16	0°–45°	260 × 200	5.5 GHz	NO
[16]	0.34λ × 0.34λ	48, 64	2	U and V band	20	0°–30°	95 × 95	30 GHz (40–70)	NO
[24]	56 × 56 × 25	1.86, 3.10	2	L and S band	20	0°–45°	280 × 280	660 MHz	YES
[56]	7 × 7 × 1.6	5	1	WLAN	20	0°–60°	30 × 30	400 MHz	NO
[57]	8.8 × 8.8 × 0.762	8.47, 10.45	2	Wideband X-band	–	0°–60°	–	–	NO
[58]	10 × 10 × 1.6	3.5, 5.2, 10.2	3	WLAN WiMAX, X-band	60	0°–30°	–	0.5 GHz (3.1–3.7), 1.9 GHz (4.1–6), 4.1 GHz (8–12.1)	NO
[59]	15 × 15 × 1.36	3.5, 5, 5.8	3	WiMAX, WLAN, and ISM/WiMAX	40	–	15 × 300	–	YES

- Metamaterial-inspired FSS for satellite communication: FSS influenced by metamaterial concepts for satellite communication.
- FSS for earth observation and satellite communication: FSS applied in earth observation and satellite communication.
- FSS for frequency shielding: FSS used for selective frequency shielding in critical locations like airports and military communications.
- FSS for radiation isolation: FSS isolating harmful radiation in settings such as hospitals, schools, and households, particularly in microwave low-frequency bands (L, S).
- Chipless wireless sensors using FSS: FSS-based passive sensors for monitoring structural integrity.
- Plasma shell-based switchable FSS: FSS employing plasma shells for aerodynamics and terrestrial electromagnetic protection systems.

In the years ahead, the developments in FSS technology are poised to encompass these diverse applications, driven by ongoing research and innovation.

6.8 Conclusion

This chapter offers an in-depth exploration of various configurations of frequency selective surfaces. It delves into a range of applications for FSSs, unveiling their capabilities beyond mere filtering. FSSs prove valuable in augmenting antenna gain and addressing mutual coupling challenges within MIMO antenna systems. Moreover, the utilization of miniaturized FSS unit cells and arrays contributes to improved angular stability. The chapter also provides an overview of three-dimensional FSS structures. Notably, FSSs facilitate frequency tuning, further expanding its utility across diverse fields.

This comprehensive review underscores the adaptability of FSS-based designs, showcasing their practical applicability across a broad frequency range, encompassing microwave, millimeter, and terahertz wavelengths. The outlook for the near future anticipates the emergence of novel FSS applications, which could present innovative avenues for researchers to explore.

Conflicts of interest

The authors declare no conflict of interest.

References

[1] Anwar R S, Mao L and Ning H 2018 Frequency selective surfaces: a review *Appl. Sci.* **8** MDPI AG, Sep. 18

[2] Munk B A 2000 *Frequency Selective Surfaces: Theory and Design* **Vol 29** (NJ)

[3] Costa F, Monorchio A and Manara G 2014 An overview of equivalent circuit modeling techniques of frequency selective surfaces and metasurfaces *Appl. Comput. Electromagn. Soc. J.* **29** 960–76

[4] Yang F and Rahmat-Samii Y 2008 *Electromagnetic Band Gap Structures in Antenna Engineering* (Cambridge: Cambridge University Press)

[5] Ferreira D, Caldeirinha R F S, Cuinas I and Fernandes T R Sep. 2015 Square loop and slot frequency selective surfaces study for equivalent circuit model optimization *IEEE Trans. Antennas Propag.* **63** 3947–55

[6] Mohan J and Gupta A 2019 Jaypee Institute of Information Technology University, Institute of Electrical and Electronics Engineers. Uttar Pradesh Section. SP/CS Joint Chapter, and Institute of Electrical and Electronics Engineers 2019 Int. Conf. on Signal Processing and Communication (ICSC) : 07–09 (NOIDA: *Jaypee Institute of Information Technology*)

[7] Kapoor A, Mishra R and Kumar P Jun. 2022 Frequency selective surfaces as spatial filters: fundamentals, analysis and applications *Alex. Eng. J.* **61** 4263–93

[8] Lv X, Ako R T, Bhaskaran M, Sriram S, Fumeaux C and Withayachumnankul W May 2022 Frequency-selective-surface-based mechanically reconfigurable terahertz bandpass filter *IEEE Trans. Terahertz Sci. Technol.* **12** 257–66

[9] Nauman M, Saleem R, Rashid A K and Shafique M 2016 A miniaturized flexible frequency selective surface for x-band applications *IEEE Trans. Electromagn. Compat.* **58** 419–28

[10] Yang G, Zhang T, Li W and Wu Q 2010 A novel stable miniaturized frequency selective surface *IEEE Antennas Wirel. Propag. Lett.* **9** 1018–21

[11] Farooq U, Shafique M F and Mughal M J Feb. 2020 Polarization insensitive dual band frequency selective surface for RF shielding through glass windows *IEEE Trans. Electromagn. Compat.* **62** 93–100

[12] Al-Gburi A J A, Ibrahim I B M, Zakaria Z, Ahmad B H, Bin Shairi N A and Zeain M Y 2022 High gain of UWB planar antenna utilising FSS reflector for UWB applications *Comput., Mater. Contin.* **70** 1419–36

[13] Yahya R, Nakamura A, Itami M and Denidni T A Jul. 2017 A novel UWB FSS-based polarization diversity antenna *IEEE Antennas Wirel. Propag. Lett.* **16** 2525–8

[14] Pirhadi A, Bahrami H and Mallahzadeh A R Nov. 2012 Electromagnetic band gap (EBG) superstrate resonator antenna design for monopulse radiation pattern *Appl. Comput. Electromagn. Soc. J.* **27** 908–17

[15] Sharma A *et al* 2019 Wideband high-gain circularly-polarized low RCS dipole antenna with a frequency selective surface *IEEE Access* **7** 156592–602

[16] Kesavan A, Karimian R and Denidni T Jan. 2016 A novel wideband frequency selective surface for millimeter-wave applications *IEEE Antennas Wirel. Propag. Lett.* **15** 1

[17] Saleem R, Bilal M, Chattha H T, Ur Rehman S, Mushtaq A and Shafique M F 2019 An FSS based multiband MIMO system incorporating 3D antennas for WLAN/WiMAX/5G cellular and 5G Wi-Fi applications *IEEE Access* **7** 144732–40

[18] Syed I S, Ranga Y, Matekovits L, Esselle K and Hay S G Dec. 2014 A single-layer frequency-selective surface for ultrawideband electromagnetic shielding *IEEE Trans. Electromagn. Compat.* **56** 1404–11

[19] Yang S, Liu P, Yang M, Wang Q, Song J and Dong L Feb. 2016 From flexible and stretchable meta-atom to metamaterial: a wearable microwave meta-skin with tunable frequency selective and cloaking effects *Sci. Rep.* **6** 21921

[20] Liu P *et al* Jul. 2015 Tunable meta-atom using liquid metal embedded in stretchable polymer *J. Appl. Phys.* **118** 014504

[21] Rashid A K, Li B and Shen Z 2014 An overview of three-dimensional frequency-selective structures

[22] Álvarez H F *et al* Dec. 2021 3D conformal bandpass millimeter-wave frequency selective surface with improved fields of view *Sci. Rep.* **11** 12846

[23] Zhou Q, Guo M, Moghadas H, Wu Z, Liu P and Daneshmand M Nov. 2019 A frequency selective rasorber with three transmission bands and three absorption bands *IEEE Access* 1

[24] Sivasamy R and Kanagasabai M Oct. 2017 Novel reconfigurable 3-D frequency selective surface *IEEE Trans. Compon. Packag. Manuf. Technol.* **7** 1678–82

[25] Ghosh S and Lim S Jul. 2018 Fluidically reconfigurable multifunctional frequency-selective surface with miniaturization characteristic *IEEE Trans. Microw. Theory Tech.* 1–9

[26] Li W, Wang Y, Sun S and Shi X 2020 An FSS-backed reflection/transmission reconfigurable array antenna *IEEE Access* **8** 23904–11

[27] Ghebrebrhan M *et al* Nov. 2017 Textile frequency selective surface *IEEE Microw. Wirel. Compon. Lett.* **27** 989–91

[28] Tennant A, Hurley W and Dias T Oct. 2012 Experimental knitted, textile frequency selective surfaces *Electron. Lett.* **48** 1386–8

[29] Seager R D, Chauraya A, Bowman J, Broughton M, Philpott R and Nimkulrat N Nov. 2013 Fabric based frequency selective surfaces using weaving and screen printing *Electron. Lett.* **49** 1507–9

[30] Xia B G, Zhang D H, Meng J, Huang J, Yao C F and Zhang J S Oct. 2013 Terahertz FSS for space borne passive remote sensing application *Electron. Lett.* **49** 1398–9

[31] Edalati A and Sarabandi K Sep. 2014 Reflectarray antenna based on grounded loop-wire miniaturised-element frequency selective surfaces *IET Microw. Antennas Propag.* **8** 973–9

[32] Edalati A and Sarabandi K 2014 Wideband, wide angle, polarization independent RCS reduction using nonabsorptive miniaturized-element frequency selective surfaces *IEEE Trans. Antennas Propag.* **62** 747–54

[33] Li M, Al-Joumayly M A and Behdad N 2013 Broadband true-time-delay microwave lenses based on miniaturized element frequency selective surfaces *IEEE Trans. Antennas Propag.* **61** 1166–79

[34] Wang Y, Yang B, Tian Y, Donnan R S and Lancaster M J 2014 Micromachined thick mesh filters for millimeter-wave and terahertz applications *IEEE Trans. Terahertz Sci. Technol.* **4** 247–53

[35] Ebrahimi A *et al* Sep. 2015 Second-order terahertz bandpass frequency selective surface with miniaturized elements *IEEE Trans. Terahertz Sci. Technol.* **5** 761–9

[36] Sanz-Izquierdo B and Parker E A 2013 3D printing technique for fabrication of frequency selective structures for built environment *Electron. Lett.* **49** 1117–8

[37] Sanz-Izquierdo B and Parker E A 2014 Dual polarized reconfigurable frequency selective surfaces *IEEE Trans. Antennas Propag.* **62** 764–71

[38] Sanz-Izquierdo B and Parker E A Dec. 2014 3-D printing of elements in frequency selective arrays *IEEE Trans. Antennas Propag.* **62** 6060–6

[39] Bossard J A, Werner D H, Mayer T S and Drupp R P 2005 A novel design methodology for reconfigurable frequency selective surfaces using genetic algorithms *IEEE Trans. Antennas Propag.* **53** 1390–400

[40] Sheng X, Zhao X, Liu N and Gao X 2021 Design of miniaturized FSS using equivalent circuit model and multi-objective particle swarm optimization *J. Phys. D: Appl. Phys.* **54** 405001

[41] Shrivastava K, Das A K and Saha U K 2023 A neural network based design of a planar double divergent nozzle 25th AIAA Int. Space Planes and Hypersonic Systems and

Technologies Conf., *in Int. Space Planes and Hypersonic Systems and Technologies Conferences* (American Institute of Aeronautics and Astronautics)

[42] Gu Z *et al* 2023 A solution to the dilemma for FSS inverse design using generative models *IEEE Trans. Antennas Propag.* **71** 5100–9

[43] La Rubia V D, Parellada I, Molero C and Padilla P 2023 EM-based design of all metal frequency selective surfaces with finite transmission zeros *2023 17th European Conf. on Antennas and Propagation (EuCAP)* pp 1–5

[44] Zou Y-J, Zhao X-W, Guo C, Zhu M-D and Zhang Y 2023 An integral equation method using periodic contact-region modeling for analyzing infinitely biperiodic composite structures *IEEE Trans. Antennas Propag.* **71** 1

[45] Liu C, Feng J and Fang M 2022 Heterogeneous CPU-GPU accelerated parallel subgridding FDTD algorithm *2022 Int. Applied Computational Electromagnetics Society Symp. (ACES-China)* pp 1–3

[46] Barton J H, Garcia C R, Berry E A, May R G, Gray D T and Rumpf R C 2014 All-dielectric frequency selective surface for high power microwaves *IEEE Trans. Antennas Propag.* **62** 3652–6

[47] Silva P, Cruz R and D'Assunção A Sep. 2010 Blending PSO and ANN for optimal design of FSS filters with Koch island patch elements *Magn., IEEE Trans.* **46** 3010–3

[48] Silva M, Nóbrega C, Silva P and D'Assunção A Apr. 2014 Optimization of FSS with Sierpinski island fractal elements using population-based search algorithms and MLP neural network *Microw. Opt. Technol. Lett.* **56** 827–31

[49] Bardi I, Tharp J and Petersson L E R Feb. 2014 Homogenization of periodic structures using the FEM *Magn., IEEE Trans.* **50** 157–60

[50] Pandhare R A, Abegaonkar M P and Dhote C 2023 High gain compact dual band reconfigurable antenna using multilayer FSS for WLAN and RADAR applications *Prog. Electromagn. Res. C.* **130** 27–240

[51] Jiang H, Liao S and Xue Q 2023 Hybrid capacitive and inductive coupling paths for implementing quasi-elliptical high-order bandpass FSS with high out-of-band suppression *IEEE Trans. Antennas Propag.* **71** 3294–303

[52] Violi V, Brizi D, Mishra V and Monorchio A 2022 A compact wide-band absorber based on resistive FSS and magnetic materials *2022 IEEE Int. Symp. on Antennas and Propagation and USNC-URSI Radio Science Meeting (AP-S/URSI)* pp 1086–7

[53] Mili S, Aguili C and Aguili T Jan. 2011 Study of fractal-shaped structures with PIN diodes using the multi-scale method combined to the generalized equivalent circuit modeling *Prog. Electromagn. Res.* B **27** 213–33

[54] Wang Z X, Zhao P C, Zong Z Y, Wu W and Fang D G Jan. 2023 An extension of partial element equivalent circuit method for frequency selective surface analysis *IEEE Trans. Antennas Propag.* **71** 1141–6

[55] El Hani R and Laurin J-J 2013 Numerical diffraction of array-guided surface waves on finite frequency selective surfaces *IEEE Trans. Antennas Propag.* **61** 3582–8

[56] Natarajan R, Kanagasabai M, Baisakhiya S, Sivasamy R, Palaniswamy S and Pakkathillam J K 2013 A compact frequency selective surface with stable response for WLAN applications *IEEE Antennas Wirel. Propag. Lett.* **12** 718–20

[57] Unaldi S, Cimen S, Cakir G and Ayten U E 2017 A novel dual-band ultrathin FSS with closely settled frequency response *IEEE Antennas Wirel. Propag. Lett.* **16** 1381–4

[58] Bashiri M, Ghobadi C, Nourinia J and Majidzadeh M Nov. 2017 WiMAX, WLAN, and X-band filtering mechanism: simple-structured triple-band frequency selective surface *IEEE Antennas Wirel. Propag. Lett.* **16** 3245–8

[59] Sivasamy R, Moorthy B, Kanagasabai M, Samsingh V R and Alsath M G N 2018 A wideband frequency tunable FSS for electromagnetic shielding applications *IEEE Trans. Electromagn. Compat.* **60** 280–3

IOP Publishing

Lulu Wang and Muharrem Karaaslan

Chapter 7

Metasurfaces

Yunus Kaya

Symbols

λ	Wavelength
λ_0	Wavelength of free space
ε	Permittivity
μ	Permeability
α	Angle of incidence
n_1	Refractive indices of medium 1
n_2	Refractive indices of medium 2
$\vec{E}$	Electric field intensity
$\vec{E_1}$	Electric field intensity of the 1st region
$\vec{E_2}$	Electric field intensity of the 2nd region
E_1	Amplitude of the wave linearly polarized in the x-direction
E_2	Amplitude of the wave linearly polarized in the y-direction
$E_x(t)$	The time–space expression of the electric field intensity in the x-directions
$E_y(t)$	The time–space expression of the electric field intensity in the y-directions
$\vec{H}$	Magnetic field intensity
$\vec{H_1}$	Magnetic field intensity of the 1st region
$\vec{H_2}$	Magnetic field intensity of the 2nd region
$\hat{n}$	Unit vector of the surface normal
$\vec{J_S}$	Surface electric current
$\vec{M_S}$	Surface magnetic current
ω	Angular frequency
k	Wave number
φ	Phase difference between E_x and E_y
φ_x	Phase of E_x
φ_y	Phase of E_y

doi:10.1088/978-0-7503-5754-8ch7

t	Thickness
R_{xx}	Co-polarization reflection coefficient
R_{yy}	Co-polarization reflection coefficient
R_{xy}	Cross-polarization reflection coefficient
R_{yx}	Cross-polarization reflection coefficient

7.1 Introduction

It is a commonly known fact that when an electromagnetic (EM) wave hits composite media, magnetic and electric dipole moments are induced in inclusions (or resonators) located in the medium. These magnetic and electric dipole moments are intimately associated with the effective permeability and permittivity, which are the unique EM properties of composite media. Because the dimension, density, form, and direction of inclusions present/designed in the medium can be controlled by researchers, materials that exhibit an EM response as desired can be easily designed. These materials, which are not found in nature and are artificially designed, are called metamaterials (MMs). MMs are synthetic materials and consist of artificial EM materials that have periodic regulations of metallic structures that are smaller than the wavelength (λ) of the incoming EM wave [1]. MMs consist of periodic subwavelength metal/dielectric structures (i.e. from meta-atoms) that are resonantly connected to the magnetic or electric or both components of the incoming EM fields, exhibiting effective magnetic (the magnetic permeability is presented by μ) and/or electrical (the electric permittivity is presented by ε) properties not found in Nature [2]. These micro- and nano-structured artificial medium class MMs, which are three-dimensional (3D), usually periodic, artificial materials consisting of metals and/or dielectrics, have been extensively studied by researchers during the past 15 years because of their matchless interactions with EM waves that surpass the abilities of homogeneous or naturally occurring materials [3, 4]. The extraordinary ability of MMs to manipulate waves is owing to their strong interactions with magnetic and/or electric fields, which are characteristically provided by resonance effects controlled with the geometry of the unit cells. Additionally, MMs have exceptional magnetic and electrical properties not prevalently found in nature. These capabilities give rise to wide applications, such as backward propagation [5], Vavilov–Cerenkov effect [6], reverse Doppler effect [7], antenna performance improvement [8], excellent absorbers [9], negative refraction [10], diffraction-boundary refraction imaging [11], cloaking [12], super lenses [13], scattering reduction [14] and energy harvesting [15], etc, as well as other applications at both microwave and optical frequencies. MMs can benefit from the beams of EM waves in an improbable manner [16]. In addition, MMs may powerfully localize and concentrate fields, which makes them effective in improving sensor selectivity in detecting extremely small analyzed sample concentrations and nonlinear chemicals [17].

MMs with resonant individual elements may have negative relative permeability and permittivity values. These left-handed materials afterward make the angle of refraction negative, according to Snell's law of refraction, thereby causing the incoming and refracted wave to be located on the same side of the normal. This fact is also known as negative refraction. Because of this unusual feature, MMs present potential applications that would not be possible using only naturally occurring materials, as listed above [3]. However, many promising potential applications of MMs are practically hindered by the high losses and difficulties in 3D manufacturing, the strong dispersion and high losses connected with the use of metallic structures and resonant responses, especially at the micro and nanoscales [18].

Metasurfaces (MSs) are the two-dimensional (2D) or surface/planar counterparts of MMs with a subwavelength thickness [19]. It is possible to characterize MSs', just like MMs', reaction through their magnetic and electrical polarizability. MSs are also known as metafilms in the literature [20]. MMs control the spread of light because of arbitrarily adjustable permeability and permittivity values; on the other hand, they nevertheless utilize the propagation effect to manipulate EM waves. While this case may result in a complex and comparatively cumbersome structure, MSs attempt to manipulate the wave on a single, highly thin layer [21]. The 2D designable nature of MSs therefore makes them less bulky and lightweight, while also allowing the formation of structures with lower loss [19]. Because of their 3D nature, MMs are also difficult to manufacture, in which MSs present a highly encouraging alternative. Owing to their planar 2D design structure, MSs have been extensively researched and applied in EM applications, as they may be comfortably produced using planar production tools/methods (e.g. traditional board fabrication method) [22]. The planar production operation is also very cost-effective compared to the difficult and complex production of 3D MMs [23]. Therefore, MSs, which are 2D materials, can not only be smoothly integrated into other devices and/or systems, which could make them a prominent property for nanophotonic circuits, but also permit them to be a part of 'laboratory-on-a-chip' photonics [24].

The extraordinary negative index of MMs is because of the resonance of individual (specific) meta-atoms [25]. This negative index feature makes MMs inherently dispersive and hence the EM features of such materials are also quite critical to alterations in operating frequency. Therefore, this limits the bandwidth of such materials. Jiang *et al* have shown that dispersion properties can be designed using highly thin MSs with deep subwavelength notches in a two-layer fishing net structure [26]. This technique was later used to make a wideband MS filter. Rejection and transmission for out-band and in-band, respectively, are achieved by matching and incompatibility of this MS impedance in free space, respectively. In addition, the dispersion properties were also controlled by adapting plasma wavelengths and primary (and secondary) magnetic resonances for permittivity. Of course, both of these properties were considerably connected with the design of deep notches at the lower wavelength of MS. The design was optimized by means of a genetic algorithm, and this wideband MS also had a very low insertion loss in the frequency range of transmission [26]. In this context, due to the various advantages that MSs offer compared to MMs, many researchers and scientific communities have recently shown great attention to this field. This, in turn, has caused quick development in the fundamentals, physics, and engineering that manage the behavior and potential applications of MS.

Surface waves may be well controlled by designing MS unit cells with the necessary impedance to manipulate group velocity or phase [27]. MSs are modeled in ways that may direct or separate waves in specific directions and/or are used in scattering control applications [28]. By controlling the sizes and shapes of MS unit cells, different effective surface refractive indices may be obtained and surfaces may be modeled to supply various functions.

Different MSs show different abilities and Acouri *et al* have extensively studied MS applications for the manipulation of plane waves in free space, as well as surface waves [29]. The reflection phase of EM waves striking the MS is controlled for applications such as ultra-thin polarization converters [30]. Wave front engineering is also widely explored in wave front engineering such as beam shaping including refraction, reflection and focusing [31]. MSs are coated on curved and planar surfaces to decrease scattering and applied at microwave and far-infrared frequency bands for cloaking and other relevant applications [32, 33]. MS-based absorbers may also offer advantages over conventional absorbers, especially when combined with active circuits or nonlinear devices [34].

In addition to manipulating the scattering or absorption of EM waves, MSs can also be used as leaky-wave antennas to propagate waves into free space by transforming surface waves into radiation modes [35]. Also, other applications of MSs at microwave and optical frequencies have recently attracted attention to a great variety of applications such as increasing antenna efficiency [36], ideal absorbers [9, 37], super lenses [38], cloaking [39], scattering elimination [14] and energy harvesting [40]. Furthermore, when nonlinear and active electronic circuit elements such as transistors [41], diodes [42] and varactors [43] are added to MSs, new switching and tunability capabilities in addition to the elements become possible.

This chapter of the book provides an overview of the basic information about MSs. Therefore, the rest of the chapter is organized as follows. In section 7.2, general information about MSs is given, while in the subsections, information about the generalized refraction law, Huygens' MSs, and MSs based on the Pancharatnam–Berry phase is given. In section 7.3, MS-based polarization converters from the application areas of MSs are examined in general terms. In the last section, important points about MSs are emphasized once again.

7.2 About MSs

MSs are identified as periodic (or non-periodic) structures in which the periodicity and thickness of the individual elements are small in comparison with the operating wavelength [19]. MSs can also be defined as a subwavelength resonant scattering array that controls the EM reply of the surface (symmetric or asymmetric) [21]. The distribution of individual scatterers is very important in specifying the reply of a surface, and this feature distinguishes MSs from conventional frequency selective surfaces (FSSs). A conventional FSS has individual elements (periodicity) of the order of the operating wavelength (this is usually $\lambda/2$). Another distinction between MSs and FSSs is closely associated with the subwavelength nature of the individual elements of an MS. The subwavelength nature of the building blocks of MSs permits them to be dealt with as a classical homogeneous structure, so that their response is characterized by effective parameters. For an MS, the areas and polarizations must be the mean of the surface areas, since the surface is thin (or ultra-thin). Albooyeh *et al* have given an analytical approach based on the transmission and reflection duality [44]. Generalized layer transition conditions permit the complex distribution of meta-elements to be modified with boundary conditions using magnetic and electric surface susceptibility [45]. This homogenization technique has been used to design leaky-wave antennas [46]. It is worth mentioning that non-periodic arrangement of elements is carried out in the design of reflection arrays. In order for a reflected beam to have a specific direction, each element is designed to supply the desired phase shift [47].

Since the response of the MS layer depends on the localized behavior of each unit, spatial modification of each unit cell is permitted [21]. Such variations may be used to shape wavefronts and design MS lenses. The miniature individual 'meta-atoms' of MSs allow subwavelength resolution of wavefronts ranging from microwave to optical frequency bands. This feature of MSs also permits them to behave as homogeneous or almost homogeneous layers [48].

Another perspective for defining an MS is through Huygens' principle [49]. Huygens' principle expresses that "every point on a wave front acts as a secondary source" [50], therefore permitting control of wavefronts beyond the source. When an EM wave hits a surface, it induces magnetic and electric dipole moments, and these moments manufacture surface currents. In regard to Schelkuno's principle, these surface currents are equivalent to the tangential magnetic and electric fields that control the EM reply of a surface [51]. A surface can only be characterized as an MS when each element can be polarized and subwavelength, which is acquired by attentive design of the individual element(s) [48].

An MS is capable of adapting both the magnetic and electric field component of EM waves. Therefore, since its impedances may be matched by adapting the magnetic and electric polarizability, an MS can have negligible mismatch losses [52]. MSs without reflection losses are defined as Huygens MSs [31].

7.2.1 The generalized law of refraction

Conventional devices such as holograms and lenses, shape wavefronts along a greater distance than the working wavelength [53]. Lenses take advantage of their shape and material features (e.g., refractive index) to focus energy by gradually changing the phase of an EM wave, and holograms also produce images in the far field by the way of constructive interference [54]. The same thing is true for double negative (DNG) medium materials whose phase is gradually tempered over a certain distance. MSs offer an encouraging alternative to this approach because an EM wave passing through an MS undergoes a phase jump. The reason for this is that the phase is discontinuous in an MS [55]. This anomalous behavior reduces the dependence on wave propagation and permits wavefronts to be shaped and focus energy over subwavelength distances. Therefore, MSs could potentially replace heavy and bulky devices traditionally used for such purposes.

This phase discontinuity (or skip/jump) owes its existence to the resonance behavior of miniature (subwavelength) MS building blocks. When an EM wave hits a single element of an MS, it induces a surface EM wave and gives rise to the charge contained in this element to oscillate. This entire fact is remembered as surface plasmon [56]. The interaction of the surface plasmon and the impinging EM wave causes phase discontinuity throughout the MS. Because of this discontinuity, Snell's law has been displaced by a 'generalized law' for MSs [57].

The generalized law is created based on Fermat's principle of the shortest time as a stationary phase [57]. Fermat's principle expresses that light travels in the least time-consuming path of all possible paths between two points [58]. If we think of a plane wave at angle of incidence α hitting an MS, two potential routes the wave may take are seen in figure 7.1. Since an MS gives rise to a sudden alteration in phase (i.e. phase discontinuity), this alteration at point X is represented by β, while X' is represented by $\beta + d\beta$. It is assumed that the distance between X and X' is dx. In addition, β is supposed to be a continuous function of x. It is supposed that the two routes are infinitely close such that the phase difference between them is zero in order to reach point X. In this case, it also gives rise to the generalized law of refraction given by the following equation (7.1) [57].

$$n_1 \sin(\alpha) - n_2 \sin(\alpha') = \frac{\lambda_0}{2\pi} \frac{dx}{d\beta} \tag{7.1}$$

where n_1 and n_2 represent the refractive indices of the 1st and 2nd medium, respectively, as seen in figure 7.1, and λ_0 is the wavelength of free space.

Likewise, the generalized law of reflection for MSs is given by equation (7.2), where α_{refl} is the angle of reflection.

$$\sin(\alpha) - \sin(\alpha_{\text{refl}}) = \frac{\lambda_0}{2\pi n} \left| \frac{dx}{d\beta} \right| \tag{7.2}$$

The nonlinear relationship between incidence and reflection is remarkable, as the angles of incoming and reflection for an MS are no longer equal. If there is no phase discontinuity along the surface, that is, if $d\beta/dx = 0$, equations (7.1) and (7.2) may be smoothly degraded to the original Snell's law of reflection and refraction. The generalized reflection and refraction laws are illustrated by reflection arrays and V-shaped antennas, respectively [59].

7.2.2 Huygens' MS

Since the phase-controlled insufficient coupling efficiency of single-layer plasmonic antennas significantly reduces the performance of transmitted type MSs, another approximation based on the surface equivalence principle is suggested to minimize reflection by simultaneously adapting the magnetic and electric polarizations at the interface. In order to generate the aspired field distributions in two areas with

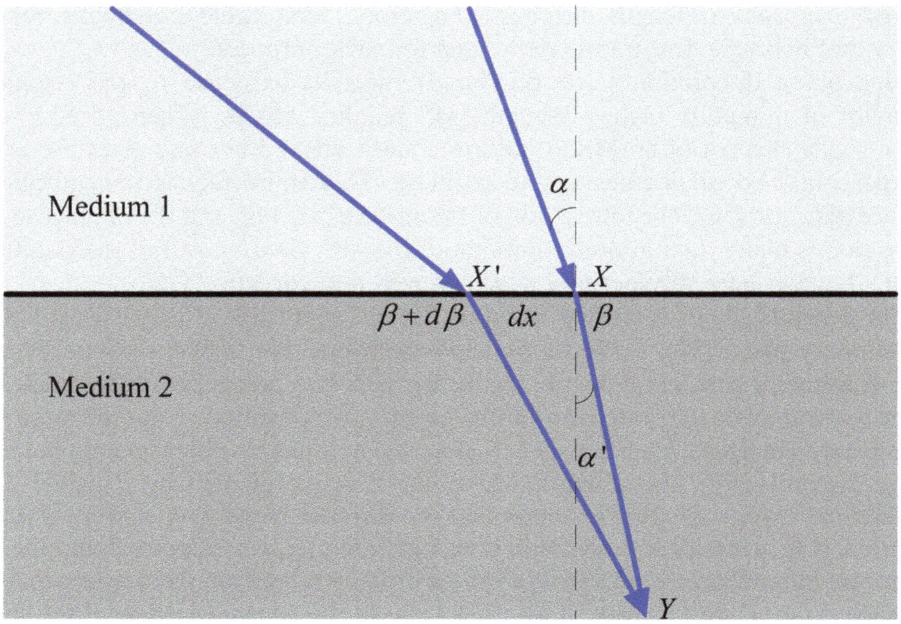

Figure 7.1. An example of the generalized law of refraction. β and $\beta + d\beta$ are phase discontinuities at X and X', respectively, and dx is the distance between them.

independent EM features (the electric field intensities of the 1st and 2nd regions are $\overrightarrow{E_1}$ and $\overrightarrow{E_2}$, respectively, the magnetic field intensities are $\overrightarrow{H_1}$ and $\overrightarrow{H_2}$, respectively, and the unit vector of the surface normal is $\hat{n}$), the surface magnetic and electric currents ($\overrightarrow{M_S}$ and $\overrightarrow{J_S}$, respectively) at the interface must provide the following relationship to compensate for the field discontinuities [31]:

$$\overrightarrow{J_s} = \hat{n} \times \left(\overrightarrow{H_2} - \overrightarrow{H_1}\right) \text{ and } \overrightarrow{M_S} = -\hat{n} \times \left(\overrightarrow{E_2} - \overrightarrow{E_1}\right) \tag{7.3}$$

With a random pattern of hitting and aspired waves, the design of Huygens' MSs can include non-periodic or multilayer structures to locally design surface impedances. For instance, Pfeiffer and Grbic (2013) used a supercell made up of 12 patterned elements to carry out a beam deflector with 86% transmittance in microwave frequency bands [31]. Monticone *et al* (2013) also suggested an MS stack consisting of three planar arrays arranged on top of each other, consisting of a combination of dielectric and plasmonic nano blocks functionalized as nano-inductors and nanocapacitors (LC nanocircuit elements), respectively, for bending light in the near-infrared frequency band with a theoretical efficiency of 75% based on optical nanocircuit concepts [52].

7.2.3 MSs based on the Pancharatnam–Berry phase

MSs based on the Pancharatnam–Berry phase provide full phase control by setting the orientation angle of antennas having the same geometry [60, 61]. These MS structures, firstly, offer an exactly different approach to introduce phase jumps and secondly, they are associated with polarization change. If we consider an anisotropic nanostructure under normal incidence, the t_0 and t_e terms, respectively, indicate complex transmission coefficients when the polarization of incident light is along the two major axes of the nanostructure. Since the nanoresonator is rotated with a θ angle from the x-axis in the cartesian coordinate system, the transmission coefficients for this rotated system may be extracted by the working of the Jones matrix to be the $R(\theta)$ rotation matrix as follows [62]:

$$\hat{t}(\theta) = R(-\theta)\begin{pmatrix} t_0 & 0 \\ 0 & t_e \end{pmatrix}R(\theta) = \begin{bmatrix} \cos\theta & -\sin\theta \\ \sin\theta & \cos\theta \end{bmatrix}\begin{bmatrix} t_0 & 0 \\ 0 & t_e \end{bmatrix}\begin{bmatrix} \cos\theta & \sin\theta \\ -\sin\theta & \cos\theta \end{bmatrix} \tag{7.4}$$

$$\hat{t}(\theta) = \begin{bmatrix} t_0\cos^2\theta + t_e\sin^2\theta & (t_0 - t_e)\cos\theta\sin\theta \\ (t_0 - t_e)\cos\theta\sin\theta & t_0\sin^2\theta + t_e\cos^2\theta \end{bmatrix} \tag{7.5}$$

When the incoming EM wave is circularly polarized, the transmitted electric field ($E^t_{L/R}$) may be reproduced by applying the matrix product of equation (7.5) and Jones vectors of left-handed circularly polarized or right-handed circularly polarized light $\hat{e}_{L/R} = (\hat{e}_x \pm i\hat{e}_y)/\sqrt{2}$ and expressed as [63]:

$$E^t_{L/R} = \hat{t}(\theta)\cdot\hat{e}_{L/R} = \frac{t_0 + t_e}{2}\hat{e}_{L/R} + \frac{t_0 - t_e}{2}e^{\pm i2\theta}\hat{e}_{R/L} \tag{7.6}$$

The first term in equation (7.6) indicates circularly polarized scattered waves with the same spirality as the incident wave, while the second term states that the opposite spirality circularly polarized scattered waves carry an additional Pancharatnam–Berry phase ($\pm i2\theta$). Therefore, a phase shift from 0 to 2π can be adjusted for the inverse radiation when rotating the nanoresonator from 0 to π [64]. Because this method uses oriented copies of a single antenna design for phase modulation, it has a number of advantages such as higher manufacturing tolerance, easy design and is useful for wideband applications.

7.3 Applications of MSs

MSs are successfully used in many areas such as designing a high-gain antenna with improved bandwidth, improving the response of FSSs, protecting sensitive electronics and reducing interference on a shared platform, manufacturing ultra-thin lenses, local phase manipulation and shaping the wavefronts of the reflected wave, increasing polarization insensitivity and improving absorption bandwidth. Another area where MSs have been used frequently in recent times is in MS-based polarization converters. In this section, information about the concept of polarization, polarization types, MS-based polarization converters, basic MS-based polarization converter design and unit cell configuration, polarization conversion parameters and MS-based polarization converter studies in the literature will be given.

7.3.1 Polarization

Polarization is a fundamental parameter of the EM wave, which mentions the oscillation direction of the electric field in a plane perpendicular to the wave propagation direction, and is one of the most important parameters of the EM wave [65]. While polarization defines the orientation and change of the electric field, it can be more clearly stated as the shape formed by the observer looking from the opposite direction as the wave oscillates with the expression $\overrightarrow{E}$ (electric field intensity) and propagates in the z-direction [66]. Or, polarization can also be defined as the path followed by $\overrightarrow{E}$ and $\overrightarrow{H}$ (magnetic field intensity) perpendicular to the direction of propagation in the plane as a function of time at a certain point [67]. Three types of polarization can be mentioned: Linear polarization (LP), circular polarization (CP) and elliptical polarization (EP). There are two types of circular and elliptical polarization, right-handed and left-handed. The shape formed according to the observer, that is, as mentioned before, the path followed by $\overrightarrow{E}$ and $\overrightarrow{H}$ are straight lines in LP, circular in CP and elliptical in EP. Accordingly, the wave propagations required to model the properties of polarization types as EM, the time–space expressions of the electric field intensity ($E_x(t)$ and $E_y(t)$, respectively) in the x- and y-directions are given by the following equations.

$$E_x(t) = E_1 \cos(\omega t - kz) \tag{7.7}$$

$$E_y(t) = E_2 \cos(\omega t - kz + \varphi) \tag{7.8}$$

Equations (7.7) and (7.8) express the electric field components in the x- and y-directions of the wave propagating in the z-direction, respectively, where E_1 and E_2 are the amplitudes of the waves linearly polarized in the x- and y-directions, respectively, k is the wave number, ω is the angular frequency and φ is the phase difference between E_x and E_y (that is, φ_x and φ_y are the phase difference in the x- and y-directions, respectively, and $\varphi = \varphi_x - \varphi_y$) [68, 69].

7.3.1.1 Linear polarization

If the phase difference is $\varphi = \pm m\pi$ ($m = 0, 1, 2, 3, \ldots$), the LP situation is realized. If $E_1 = 0$, LP occurs in the y-direction, and if $E_2 = 0$, LP occurs in the x-direction (as seen in figure 7.2) [68].

7.3.1.2 Circular polarization

If the phase difference is $\varphi = \pm m\frac{\pi}{2}$ ($m = 1, 3, 5, 7, \ldots$) and $E_1 = E_2$, the CP situation is realized. As mentioned earlier, there are two types of CP, right-handed and left-handed. In the case of phase difference, $\varphi = -m\frac{\pi}{2}$, the right-handed circular polarization (RHCP) situation occurs and $\overrightarrow{E}$ occurs counterclockwise (as seen in figure 7.3(a)) while in the event of phase difference $\varphi = +m\frac{\pi}{2}$, the left-handed circular polarization (LHCP) situation occurs and $\overrightarrow{E}$ occurs clockwise (as shown in figure 7.3(b)) [68].

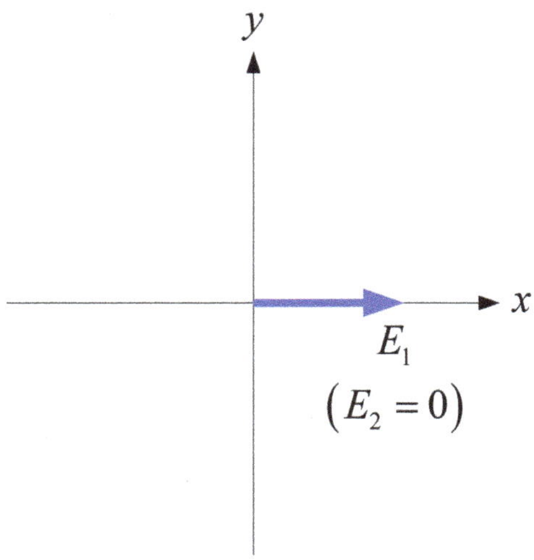

Figure 7.2. LP status.

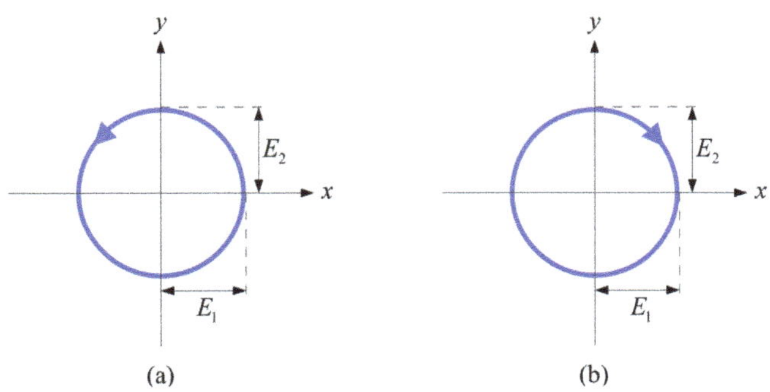

Figure 7.3. (a) RHCP and (b) LHCP situations.

7.3.1.3 Elliptical polarization

If $E_1 \neq E_2$, the EP situation is realized. As mentioned earlier, there are two types of EP, right-handed and left-handed, and whether the polarization is right-handed or left-handed depends on the φ angle. EP is the general case of CP where the phase difference between E_1 and E_2 is any angle that is not necessarily $\varphi = \pm m\frac{\pi}{2}$ ($m = 1, 3, 5, 7, \ldots$) (figure 7.4) [67, 68].

7.3.2 MS-based polarization converters

The polarization case of the EM wave has attracted the interest of more and more researchers due to the variety of polarization-sensitive devices and applications. Polarization converters among the devices exhibiting polarization conversion control have been extensively studied for a variety of applications such as improvement of antenna gain [70], reduction of interference and radar cross section (RCS) [71, 72], satellite communication [73] and EM shielding [74]. Conventional polarization converters using the Faraday effect and the optical activity of crystals,

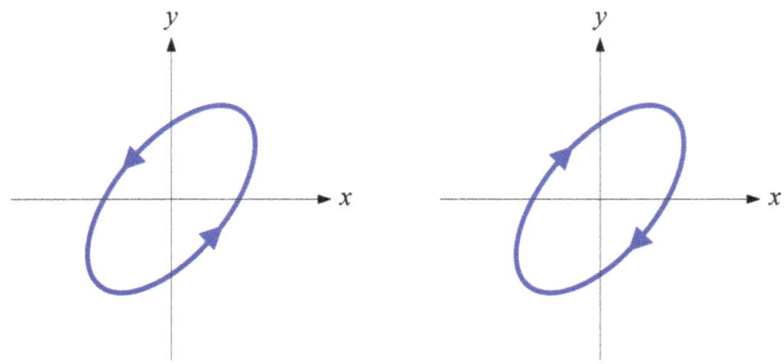

Figure 7.4. (a) Right-handed and (b) left-handed EP situations [67, 68].

because of their large volume, narrow bandwidth, require much greater thickness compared to the operating wavelength and low dependence under oblique angle; they are exceedingly unfavorable to integrate with ultra-thin devices such as advanced sensors and nanophotonic devices and are incompatible for real-world applications [75]. Therefore, it is imperative to design polarizing converters with simple structures, compact size, wide band gap, wide angular stability and good performance. In recent years, a large number of polarization converters have been developed based on a 2D planar MM structure and artificial materials called MSs. These polarization converters provide a potential approximation to manipulate the polarization case of EM waves by adjusting the material parameters or altering the size and geometry of the MS [76]. These MS-based structures have shown the capability to control polarization in different frequency bands of the EM wave regime, such as microwave [77], terahertz [78], infrared [79] and visible [80] in transmission and reflection modes.

7.3.2.1 Design of a basic MS-based polarization converter and unit cell configuration

The unit cell of a suggested basic MS-based polarization converter design in metallic–dielectric–metallic configuration with a single-layer is shown in figure 7.5. As seen in figure 7.5, the back part of the MS-based polarization converter structure is terminated with metal material with a certain conductivity and thickness (t). The MS on the upper part of the structure is also formed with the same metal material (or it can be formed with another alloy or metal depending on the design). In the middle part of the structure, dielectric substrate material with a certain relative dielectric constant and loss tangent value was used.

7.3.2.2 Polarization conversion parameters

From Stoke's parameters, S_0 is the density of the total spreading area, S_1 is linear horizontal polarization and linear vertical polarization dominance, S_2 is linear +45° polarization and linear −45° dominance, and S_3 is right-handed and left-handed CP; these parameters are as follows [81]:

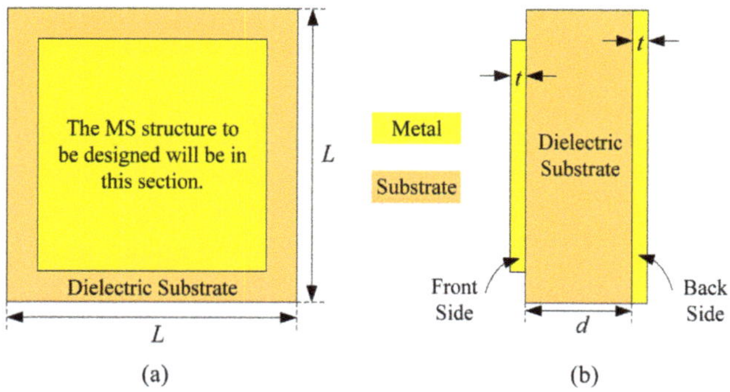

Figure 7.5. Design of a basic MS-based polarization converter and the unit cell configuration geometry of this design: (a) front and (b) side view.

$$S_0 = E_1^2 + E_2^2 \tag{7.9}$$

$$S_1 = E_1^2 - E_2^2 \tag{7.10}$$

$$S_2 = 2E_1E_2 \cos \varphi \tag{7.11}$$

$$S_3 = 2E_1E_2 \sin \varphi \tag{7.12}$$

Also, the relationship between these Stoke's parameters is as follows:

$$S_0^2 = S_1^2 + S_2^2 + S_3^2 \tag{7.13}$$

The ellipticity value obtained from Stoke's parameters is a parameter that gives information about the CP of the EM wave. This value denominates that the EM wave has left-handed or right-handed CP. If the ellipticity value is −1, the EM wave shows the LHCP feature, while if this value is +1, the EM wave shows the RHCP feature. The ellipticity value is calculated using Stoke's parameters S_0 and S_3, as follows [82]:

$$\text{Ellipticity} = \frac{S_3}{S_0} \tag{7.14}$$

The polarization conversion ratio (PCR) expresses the efficiency ratio of the LP conversion. As mentioned before, the LP conversion is a polarization conversion in which the x- or y-polarized incoming wave is reflected or transmitted as y- or x-polarized [83]. R_{xx} and R_{yy} are co-polarization reflection coefficients and R_{xy} and R_{yx} are cross-polarization reflection coefficients. Here, R_{xx}, R_{yy}, R_{yx} and R_{xy} express the x-polarized reflection coefficient of the x-polarized incoming wave, the y-polarized reflection coefficient of the y-polarized incoming wave, the y-polarized reflection coefficient of the x-polarized incoming wave, and the x-polarized reflection coefficient of the y-polarized incoming wave, in the given order, and these reflection coefficients are as follows [84].

$$R_{xx} = \frac{E_{rx}}{E_{ix}},\ R_{yy} = \frac{E_{ry}}{E_{iy}},\ R_{xy} = \frac{E_{rx}}{E_{iy}},\ \text{and}\ R_{yx} = \frac{E_{ry}}{E_{ix}} \tag{7.15}$$

In equation (7.15), the subscripts i and r in electric fields state the incoming and reflected wave, respectively, while x and y state the direction of polarization. Accordingly, using the co- and cross-polarization reflection coefficients, the PCR value can be calculated with the help of the following equation [82, 84].

$$PCR = \frac{R_{yx}^2}{R_{yx}^2 + R_{xx}^2}\ \text{or}\ PCR = \frac{R_{xy}^2}{R_{xy}^2 + R_{yy}^2} \tag{7.16}$$

Equation (7.16) demonstrates that the incoming wave in x- or y-polarized is reflected as y- or x-polarized. That is, the closer the PCR value is to one, the more linear the conversion.

7.3.3 MS-based polarization converter studies

In this section, selected studies of MS-based polarization converters from 2015 to the present are given. By examining the geometries, shapes, materials, conversion performances, conversion types and operating ranges of these studies, more information in the literature about MS-based polarization converters has been attempted to be found.

Gao *et al* (2015) suggested a double V-shaped MS-based LP converter. With their suggested polarization converter, they achieved a PCR rate of over 90% in the ultra-wideband in the frequency band of 12.4–27.96 GHz. It has been reported that the suggested polarization converter may be used in a lot of applications such as imaging systems, reflector antennas, radiometers and remote sensors due to its simple geometry and operation in ultra-wideband [76].

Liu *et al* (2016) suggested an MS-based converter consisting of reflective mode, double-layer and metallic L-shapes. According to both simulated and measured results, a three-band polarization has been confirmed for the angle of incidence up to 30°. The suggested polarization converter first shows RHCP conversion for the transverse magnetic (TM) wave in the frequency band of 9.1–16.5 GHz, and LHCP conversion for transverse electric (TE) wave in the same frequency band. Second, the converter shows LHCP conversion for the TM wave in the frequency band of 20–25.4 GHz, while it also shows the RHCP conversion for the TE wave in the same frequency band. Thirdly, the converter shows LP conversion feature in the frequency band of 17.4–18.9 GHz for both TE and TM waves [85].

Jiang *et al* (2017) suggested an MS-based CP converter design with ultra-broadband and high efficiency. The structure consisting of a double split resonant square ring MS, dielectric layer and completely gold layer shows LHCP conversion performance in the frequency range of 0.6–1.41 THz and RHCP in the frequency range of 1.48–1.54 THz. In addition, this polarization converter design has potential applications such as cloaking technology, antenna design and EM measurement [78].

Xu *et al* (2018), using a polarization converter design consisting of a metal–dielectric–metal arrangement using a compact H-shaped MS, have obtained over 90% PCR efficiency in the frequency band of 7–19.5 GHz, and an average PCR efficiency at 80% up to 41.5° incidence angle in the aforementioned frequency range [86].

Huang *et al* (2019) obtained LP conversion in the frequency ranges of 4.40–5.30 GHz and 9.45–13.60 GHz and CP conversion in the frequency ranges of 4.47–5.35 GHz and 9.57–13.57 GHz with a metal–dielectric–metal sandwich structure using a FR-4 dielectric substrate with 3 mm thickness. With the experimentally confirmed ultra-thin polarization converter, it is thought to provide an essential stepping stone for the future design of other dual-band MS devices and can be extended to higher frequency bands [87].

Qi *et al* (2020) suggested an MS-based, high-efficiency and polarization converter operating in the THz region, similar to a sandwich structure composed of an upper double split-ring resonator (DSRR), dielectric interlayer and bottom metallic layer. The suggested converter shows LP conversion performance with a PCR efficiency of more than 90% in the frequency range of 2.04–5.33 THz [88].

Pouyanfar *et al* (2021) suggested a multifunctional, single-layer, thin, simple and MS-based polarization converter. The converter they propose performs LP conversion in the frequency range of 15.5–16.5 GHz, while RHCP and LHCP conversions are performed at 13 GHz and 18 GHz, respectively. In addition, the structure has angular stability up to 60° [89].

Tutar and Ozturk (2022) designed and implemented an MS-based polarization converter showing LP and CP conversion features with reflection mode operating in the microwave C- and X-bands. The converter they propose shows LP conversion performance with PCR efficiency over 90% in two frequency bands, 4.61–4.85 GHz and 6.86–7.78 GHz, while RHCP in the frequency range of 5.35–5.94 GHz and LHCP in the frequency range of 8.88–12.03 GHz. In addition, their design has a PCR value of 80% for angle of incidence up to 45° in the LP conversion frequency range [90].

Kaya (2023) designed and implemented an ultra-thin, MS-based polarization converter with both LP and CP conversion performance in the microwave C-, X- and Ku-bands. The design, which consists of MS metallic patterns–FR-4 substrate–completely metallic surface coating, shows the LP conversion feature in three frequency ranges (PCR efficiency is over 90% in the frequency ranges of 6.46–6.78, 10.52–11.85 and 16.49–17.37 GHz) and CP conversion feature in two frequency ranges (LHCP in the frequency range 7.28–9.4 GHz and RHCP conversion in the frequency range of 13.38–15.19 GHz). In addition, the PCR value is around 90% in the LP conversion frequency band under the angle of incidence of 45° [91].

7.4 Conclusion

The purpose of this book chapter is to ensure an extensive and appropriate definition of MSs, to determine their outstanding features, and also to emphasize some of the advantages that may be gained by using them. A superior property of MSs is their low loss behavior in contrast to MMs. Therefore, MSs permit us to use the extraordinary properties of MMs such as refraction and reflection without incurring production complications and the losses related to MMs.

Although very successful studies have been made with MSs in the microwave and optical frequency region, the inclusion of MSs in the design of imaging and telecommunication devices in the future has great potential for researchers and technological developments. In addition, the ease of fabrication and cost-effectiveness of MS-based designs will also play an important role in their incorporation into technology.

For this reason, an overview of the basic information about MSs is presented first in the study. Secondly, while giving general information about MSs, information about generalized refraction law, Huygens' MSs and MSs based on the Pancharatnam–Berry phase are given in the subsections. Thirdly, MS-based polarization converters, polarization (LP, CP and EP), basic MS-based polarization converter design and unit cell configuration, polarization conversion parameters, and MS-based polarization converter studies, which are the application areas of MSs, are examined.

In this context, it is hoped that this book chapter will provide strategic guidance for other researchers and also create an infrastructure for potential future studies and research.

References

[1] Collin R E 1991 *Field Theory of Guided Waves* (Wiley-IEEE Press)

[2] Aydin K, Li Z F, Hudlicka M, Tretyakov S A and Ozbay E 2007 Transmission characteristics of bianisotropic metamaterials based on omega shaped metallic inclusions *New J. Phys.* **9** 326

[3] Engheta N and Ziolkowski R W 2006 *Metamaterials: Physics and Engineering Explorations* (Wiley-IEEE Press)

[4] Smith D R, Padilla W J, Vier D C, Nemat-Nasser S C and Schultz S 2000 Composite medium with simultaneously negative permeability and permittivity *Phys. Rev. Lett.* **84** 4184–7

[5] Motcheyo A B T, Kanaa T and Essiane S N 2021 Backward-energy propagation in nonlinear magnetic metamaterials *J. Supercond. Novel Magn.* **34** 2619–26

[6] Tyukhtin A V and Vorobev V V 2013 Cherenkov radiation in a metamaterial comprised of coated wires *J. Opt. Soc. Am. B: Opt. Phys.* **30** 1524–31

[7] Zhai S L, Zhao X P, Liu S, Shen F L, Li L L and Luo R 2016 Inverse Doppler effects in broadband acoustic metamaterials *Sci Rep.* **6** 32388

[8] Alici K B and Ozbay E 2007 Radiation properties of a split ring resonator and monopole composite *Phys. Status Solidi* B **244** 1192–6

[9] Landy N I, Sajuyigbe S, Mock J J, Smith D R and Padilla W J 2008 Perfect metamaterial absorber *Phys. Rev. Lett.* **100** 207402

[10] Valentine J, Zhang S, Zentgraf T, Ulin-Avila E, Genov D A, Bartal G and Zhang X 2008 Three-dimensional optical metamaterial with a negative refractive index *Nature* **455** 376

[11] Cheng B H, Lan Y C and Tsai D P 2013 Breaking optical diffraction limitation using optical hybrid-super-hyperlens with radially polarized light *Opt. Express* **21** 14898–906
[12] Cai W S, Chettiar U K, Kildishev A V and Shalaev V M 2007 Optical cloaking with metamaterials *Nat. Photonics* **1** 224–7
[13] Fang N, Lee H, Sun C and Zhang X 2005 Sub-diffraction-limited optical imaging with a silver superlens *Science* **308** 534–7
[14] Modi A Y, Balanis C A, Birtcher C R and Shaman H N 2017 Novel design of ultra-broadband radar cross section reduction surfaces using artificial magnetic conductors *IEEE Trans. Antennas Propag.* **65** 5406–17
[15] Qi S B, Oudich M, Li Y and Assouar B 2016 Acoustic energy harvesting based on a planar acoustic metamaterial *Appl. Phys. Lett.* **108** 263501
[16] Orazbayev B, Beruete M and Khromova I 2016 Tunable beam steering enabled by graphene metamaterials *Opt. Express* **24** 8848–61
[17] Jaksic Z, Vukovic S, Matovic J and Tanaskovic D 2011 Negative refractive index metasurfaces for enhanced biosensing *Materials* **4** 1–36
[18] Hao J M, Wang J, Liu X L, Padilla W J, Zhou L and Qiu M 2010 High performance optical absorber based on a plasmonic metamaterial *Appl. Phys. Lett.* **96** 251104
[19] Holloway C L, Kuester E F, Gordon J A, O'Hara J, Booth J and Smith D R 2012 An overview of the theory and applications of metasurfaces: the two-dimensional equivalents of metamaterials *IEEE Antennas Propag. Mag.* **54** 10–35
[20] Kuester E F, Mohamed M A, Piket-May M and Holloway C L 2003 Averaged transition conditions for electromagnetic fields at a metafilm *IEEE Trans. Antennas Propag.* **51** 2641–51
[21] Yu N F and Capasso F 2014 Flat optics with designer metasurfaces *Nat. Mater.* **13** 139–50
[22] Soukoulis C M and Wegener M 2011 Past achievements and future challenges in the development of three-dimensional photonic metamaterials *Nat. Photonics* **5** 523–30
[23] Meinzer N, Barnes W L and Hooper I R 2014 Plasmonic meta-atoms and metasurfaces *Nat. Photonics* **8** 889–98
[24] Kildishev A V, Boltasseva A and Shalaev V M 2013 Planar photonics with metasurfaces *Science* **339** 1232009
[25] Yang S M, Liu P, Yang M D, Wang Q G, Song J M and Dong L 2016 From flexible and stretchable meta-atom to metamaterial: a wearable microwave meta-skin with tunable frequency selective and cloaking effects *Sci Rep.* **6** 21921
[26] Jiang Z H, Yun S, Lin L, Bossard J A, Werner D H and Mayer T S 2013 Tailoring dispersion for broadband low-loss optical metamaterials using deep-subwavelength inclusions *Sci Rep.* **3** 1571
[27] Quarfoth R and Sievenpiper D 2013 Artificial tensor impedance surface waveguides *IEEE Trans. Antennas Propag.* **61** 3597–606
[28] Maksimenko A V, Shcherbinin V I and Tkachenko V I 2019 Coupled-mode theory of an irregular waveguide with impedance walls *J. Infrared, Millimeter, Terahertz Waves* **40** 620–36
[29] Achouri K, Lavigne G, Salem M A and Caloz C 2016 Metasurface spatial processor for electromagnetic remote control *IEEE Trans. Antennas Propag.* **64** 1759–67
[30] Shi Z J, Zhu A Y, Li Z Y, Huang Y W, Chen W T, Qiu C W and Capasso F 2020 Continuous angle-tunable birefringence with freeform metasurfaces for arbitrary polarization conversion *Sci. Adv.* **6** eaba3367

[31] Pfeiffer C and Grbic A 2013 Metamaterial Huygens' surfaces: tailoring wave fronts with reflectionless sheets *Phys. Rev. Lett.* **110** 197401

[32] Orazbayev B, Estakhri N M, Alu A and Beruete M 2017 Experimental demonstration of metasurface-based ultrathin carpet cloaks for millimeter waves *Adv. Opt. Mater.* **5** 1600606

[33] Iwaszczuk K, Strikwerda A C, Fan K B, Zhang X, Averitt R D and Jepsen P U 2012 Flexible metamaterial absorbers for stealth applications at terahertz frequencies *Opt. Express* **20** 635–43

[34] Eleftheriades G V 2014 Protecting the weak from the strong *Nature* **505** 490–1

[35] Fong B H, Colburn J S, Ottusch J J, Visher J L and Sievenpiper D F 2010 Scalar and tensor holographic artificial impedance surfaces *IEEE Trans. Antennas Propag.* **58** 3212–21

[36] Alici K B, Serebryannikov A E and Ozbay E 2010 Radiation properties and coupling analysis of a metamaterial based, dual polarization, dual band, multiple split ring resonator antenna *J. Electromagn. Waves Appl.* **24** 1183–93

[37] Yao Y, Shankar R, Kats M A, Song Y, Kong J, Loncar M and Capasso F 2014 Electrically tunable metasurface perfect absorbers for ultrathin mid-ınfrared optical modulators *Nano Lett.* **14** 6526–32

[38] Liu Z W, Lee H, Xiong Y, Sun C and Zhang X 2007 Far-field optical hyperlens magnifying sub-diffraction-limited objects *Science* **315** 1686 1686

[39] Jiang L, Chu C F, Fang B, Zhang M N, Gan H Y, Li C X, Jing X F and Hong Z 2020 Multi-wavelength carpet cloaking based on an ultrathin single layer metamaterial microstructure *Laser Phys. Lett.* **17** 066202

[40] Lee T G, Jo S H, Seung H M, Kim S W, Kim E J, Youn B D, Nahm S and Kim M 2020 Enhanced energy transfer and conversion for high performance phononic crystal-assisted elastic wave energy harvesting *Nano Energy* **78** 105226

[41] Li A B, Kim S, Luo Y, Li Y B, Long J and Sievenpiper D F 2017 High-power transistor-based tunable and switchable metasurface absorber *IEEE Trans. Microwave Theory Tech.* **65** 2810–8

[42] Kim S, Wakatsuchi H, Rushton J J and Sievenpiper D F 2016 Switchable nonlinear metasurfaces for absorbing high power surface waves *Appl. Phys. Lett.* **108** 041903

[43] Habib M A, Barkat M, Aissa B and Denidni T A 2008 Ca-CFAR detection performance of radar targets embedded in 'Non Centered Chi-2 Gamma' clutter *Prog. Electromagn. Res.* **88** 135–48

[44] Albooyeh M, Simovski C and Tretyakov S 2016 Homogenization and characterization of metasurfaces: general framework *2016 10th European Conf. on Antennas and Propagation (EUCAP) (Davos, Switzerland)* pp 10–5

[45] Holloway C L and Kuester E F 2016 A homogenization technique for obtaining generalized sheet-transition conditions for a metafilm embedded in a magnetodielectric interface *IEEE Trans. Antennas Propag.* **64** 4671–86

[46] Fuscaldo W, Tofani S, Zografopoulos D C, Baccarelli P, Burghignoli P, Beccherelli R and Galli A 2018 Systematic design of THz leaky-wave antennas based on homogenized metasurfaces *IEEE Trans. Antennas Propag.* **66** 1169–78

[47] Huang J and Pogorzelski R J 1998 A Ka-band microstrip reflectarray with elements having variable rotation angles *IEEE Trans. Antennas Propag.* **46** 650–6

[48] Glybovski S B, Tretyakov S A, Belov P A, Kivshar Y S and Simovski C R 2016 Metasurfaces: from microwaves to visible *Phys. Rep.-Rev. Sect. Phys. Lett.* **634** 1–72

[49] Huygens C 1690 *Traité De La Lumière (A Treatise on Light)* (Leiden, Netherlands: Pieter van der Aa)
[50] Subrahmanyam N 2012 *A Textbook of Optics* (New Delhi: S. Chand Publishing)
[51] Rengarajan S R and Rahmat-Samii Y 2000 The field equivalence principle: illustration of the establishment of the non-intuitive null fields *IEEE Antennas Propag. Mag.* **42** 122–8
[52] Monticone F, Estakhri N M and Alu A 2013 Full control of nanoscale optical transmission with a composite metascreen *Phys. Rev. Lett.* **110** 203903
[53] Cornbleet S 1983 Geometrical optics reviewed: a new light on an old subject *Proc. IEEE* **71** 471–502
[54] Cathey W T 1965 Three-dimensional wavefront reconstruction using a phase hologram *J. Opt. Soc. Am.* **55** 457
[55] Zhang Y, Liang L J, Yang J, Feng Y J, Zhu B, Zhao J M, Jiang T, Jin B B and Liu W W 2016 Broadband diffuse terahertz wave scattering by flexible metasurface with randomized phase distribution *Sci Rep.* **6** 26875
[56] Zheng G X, Muehlenbernd H, Kenney M, Li G X, Zentgraf T and Zhang S 2015 Metasurface holograms reaching 80% efficiency *Nat. Nanotechnol.* **10** 308–12
[57] Yu N F, Genevet P, Kats M A, Aieta F, Tetienne J P, Capasso F and Gaburro Z 2011 Light propagation with phase discontinuities: generalized laws of reflection and refraction *Science* **334** 333–7
[58] Feynman R P, Leighton R B, Sands M and Hafner E M 1965 The Feynman lectures on physics; Vol. I *Am. J. Phys.* **33** 750–2
[59] Yu N F, Aieta F, Genevet P, Kats M A, Gaburro Z and Capasso F 2012 A broadband, background-free quarter-wave plate based on plasmonic metasurfaces *Nano Lett.* **12** 6328–33
[60] Huang L L, Chen X Z, Muehlenbernd H, Li G X, Bai B F, Tan Q F, Jin G F, Zentgraf T and Zhang S 2012 Dispersionless phase discontinuities for controlling light propagation *Nano Lett.* **12** 5750–5
[61] Jiang S C, Xiong X, Hu Y S, Jiang S W, Hu Y H, Xu D H, Peng R W and Wang M 2015 High-efficiency generation of circularly polarized light via symmetry-induced anomalous reflection *Phys. Rev.* B **91** 125421
[62] Cong L Q, Xu N N, Zhang W L and Singh R 2015 Polarization control in terahertz metasurfaces with the lowest order rotational symmetry *Adv. Opt. Mater.* **3** 1176–83
[63] Wen D D *et al* 2015 Metasurface for characterization of the polarization state of light *Opt. Express* **23** 10272–81
[64] Ding X M, Monticone F and Zhang K *et al* 2015 Ultrathin Pancharatnam-Berry metasurface with maximal cross-polarization efficiency *Adv. Mater.* **27** 1195–200
[65] Beruete M, Navarro-Cia M, Sorolla M and Campillo I 2008 Polarization selection with stacked hole array metamaterial *J. Appl. Phys.* **103** 053102
[66] Ozbay M and Ozkazanc Y 2018 Decomposition of optical reflections via polarization techniques *2018 26th Signal Processing and Communications Applications Conf. (SIU) (İzmir, Turkey)* pp 2–5
[67] Sadiku M 2014 *Elements of Electromagnetics* (Oxford: Oxford University Press)
[68] Kraus J D and Carver K R 1973 *Electromagnetics* (New York: McGraw-Hill)
[69] Popovic Z and Popovic B D 1999 *Introductory Electromagnetics* (Prentice Hall)
[70] Li H P, Wang G M, Xu H X, Cai T and Liang J G 2015 X-band phase-gradient metasurface for high-gain lens antenna application *IEEE Trans. Antennas Propag.* **63** 5144–9

[71] Jha P K, Ni X J, Wu C H, Wang Y and Zhang X 2015 Metasurface-enabled remote quantum interference *Phys. Rev. Lett.* **115** 025501
[72] Liu Y, Hao Y W, Li K and Gong S X 2016 Wideband and polarization-independent radar cross section reduction using holographic metasurface *IEEE Antennas Wirel. Propag. Lett.* **15** 1028–31
[73] Wu Z, Li L, Li Y J and Chen X 2016 Metasurface superstrate antenna with wideband circular polarization for satellite communication application *IEEE Antennas Wirel. Propag. Lett.* **15** 374–7
[74] Huang F C, Chiu C N, Wu T L and Chiou Y P 2015 A circular-ring miniaturized-element metasurface with many good features for frequency selective shielding applications *IEEE Trans. Electromagn. Compat.* **57** 365–74
[75] Zhao Y and Alu A 2011 Manipulating light polarization with ultrathin plasmonic metasurfaces *Phys. Rev.* B **84** 205428
[76] Gao X, Han X, Cao W P, Li H O, Ma H F and Cui T J 2015 Ultrawideband and high-efficiency linear polarization converter based on double V-shaped metasurface *IEEE Trans. Antennas Propag.* **63** 3522–30
[77] Ozturk G, Hasar U C, Ertugrul M, Tutar F, Corapsiz M F, Kurt M, Donmez K, Ismail I and Alfaqawi M S S 2023 An efficient cost effective wide-angle metasurface-based linear and circular polarization converter for X-, Ku- and K-band applications *Opt. Laser Technol.* **163** 109404
[78] Jiang Y N, Wang L, Wang J, Akwuruoha C N and Cao W P 2017 Ultra-wideband high-efficiency reflective linear-to-circular polarization converter based on metasurface at terahertz frequencies *Opt. Express* **25** 27616–23
[79] Guo Z Y, Tian L H, Shen F, Zhou H P and Guo K 2017 Mid-infrared polarization devices based on the double-phase modulating dielectric metasurface *J. Phys. D: Appl. Phys.* **50** 254001
[80] Qin F *et al* 2016 Hybrid bilayer plasmonic metasurface efficiently manipulates visible light *Sci. Adv.* **2** e1501168
[81] Papas C H 2011 *Theory of Electromagnetic Wave Propagation* (Dover Publications)
[82] Khan M I, Khalid Z and Tahir F A 2019 Linear and circular-polarization conversion in X-band using anisotropic metasurface *Sci Rep.* **9** 4552
[83] Khan M I and Tahir F A 2018 A broadband cross-polarization conversion anisotropic metasurface based on multiple plasmon resonances *Chin. Phys.* B **27** 014101
[84] Fu C F, Sun Z J, Han L F, Liu C, Sun T and Chu P K 2019 High-efficiency dual-frequency reflective linear polarization converter based on metasurface for microwave bands *Appl. Sci.* **9** 1910
[85] Liu X B, Zhang J S, Liu W, Lu R, Li L M, Xu Z and Zhang A X 2017 Three-band polarization converter based on reflective metasurface *IEEE Antennas Wirel. Propag. Lett.* **16** 924–7
[86] Xu J, Li R Q, Qin J, Wang S Y and Han T C 2018 Ultra-broadband wide-angle linear polarization converter based on H-shaped metasurface *Opt. Express* **26** 20913–9
[87] Huang X J, Yang H L, Zhang D H and Luo Y 2019 Ultrathin dual-band metasurface polarization converter *IEEE Trans. Antennas Propag.* **67** 4636–41
[88] Qi Y P, Zhang B H, Liu C Q and Deng X Y 2020 Ultra-broadband polarization conversion meta-surface and its application in polarization converter and RCS reduction *IEEE Access* **8** 116675–84

[89] Pouyanfar N, Nourinia J and Ghobadi C 2021 Multiband and multifunctional polarization converter using an asymmetric metasurface *Sci Rep.* **11** 9306
[90] Tutar F and Ozturk G 2022 An effective metasurface-based linear and circular polarization converter for C- and X-band applications *Opt. Mater.* **128** 112355
[91] Kaya Y 2023 Cheaper, wide-band, ultra-thin, and multi-purpose single-layer metasurface polarization converter design for C-, X-, and Ku-band applications *Symmetry* **15** 442

Chapter 8

Flexible metamaterials

Erkan Tetik

Symbols

ε Permittivity
μ Permeability
n Refractive index

In recent years, comprehensive research has been carried out on using advanced materials and structures to improve the properties and functions of flexible materials. Emerging as a revolutionary advance in materials engineering compared to conventional methods, metamaterials are nominated as a remarkable candidate for flexible materials. Metamaterials, which are defined as artificial/synthetic structures, exhibit acoustic, electronic, optical, mechanical, and thermal behaviors that natural materials do not have. The bending, stretching, and rolling abilities of metamaterial structures designed using flexible materials, offer a new perspective on the manipulation of electromagnetic waves. Flexible MTM structures with features such as adjustable hardness/Poisson's ratio, portable energy source, and programmable mechanical structures inspire innovative applications as a new research area. In addition, they have many uses with the increase in portable devices. Herein, recent developments in flexible metamaterials are reviewed and their potential values are examined.

8.1 Introduction

In recent years, comprehensive research has been carried out on using advanced materials and structures to improve the properties and functions of flexible materials. With the rapid and extensive progress in micro/nano simulation and fabrication technologies, materials have been studied by modeling micro/nano structures on rigid substrates. Considering the importance of reliable supporting material, such rigid substrates are remarkably beneficial in material fabrication. However, it emerges with notable challenges, such as wrapping material around

doi:10.1088/978-0-7503-5754-8ch8

objects in cases without a flat plane, such as a fabric, a cloak, or a human body. Mechanically flexible materials step forward in terms of lightness, portability, harmonious manipulation, and functionality. In this regard, complementing the shortcomings of conventional solid materials, flexible materials have aroused extensive interest in all material science research fields and have become one of the most advanced disciplines. Emerging as a revolutionary advance in materials engineering compared to conventional techniques, metamaterial (MTM) structures are nominated as remarkable candidates for flexible materials [1–3]. MTMs, which are defined as artificial/synthetic structures, exhibit acoustic, electronic, optical, and thermal behaviors that natural materials do not have. The geometrical scalability of MTM designs has enabled free control of the electromagnetic (EM) permittivity $\varepsilon(\omega)$ and permeability $\mu(\omega)$. In this way, it is possible for these structures to have a negative refractive index. Due to such unusual properties, MTM research has increased significantly, and researchers have started to focus on many branches such as energy harvesting [4–6], sensors [7–9], super lenses [10–12], EM cloaking [13, 14], and mechanical structures [15, 16] with simulation and fabrication methods. Since the MTM structures are mechanically flexible, they provide significant benefits in many areas such as programmable flexible materials, auxetic structures, soft robotics, and reconfigurable devices, especially in THz applications, and they will pave the way for stimulating new studies. During the design stage of the MTM structures, flexible materials can be preferred for the plane, resonator, and substrate. The ability to bend, stretch, and roll of MTM structures designed on flexible substrates offers a new perspective on the manipulation of EM waves. A wide variety of materials such as polyimide (PI) [17–19], benzo cyclobutene (BCB) [20–23], polydimethylsiloxane (PDMS) [24–26], polymethylmethacrylate (PMMA) [27–30], polyethylene terephthalate (PET) [19, 31, 32], and polystyrene (PS) [27, 33, 34] may add flexibility to the MTMs. In addition, conductive materials such as pure copper polyester taffeta fabrics (PCPTF) [35, 36], Shieldit super fabric [37–39], and zelt fabric [40–42], are the preferred materials by researchers to gain flexibility to the resonator and the sub-plane. Figure 8.1 shows structures designed using flexible materials. Today, flexible displays (figure 8.1(a)), which are also used in industrial areas, can be shown as an important example of the use of flexible materials [17]. Figure 8.1(b) shows the free-standing electric MTM structures integrated on the PI substrate. Unless this MTM structure is supported on a frame, it rolls like a cylinder and can be applied to human skin (such as fingers) [18]. Figure 8.1(c) shows examples of the constant-mean-stretch MTMs exhibiting combinatorial singularities [43]. Negative Poisson's ratio can be also obtained with such MTM structures.

One of the most important areas in which MTMs are preferred can be described as flexible electronics. After the MTM structures were demonstrated experimentally, they spread rapidly to many research areas and the extraordinary characteristics of MTMs have been investigated by integrating them on flexible electronic devices. Flexible electronics, which fulfill the shortcomings of conventional rigid electronics in terms of flexibility, are of great interest in the research community and have become one of the most important disciplines. In the last two decades, considerable

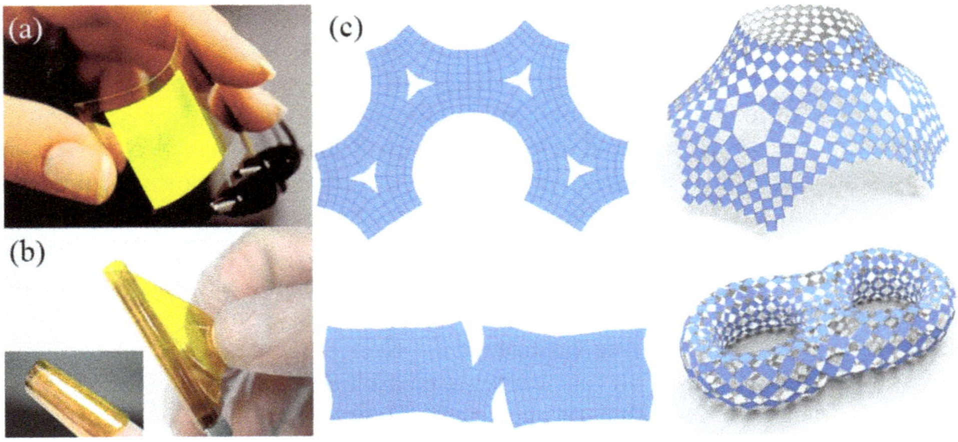

Figure 8.1. (a) Photograph of a flexible organic light emitting diode display. Reproduced from [16] with permission from the Royal Society of Chemistry.) (b) The free-standing electric MTMs fabricated on polyimide substrate. (Reproduced from [17]. © IOP Publishing Ltd. All rights reserved.) (c) The constant-mean-stretch MTMs exhibiting combinatorial singularities. (Reprinted from [43], Copyright (2022), with permission from Elsevier.)

progress has been achieved in soft materials and designs to improve flexible electronics. Flexible electronic devices have a strong impact on many fields of smart electronics such as wearable electronics, flexible touch sensors, soft robotics, healthcare devices, military textiles, and wireless communication technologies. Devices with a flexible electronic structure are in demand for use in many fields, especially in display technologies. The features of MTM structures designed with flexible materials, which cannot be obtained with conventional structures, make significant contributions to the development of flexible electronics.

Conductive and insulating flexible materials are extensively studied in wearable antenna designs. The MTM-based antennas, which are intensively preferred in conventional antennas, are also used in flexible antenna designs to increase their performance and application area. Such studies crucially increase the interest in flexible MTM structures. Integrated structures in which the MTM designs are used with antennas are preferred for different purposes in different frequency ranges. In particular, the studies with EM band gap (EBG) materials used with antennas are magnificent. With the development in wireless communication, the type and number of wearable devices also increase significantly. Although wireless devices make important contributions to human life, it is a crucial detail that these devices are in direct contact with the human body. Because wearable devices in contact with the human body for a long time are thought to cause negative effects in terms of health in the short or long term, International standards have been developed to reduce these negative effects, and the MTM-based antennas have been designed to minimize damage to the human body, considering these standards. Research shows that the use of MTM structures with antennas significantly reduces damage to the human body.

Today, with the expansion of the usage areas of wireless communication technologies, interest in portable energy sources is ascending significantly. Battery technologies are accepted as the most preferred energy source for portable and wearable devices. However, there are significant disadvantages when the size and weight of the batteries are considered. For example, the size of the battery occupies an important part of an electronic-based system. Moreover, its weight significantly limits the use of wearable devices, and their limited lifetime is another restriction. In addition to these disadvantages, developments in low power consumption in electronic circuits increase the interest in different technologies like energy harvesting, which can be an alternative to batteries. Energy harvesting is based on the development of the necessary material design and electronic infrastructure to convert energy sources such as the EM, kinetic, or thermal in the environment into electrical energy. The MTM-based EM energy harvesting applications are important candidates to overcome such disadvantages. The perfect absorption properties of the MTM structures make important contributions to the enhancement of the MTM-based EM energy harvesting studies. Although the MTM-based energy harvesting applications are being developed with rigid materials, the integration of portable devices with the human body necessitates a quite flexible design. In this respect, energy collection capacities of the MTM structures are being investigated by using flexible materials. Research shows that flexible MTMs are efficient in collecting EM energy.

Mechanical MTMs, which are similar to wave-based MTMs, create a more recent branch of research compared to research areas such as absorbers, sensors, and harvesting. When designing mechanical MTMs, the levels of parameters such as deformations, motion, and stress using mechanical energy are primarily taken into account. By means of the extraordinary properties of the MTM structures, mechanical MTMs can have characteristics and functions that cannot be realized with conventional materials. Parameters such as compressibility, Poisson's ratio or density have initially been investigated, and advanced mechanical MTM classes such as nonlinear and shape morphing have recently been developed. With the aim of obtaining these parameters and features at the expected level, the importance of flexible materials has attracted attention. When force is applied to the mechanical MTM structures created using these materials, they show extraordinary functions like shape transformations in the intended direction. In addition, these structures can have reprogrammable dispersion or stiffness characteristics.

Terahertz (THz) applications have attracted great interest from researchers since they were introduced. Moreover, they have intensively been investigated in many respects. One of these interesting fields is MTM-based THz applications. Since most natural materials exhibit only weak electrical and magnetic responses, their use for controlling radiation appears to be difficult. The positive results of the MTM structures in terms of radiation control are very important in the THz regime. Studies on the THz MTMs are taken into consideration as an important topic that can further advance applications in THz research and development. Along with these developments, it is noteworthy that flexible materials are examined together with the MTM structures for the THz frequency range. Elastomeric polymers like

PDMS are often preferred for the THz MTMs. The MTM studies are carried out on a wide variety of polymers and the properties of such flexible polymers play remarkable roles in functionality. One of the most important steps of THz applications is to tune the resonance frequency of the MTM structures, and elastically deformable flexible substrates present significant contributions to such studies.

The goal of this chapter is to examine the electronics, antennas, THz, energy harvesting and mechanically flexible MTM structures, bring them to the attention of the academic community and arouse interests among related disciplines. To this end, the review begins with an overview of the properties of flexible materials, their integrated use with the MTMs, and continues with a literature review of the progress achieved to date. Then, the material designs for the flexible MTM electronic structures in terms of parameters such as functionality, motion, and deformation are introduced. Thereafter, designed flexible structures related to antennas, THz applications, energy harvesting, and mechanical issues, which are important research areas of the MTM structures, are taken into account. Finally, a brief overview is projected toward future research on the basis of the current flexible MTM applications.

8.2 Flexible materials for MTMs

Electronic systems developed with rigid materials have been observed as insufficient, especially in terms of ease of use with the development of wearable technologies. This situation directs researchers to the areas of use of flexible materials in electronic systems and increases the interest in flexible materials. The MTMs have important applications in many branches such as energy harvesting, absorbers, sensors, antennas, THz, and mechanical materials. The flexible properties of the materials used in the structures developed in these branches provide to MTM researchers a different perspective. The MTM structures usually consist of three layers, but in some designs, they may have two, four, or more layers. The standard design consisting of three parts is created by integrating metal plate, resonator, and dielectric substrate layers. In two-layer designs, it is generally preferred to remove the metal layer, while in multi-layer structures, designs have been created by adding new dielectric structures or metal plates. Although flexible materials are used in all layers for the function of the design, they are mostly preferred in the dielectric substrate layer. Flexible substrates present remarkable applications for revealing unique characteristics in the MTMs due to some properties such as ease of movement, deformation, and stress. Flexible substrates provide extraordinary functions to the designed structure when compared to conventional structures, and it allows the rapid development of research on MTM structures created using such substrates [24, 44].

Structural changes in material design due to mechanical deformation contribute positively to the advancement of flexible substrates. For example, the mechanical deformation of flexible substrates is very effective at achieving tunability at high frequencies, thus eliminating the need to integrate an extra bias voltage or actuator

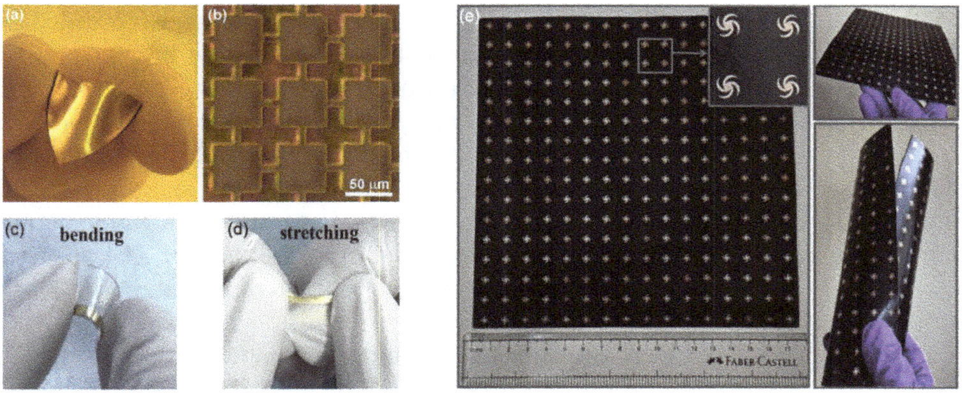

Figure 8.2. (a) Photograph of a section of the microfabricated flexible fishnet MTM and (b) optical micrograph of the unit cell of the flexible fishnet MTM. (Reprinted from [45], with the permission of AIP Publishing.) (c) Bending and (d) stretching photographs of the stretchable parabolic-shaped MTM fabricated on a PDMS substrate in the THz frequency range, [46] John Wiley & Sons. (2019). (e) Photograph of the epoxy/graphene nanoplatelets composite flexible MTM structure. (Reprinted from [47], with the permission of AIP Publishing.)

to achieve the targeted tuning. In addition, such structures show resistance to breakage due to their stretching and allow the material produced to be used in moving environments. While integrating the resonators on the flexible substrate requires some effort, the resonators designed on the flexible structure show high sensitivity to structural parameters and small changes in dimensions can easily be sensed. Moreover, the use of flexible substrates for the design of MTM structures provides the opportunity to obtain different properties as it offers a wide choice of parameters such as thickness, flexibility, and dielectric constant. Khodasevych *et al* characterized the EM properties of PDMS material, which is often preferred for flexible substrates, and used it as a free-standing substrate to design and fabricate a flexible fishnet MTM structure in the THz frequency range (figures 8.2(a) and (b)) [24]. They observed a strong magnetic resonance and the flexible MTM structure they suggested exhibits high transmission values in the passbands where the PDMS substrate has low propagation loss. They showed that the use of flexible THz MTMs is expanding, due to the properties of PDMS, such as the potential for elastic deformation or tuning with integrated microfluidics. Xu and Lin proposed a stretchable parabolic shape MTM coated on a PDMS substrate in the THz frequency range [45]. By stretching the PDMS based MTM structure they designed along different directions, they obtained polarization-dependent, ultra-narrowband, and switchable optical properties. The bending and stretching photographs of the proposed MTM structure are shown in figures 8.2(c) and (d), respectively. This MTM structure provides a highly effective approach to the creation of programmable MTM structures due to their applicability, cost effectiveness, and high flexibility. Anjali *et al* fabricated flexible polymer nanocomposites containing graphene nanoplatelets at different weight percentages in epoxy and performed their dielectric characterization [46]. They obtained the most efficient flexible substrate based on the characterization results. They designed and fabricated an MTM absorber (figure 8.2(e)). It was shown that the proposed structure has a maximum absorption of 99.8% at a frequency of 9.88GHz.

Parameters such as permittivity, permeability, loss tangent, and Young's modulus of the material that can be used as a flexible substrate directly affect the characteristics of the designed structures. Therefore, it is important to make a selection considering these parameters in order to obtain an MTM structure with the desired characteristics. The MTMs consist of periodically arrayed unit cells. The unit cells that compose the MTM should be small enough so that the radiation interacts with the material as a whole and does not scatter from its individual components. Therefore, the structure of the resonators is very important in the unit cell design. On the other hand, although the EM properties of MTMs are strongly affected by the subwavelength resonator design, the entire designed structure affects the media performance. It is very important to choose the parametric properties of the substrate correctly to maximize the performance; for example, the dielectric permittivity and refractive index are correlated with the equation $n = \sqrt{\varepsilon}$, and the refractive index is chosen at a low level to minimize the reflection losses that may occur due to the substrate. Moreover, a low-permeability substrate may be preferred to preserve the resonance power and bandwidth of the MTMs, and it is also important to provide low loss to maximize transmission/propagation.

Making MTMs flexible is attracting interest, and thus increasing the variety of flexible materials. Flexible materials are primarily used in the substrate of the MTMs. Some of the significant properties of flexible substrates often preferred for the designs of the MTMs are listed in table 8.1. Flexibility enables the MTMs to gain capabilities such as range of motion and deformation, as well as being also preferred for unusual purposes such as wrapping light/transparent MTMs as an optical cloak around a more important object. The permittivity and the loss tangent parameters are considered when defining flexible materials. Permittivity can be defined as the proportion constant that associates the electric field in a structure to the electric

Table 8.1. Properties of materials commonly used as flexible substrates.

Material	Permittivity	Loss tangent	Resistivity (Ω cm)	Young's modulus (GPa)	References
Felt	1.3	0.044			[38, 47–49]
PI	3.24, 3.4	0.031	1.7×10^{17}	2.5	[7, 17–19]
PET	2.86	0.019, 0.053, 0.072		4.0	[19, 31]
Fleece	1.25	0.0025		7.6	[50, 51]
BCB	2.65	0.001, 0.009	9×10^{19}	2.9	[20–23]
PDMS	2.35, 2,67	0.02, 0.04	2.9×10^{14}	6×10^{-4}	[24–26]
PS	2.5, 2.6, 4.4	0.02	$>10^{10}$	3.1	[27, 33, 34]
PMMA	2.25	0.02, 0.042, 0.09		4.84	[27–30]

displacement in that structure. It has effects on the spread of the electric field and is one of the most important parameters for EM and material science. The complex permeability can be expressed as a product of the relative permeability multiplied by the vacuum permittivity constant. The loss tangent, which is also known as dielectric loss or tan δ, can be expressed as the coefficient of conversion of stored energy into heat energy in a material. Materials are classified as good or poor conductors depending on whether tan δ value is greater or less than one, respectively.

Flexible materials are primarily used in substrate design. However, with the good results obtained with flexible conductive materials, they have begun to be preferred in resonator or ground plane design. Such designs offer different options and results for flexibility and deformation parameters of the material. Some of the properties of flexible conductive materials preferred for designing MTMs are listed in table 8.2. Young's modulus is a numerical constant that determines the extent to which materials deform/stretch and is expressed by the relationship between the compressive/tensile stress (force per unit area) of a material and axial strain (proportional deformation). Electrical resistivity, also known as volume resistivity or specific electrical resistance, is a fundamental quantity that determines how strongly materials resist electric current. Resistivity, which is a characteristic feature of materials, is a quantity independent of the geometric dimensions of the conductor and is only related to the properties of the material of which the conductor is made. For example, low resistivity indicates that the material easily conducts electric current. Electrical conductivity, also known as specific conductance, is the reciprocal of electrical resistance. It is a parameter that represents the ability of a material to conduct electric current, and it occurs in two forms: direct current without frequency dependence, and alternating current, which is frequency dependent. These and similar parameters can be obtained by experimental studies and defined in simulation programs and the studies can be expanded by applying them to different designs.

8.3 Electronics for flexible MTMs

Although studies on the negative index of refraction have been done before (from the 1905s to the 1968s) [59], its basis is expressed by Veselago's theoretical research [60]. It was stated by Pendry *et al* that a medium composed of periodic thin wire

Table 8.2. Properties of materials commonly used as flexible conductive materials.

Material	Resistivity (Ohm/sq)	Conductivity (S/m)	References
PCPTF	0.05	2.5×10^5	[35, 36, 52]
Rubber		2×10^4	[53, 54]
Liquid metal		5.96×10^6	[25, 55]
CNT		10^6	[56–58]
Shieldit	0.025	1.18×10^5	[37–39]
Zelt	0.01, 0.05	1.74×10^5	[40–42]

structures has the properties of homogeneous materials at the plasma frequency when the lattice constant and the diameter of the wire are smaller when compared to its wavelength at its plasma frequency [61, 62]. Later, Pendry *et al* showed that a periodic structure consisting of split-ring resonator (SRR) has effective negative permittivity and permeability in a particular frequency range [63]. A generalized SRR design and periodic structure are depicted in figures 8.3(a) and (b). The MTMs consist of materials found in nature and are periodically arranged. These materials are much larger in size than their atomic structure, but have small dimensions compared to the wavelength of the incident wave. Therefore, the resonance frequency of the designed structure can be affected by the placement of these periodic cells and can be controlled according to its dimensions. In the following years, the MTMs were investigated in one dimension by Smith *et al* [64] and in two dimensions by Shelby *et al* [65]. A medium that expresses a frequency region with simultaneously negative effective permittivity ($\varepsilon_{\text{eff}}(\omega)$) and permeability ($\mu_{\text{eff}}(\omega)$) in the value of the microwave regime has been proposed by Smith *et al.* Thus, a 'left-handed' environment was created from these artificial materials, and they took the name 'metamaterial'.

With the formation of the fundamental components of the MTM structures theoretically and experimentally, various specific study areas related to MTMs have emerged and researchers have started to focus on these specific areas. In addition, the use of flexible materials makes important contributions to the advancement of MTM research. Flexible electronic MTM structures are one of the important examples in this regard and overcome the deficiencies encountered in the designs developed with rigid materials. Besides these developments, there are also applications where the design of flexible materials is limited due to mechanical deformation problems relative to the study areas. In addition, developing the capability of flexible electronics using conventional strategies with structural adjustments can notably be challenging and costly. Studies revealing the advantages and the disadvantages show

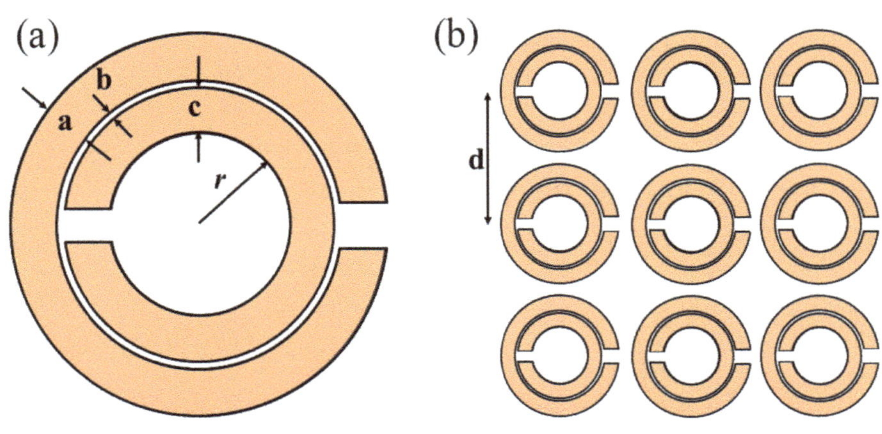

Figure 8.3. (a) A generalized SRR design and (b) its periodic sequence.

that the MTM structures have potential for flexible electronics. The MTMs have diversified applications in flexible electronics such as flexible photonics, smart skin, health monitoring, flexible sensors, soft robotic, flexible implantable electronics, and ultrabroad absorbers.

In flexible electronics, there are concepts such as EM, acoustic, and elastic MTMs, which are called wave-based MTMs, and these are used to manipulate EM, acoustic and elastic waves, respectively. The EM-MTMs can be classified as THz, photonic and optical MTMs, and acoustic and elastic MTMs can be named sonic and phononic crystals, respectively. The photonic crystals, which exhibit very similar properties to the MTM structures, indicate unusual EM responses like the MTMs [66]. Photonic crystals, which consist of periodically arranged building blocks in one, two or three dimensions, do not have a resonator like MTMs, but exhibit properties like negative refraction [67, 68]. The optical MTMs offer many features for optical devices such as negative refraction [69], cloaking [70], and super resolution image acquisition [71]. The optical MTM structures usually consist of rigid materials that are inflexible and non-configurable, as in other MTM structures, and have various limitations. The optical MTM obtained by integrating flexible materials can overcome such limitations. For example, when optical, EM, thermal stimuli are applied on optical MTMs, they acquire new optical functions due to their reconfigurable properties. While designing metamaterials on flexible substrates gives them the ability to stretch and roll, manipulating electromagnetic waves also provides remarkable advantages in the THz region. Metasurfaces (MTSs), which consist of a planar metamaterial design with one/several layers of meta-atoms, are generally preferred in flexible THz applications [72]. The MTSs can manipulate EM waves with the advantages of being ultra-thin/light, easy to configure and design. Acoustic MTMs and phononic crystals have become emerging fields in acoustics science, expanding the application properties of existing acoustic materials [73, 74]. These structures provide an important infrastructure for acoustic wave control in meta-systems/devices.

New technologies such as wearable and portable devices, flexible robotics, and flexible screens highlight the use of MTM-based flexible photonic sensors, and these sensors provide significant advantages over conventional sensors. Flexible photonic sensors, which can be integrated into curved surfaces such as textile materials or the human body, provide ease of use, flexibility, and even new functions. Moreover, high sensitivity can be achieved in these sensors. Flexible designs created by combining MTMs and photonic structures are periodically composed of micro or nano cells in arrays. These photonic MTMs are subwavelength periodic structures and allow tuning of electromagnetic waves in flexible photonic materials according to their geometric structures. By modulating the geometric structure of the unit cells or the distance between the unit cells of photonic MTMs, the interaction of the signal coming to the material can be controlled. Thus, it is possible to obtain optical performances that cannot be obtained with the materials that make up the photonic structure itself. Peng and Vu present the recent progress by evaluating the structure design, material selection, fabrication methods, sensing properties, and performance of flexible and stretchable photonic sensors [75]. The features brought by flexible MTM structures also contribute to the development of research such as electronic skin (e-skin), flexible inkjet-printed, and soft robotics applications where the MTM-based photonic sensors can be integrated into the human body. Creating soft grasping and manipulation conditions for robotic tactile sensing and applying e-skin sensors based on flexible MTMs to e-skin cells are very significant areas. Although the difficulties in sensing and data processing on this subject limit the research, flexible MTM structures that provide sensitive measurement capability according to compressive and tensile stresses constitute an important part of the studies [76]. MTM-based e-skin can be created from a periodically SRR design integrated into a flexible material. When this structure flexes, it has a wide and tuning resonance frequency range and acts as a tunable frequency-selective surface. In addition, when this MTM-based e-skin is wrapped around a curved dielectric structure, it has a cloaking feature to reduce scattering across the surface.

Yang *et al* proposed a flexible MTM structure with tunable frequency-selective and cloaking properties operating at the microwave frequency regime [77]. Flexible materials can cause quite a lot of EM energy loss according to the way they are designed, and this may manifest itself especially in MTS-based applications. By etching on the conduction structure of the MTSs, the electric field distribution can be improved, and the resulting energy loss can be minimized. The geometric parameters of the proposed meta-skin structure and the simulated surface current distribution result are shown in figure 8.4(a). In addition, figure 8.4(b) represents the fabricated designs of the unstretched and stretched mode. A dielectric nylon rod wrapped by meta-skin and a glass flask wearing meta-skin are shown in figures 8.4(c) and (d), respectively. Peng *et al* propose a flexible harvester using an inkjet-printed quasi metastructure for the microwave power transfer, and according to the results obtained, they observed that the absorption efficiency and harvesting efficiency increased significantly [78]. As shown in figure 8.4(e), each unit cell topology of their proposed flexible quasi metasurface harvester consists of five layers. The photograph of the additionally produced quasi metastructure and the measuring setup are shown in figures 8.4(f) and (g), respectively. Mechanical MTMs can be designed to exhibit unusual characteristics such as ultra-high stiffness, strength to weight ratio, negative thermal expansion, and negative Poisson's ratio. While early studies on mechanical properties have tended to create mechanical MTMs with linear elastic responses, later studies have focused on functions such as nonlinear deformations, programmable geometric shape transformations, and mechanical instabilities. Advances in technologies such as additive manufacturing have positively affected the fabrication of mechanical MTM structures of unusual complexity. In advanced soft robotics technologies, researchers can create integrated monolithic designs that can undergo complex motions programmed according to the architecture of the mechanical metamaterial. Programmable MTM designs can transform a simple pressure impulse input into complex outputs such as tension, flexion or torsion, allowing the creation of soft robotic systems. Thus, more complex functions can be performed with less input energy [79].

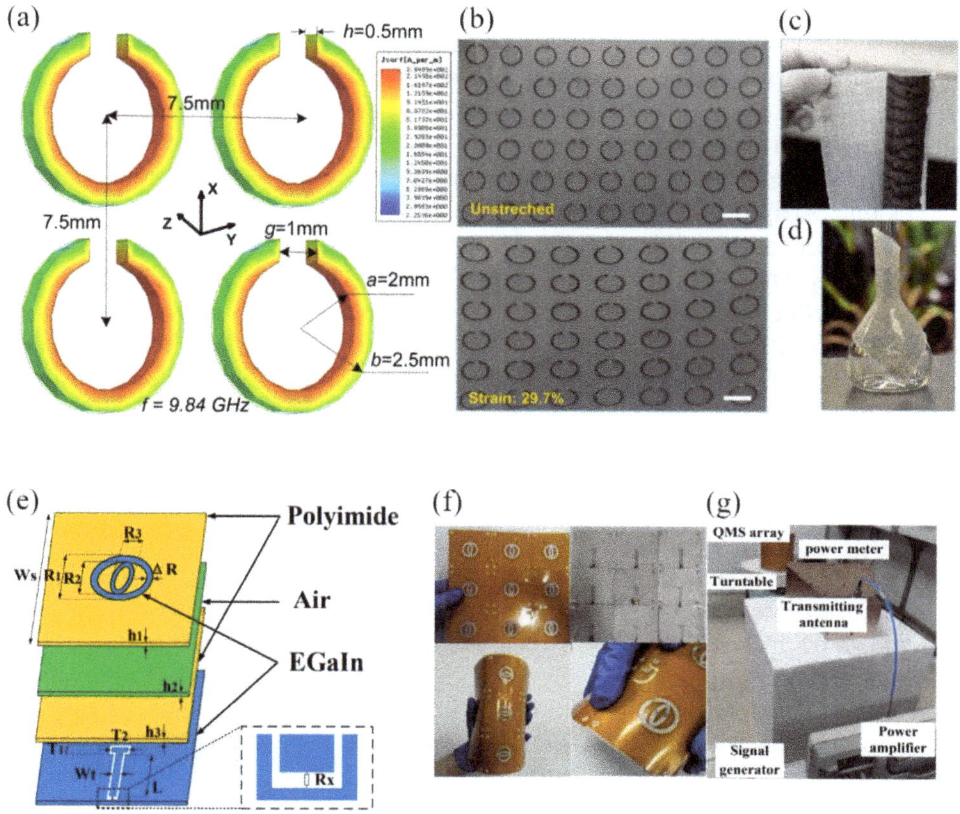

Figure 8.4. (a) The simulated surface current distribution and geometry of the meta-skin, (b) the fabricated designs of the unstretched and stretched mode (Strain: 29.7%) meta-skin, (c) a 30.48 cm long and 3.175 cm diameter dielectric nylon rod wrapped by the meta-skin, and (d) a glass flask wearing the meta-skin. (e) The topology of the quasi metasurface resonant unit, (f) the fabricated quasi metasurface, and (g) measurement system for quasi metasurface.

8.4 Antennas for flexible MTMs

Antenna-based systems are preferred in many research areas such as mobile, wireless network, image processing, communication, and radar technologies due to their useful and portability features for human life. With the increase and intensive use of wireless communication methods, antenna types and operating frequency ranges have also made significant improvements. Basically, antennas working together with a receiver and a transmitter provide the connection between the emitted EM waves and the electric current moving in conductive metal materials. With the wireless and rapid transmission of data from one medium to another, wireless communication technologies have become an integral part of daily life. These devices, which are also described as smart technology, are used to be close to or in contact with the human body. This situation limits the use of devices developed with solid materials, especially when they are in direct and prolonged contact with the body. In this respect, all or some parts of these devices, which were developed using solid materials such as telephones in the early stages, have begun to be designed with a more flexible infrastructure. Flexible displays that offer portability and extra ease of use are examples of such designs. With these advances, the development of antennas using flexible materials has become an important research area. Performance and fabrication studies are carried out by using flexible materials for many antenna types such as microstrip, array, and wire [80–82]. With the development of textile-based wearable antennas [83–86], which have a flexible structure, a technology called Wireless Body Area Network (WBAN) has developed in many important fields such as medicine, education, and military. WBAN technology stands out especially in the field of health with its applications based on monitoring the human body, as well as the advantages it provides for wireless communication [87, 88]. In these wearable systems, it is possible to monitor the parameters and transfer data quickly to the control unit by using sensors with different characteristics together with the flexible antennas. Bharadwaj *et al* have proposed a study to classify/monitor various daily living activities such as walking, running, sitting, climbing/descending stairs, and opening/closing doors, using wearable ultra-wideband (UWB) technology. In the study, compact UWB antennas act as sensors to receive measurements and data on the activities performed [89]. With wearable antennas, electronic systems become less disturbing in terms of where they are used, but it is important that these antennas are light, thin, not easily damaged, and low cost [90]. For example, properties such as performance and bandwidth of microstrip antennas are directly related to the properties and thickness of the substrate, thus allowing a light and thin design [91, 92].

Studies have shown that flexible MTM structures have been significantly preferred in wearable antenna designs for research areas such as improving bandwidth, gain, and efficiency, protecting the human body from backward radiation, and deformation performance in the last decade. In addition, MTM-based flexible antennas enable applications in 5G/6G communication technologies that improve features such as phase manipulation and subwavelength function. With the demonstration of MTMs in experimental studies, their effects on antennas have also started to attract attention by researchers. In 2001, considering that the antenna dimensions of wearable systems are not limited to the place of installation and that there are many devices using different radio systems to be considered during the design process, the photonic bandgap patch antenna design was examined [93]. The proposed study shows that sufficient bandwidth and impedance values can be obtained by using low-cost materials. Enoch *et al* investigated emission control in MTMs using antennas according to emission direction and power in 2002. They aimed to control the emission direction of a source in the material to collect all the energy in a small angular area with the MTM structure [94]. Researchers have focused on MTM-based antennas to overcome the limitations of rigid/flexible antennas such as narrow bandwidth, low-profile and conformal problems, radiation damage to the human body, limitations in wearable technologies, and performance problems. Rao *et al* proposed a flexible antenna design with dual sensing MTM load, circularly polarized and coplanar waveguide (CPW) fed using liquid crystal

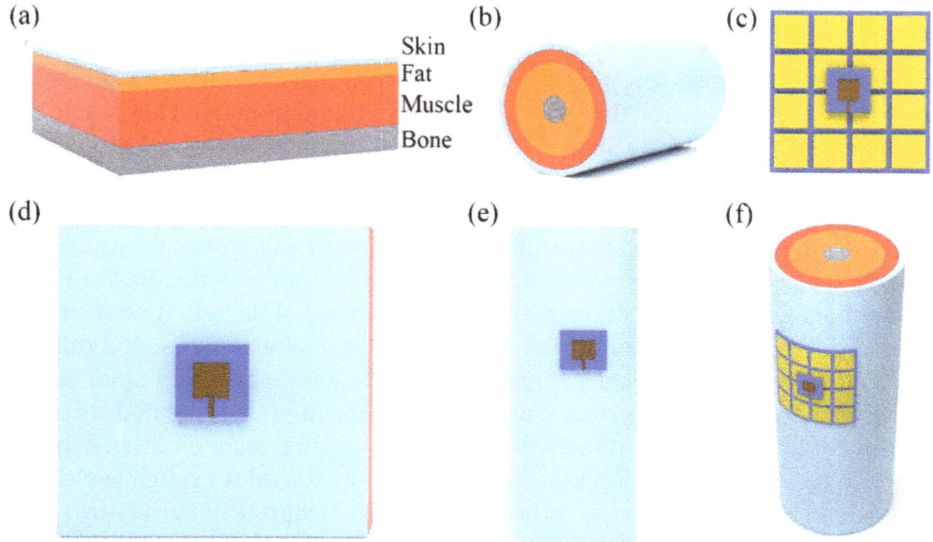

Figure 8.5. (a) A generalized flat and (b) cylindrical human model. (c) EBG structure and microstrip antenna integration. A generalized (d) flat and (e) cylindrical human model with integrated microstrip antenna. (f) Integration of antenna and EBG structure into cylindrical human model.

polymer [95]. In that study, the bending properties of the antenna designed with the MTM infrastructure were analyzed according to various bending angles. According to the calculations of the bend antenna and the planar antenna, the results such as reflection coefficient, axial ratio, and radiation patterns were confirmed with the measurement results. Yan *et al* proposed a wearable MTM loaded dual-band antenna that can operate at 2.4/5.2 GHz frequencies, which is designed from textile fabric except for the connector material used for feeding. They simulated a numerical SAR value on a human phantom, with an integrated design and a compact and reasonably simple antenna topology. The lower and upper bands were well below the limits, with a value of 0.37 and 0.69 W kg^{-1}, respectively [96].

The human body model, integrated with antenna and metamaterials, can be created with various designs. Figures 8.5(a) and (b) depict human models with flat and cylindrical structures, respectively. A flat human model can be designed in a cylindrical or semi-cylindrical form to analyze the deformations of the proposed design in case of bending. In addition, the performance of the antenna and the MTM structure can be evaluated by analyzing the data before and after bending. Figure 8.5(c) shows the antenna integrated with the EBG structure. This structure can be integrated directly into a flat human model and, as well as being an integrated structure, can be created by bending as much as the amount of bend applied to the designed human body. Figures 8.5(d) and (e) represent the antenna and the EBG structures integrated into flat and cylindrically designed human models, respectively. When the EBG integrated with the antenna is placed on the cylindrical human model by applying a slope to the structure, the radiation effect of the antenna can be reduced (figure 8.5(f)). In [52], the authors analyzed the low specific absorption rate (SAR) performance of a monopole antenna using flat and cylindrical human models. By integrating the proposed antenna with the EBG structure, they observed the change in SAR value for the straight and curved human body. They found that the EBG integrated structure significantly reduced the SAR value in both flat and bending phases.

Antennas, one of the most important components of wearable devices, have important challenges such as physical deformations, different working surfaces, and damage to the human body, contrary to the advantages they provide. Basically, EM waves are used for communication in antennas, and some of the radiation penetrates the human body depending on the antenna's location, characteristics, and duration of use. Possible damage should be considered if the antennas are in close contact with the human body for a long time. Researchers have investigated advanced design methods to overcome these challenges in wearable technologies. A perfect electric conductor (PEC), which is a useful model that provides negligible electrical resistance compared to other materials, consists of idealized materials to exhibit perfect electrical conductivity. In this regard, PEC reflectors can be used to reduce the negative effects of wearable antennas [52]. Among the proposed methods to prevent this damage, the use of EBG structures or artificial magnetic conductor (AMC) structures that have zero reflect phase, is preferred. A simple EBG structure integrated antenna design can be created by using conductive textile fabric instead of the copper material used in rigid antenna designs. Thus, with the flexibility obtained, movement restriction can be eliminated, and a low SAR value can be obtained at the same time [97]. Another important point to be considered in wearable devices is to achieve as low a value of SAR as possible without sacrificing efficiency. The amount of energy absorbed by the body per unit mass is defined as SAR (W/kg).

Antennas can be directly integrated into EBG structures that are arranged periodically. By means of this design, the radiation generated by the antenna can be avoided due to the key features of EBGs such as in-phase reflections and surface waveband gap. The EBG contributes to the gain increase of the antenna depending on the structure by bringing together the antenna and image currents in a constructive way with in-phase reflections. The EBG structures exhibit a high impedance surface behavior in the surface waveband range and thus impedance matching occurs. Thus, it contributes to the prevention of EM energy generated by the antenna from reaching to the human body. Considering that the human body cannot always have a flat surface, the resistance and performance of the integrated system consisting of antenna and the EBG structure against bending becomes important. Such parameters can be obtained through a cylindrical structure created using materials that represent the human body, such as skin, fat, muscle, and bone. The integrated structure can be deformed by the angle determined by considering the radius of the cylindrical structure created. In this way, the integrated structure can be designed to fit into the cylindrical human model. The differences between the calculations made according to the new structure and the results obtained from the previous structure are evaluated. Tetik and Tetik designed an integrated structure using a coplanar flexible antenna (figure 8.6(a)) and an EBG structure (figure 8.6(b)) [98]. Then, they combined

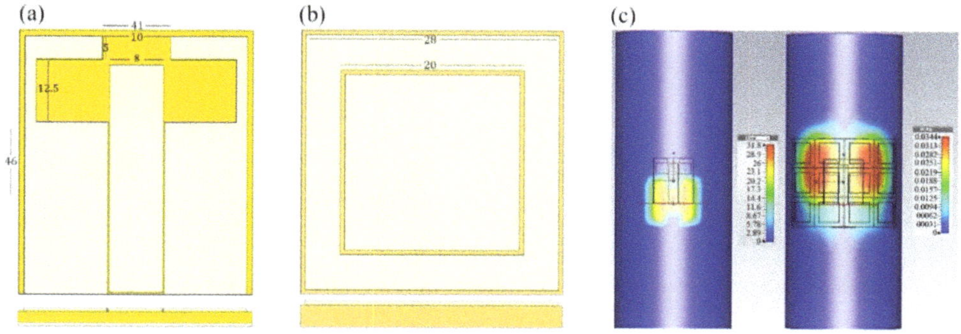

Figure 8.6. (a) The top view of the coplanar flexible antenna, (b) the top view of the MTM-based EBG structure, and (c) the SAR value obtained by direct contact and the SAR value obtained by the EBG integrated structure. (Reprinted from [100], with the permission of Journal of the Institute of Science and Technology.)

the antenna and integrated structure separately with a human model consisting of four parts. To analyze the effect of flexibility, both designs were bent considering the radius of the human model. The calculations showed that the EBG integrated structure significantly reduced the SAR value compared to direct contact of the antenna (figure 8.6(c)). It is understood that such integrated structures significantly reduce the negative effect of radiation on the human body.

AMC structures exhibit in-phase reflection properties as in the EBG structures, thus limiting the retroactive radiation of the wearable antenna reaching the human body, while maintaining the low-profile structure of the antenna, it can also contribute to the improvement of the antenna's gain. In this respect, AMC structures, which can be used as reflectors for radiated energy at certain frequencies, have become an important design method in MTS-based wearable antenna design to increase antenna performance and achieve low radiation. Di *et al* present a dual-band monopole antenna design consisting of a large ring-shaped strip for WBAN applications (figure 8.7(a)). The proposed design is integrated with a circular-shaped AMC structure to reduce the backscattered wave towards the human body. It is seen that the integrated system simulated on the phantom model, when combined with the AMC structure, improves the radiation rate, and has an average gain of 10 dB [99]. The circular-shaped AMC element and the integrated structure of the CPW-fed monopole antenna with a circular-shaped AMC reflector are shown in figures 8.7(b) and (c). Abirami *et al* proposed a printed Yagi–Uda antenna consisting of a polyester sheet and copper foil tape. This flexible antenna operated at 2.4 GHz frequency and aimed to lower the SAR value and improve the gain [100]. The top view of the proposed Yagi–Uda antenna and the EBG-backed Yagi–Uda antenna are demonstrated in figures 8.7(d) and (e), respectively. In addition, the perspective view of the Yagi–Uda antenna placed on the rectangular body model is shown in figure 8.7(f). They observed that when the antenna was backed with the EBG, the antenna gain increased by 60% and the SAR decreased from 8.55 to 0.07 W kg^{-1}.

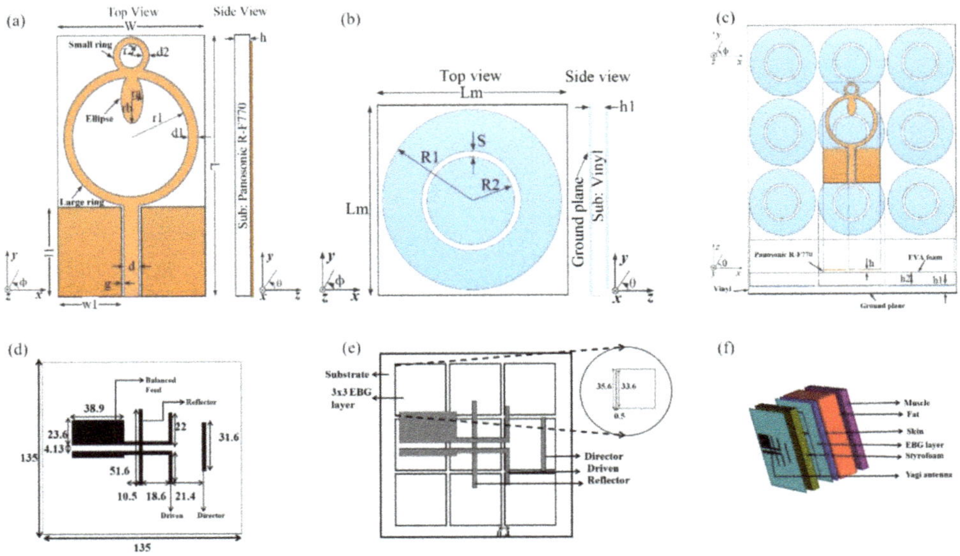

Figure 8.7. (a) The geometry of the CPW-fed monopole antenna, (b) the circular-shaped AMC element, and (c) the integrated structure of the CPW-fed monopole antenna with circular-shaped AMC reflector. (d) Top view of the printed Yagi–Uda antenna, (e) top view of the EBG-backed Yagi–Uda antenna, and (f) the Yagi–Uda antenna placed on the rectangular body model.

8.5 Energy harvesting for flexible MTMs

Since wearable technologies are directly integrated into the human body through fabric or skin, they need a continuous energy source. This energy is mostly provided by batteries of different sizes and features. Lightness is one of the priority criteria for devices that integrate with the human body to be useful. The flexible materials that make up these devices are quite light when compared to common batteries used as energy sources. Although they provide advantages, wearable technologies also cause an important problem like energy source when the battery problem is considered. In addition to its weight, batteries take up more space when compared to other circuit elements and the wearable device occupies more space. With the increase in the use of wearable devices, researchers are leading the search for different energy sources, and research areas such as low-power designs and energy transfer directly to the electronic structure have begun to attract attention. Studies on electronic circuits indicate that the power consumption of electronic circuits becomes increasingly smaller, and these advances reveal the possibility of energy harvesting from sources of environments for power supplied to the electronic device. Energy harvesting can be defined as the conversion of media sources such as EM, thermal, and kinetic energy into electrical energy. The efficiency of the energy harvesting application largely depends on the conversion structure. In conventional energy harvesting applications, the energy conversion structure is limited by the characteristics of the materials used. The way to overcome the limitations of conventional methods is to develop innovative energy harvesting methods with configurable materials.

Energy is an important requirement, and the concept of energy harvesting is a very old idea. Many energy sources such as thermal, mechanical, solar, and EM have been used from past to present. Figure 8.8 shows a general energy harvesting process that harvests the surrounding energy using a specific conversion mechanism relative to the source. Today, one of the most important energy harvesting mechanisms relies on using sources of environments to power low-power systems such as mobile and wireless devices, electronic devices, and wearable technologies. Since the amount of energy collected in the energy collection mechanism, which has a very complex structure, is generally low, the conversion system and the efficiency obtained become important. In this respect, the focus is on innovative structures that can be adjusted and can absorb energy at a high rate instead of conventional materials. Research in the last decade showed that the EM energy harvesting potential of structures designed using the unusual properties of MTMs, which are periodical artificial materials, is a remarkable alternative [101].

After MTMs were demonstrated experimentally, the structures were investigated for EM waves and classified as EM-MTMs. Studies in the following years have revealed new research areas for MTMs. The remarkable properties of MTMs such as negative refractive index and perfect absorption can provide solutions to achieve higher performance in EM energy harvesting research [102–104]. In figure 8.9, the paths that the EM wave follows on a MTM structure are depicted [48]. It is seen that some of the incoming wave (Power Outgoing) is reflected, and some of it (Power Accepted) penetrates the MTM structure. MTM can transfer the incident EM wave to the gap region due to its geometric structure. The MTM designs generally consist of electrically small-sized periodic unit cells, which are composed of a substrate, plane, and resonator. The fact that the unit cells—the working principle of which largely depends on the design of the resonators—resonate individually to create a bulk material behavior according to the constitutive parameters, indicates that

Figure 8.8. Schematic representation of a generalized energy harvesting mechanism.

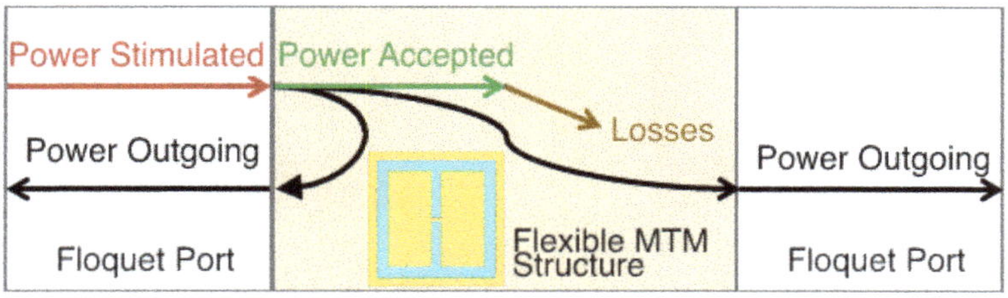

Figure 8.9. Schematic representation of generalized power and loss parameters. (Reprinted by permission from Springer Nature [49], Copyright (2022).)

MTMs have energy harvesting potential. The most familiar design of resonators is the so-called SRR, which consists of one or more broken-loop designs arranged as concentric loops. Ramahi *et al* showed that the designed MTM structure can be used as an energy collector when a resistive load is placed in the gap/gaps of the resonator [105]. If the SRR element of the designed MTM structure is excited by an external field at a specific frequency, it is observed that the structure has resonance. The storage of energy in the cavity of the resonator together with the resonance at the excited frequency means that it oscillates between the electric and magnetic field, quite similar to the resonance system in an RLC circuit. This oscillation, which is significantly influenced by the resonator design, allows the generation of electric field in the gap region(s) of the SRR structure. In this way, the generated electric field can be transferred to mechanisms such as electronic load vehicles and storage units with various energy conversion methods.

A generalized SRR design, one of the first MTMs, is depicted in figure 8.10(a). Like the basic features of MTMs, energy harvesting characteristics can also be reconfigured according to the resonator design and gap region dimensions. The EM energy harvesting mechanism can be created by integrating a circuit element such as diode and resistor in the gap region of the resonator. The top view and left view of the circuit element integrated resonator design are shown in figures 8.10(b) and (c), respectively. Hawkes *et al* presented design, simulation, and experimental application of a functional MTM-based energy harvester operating at the 900 MHz regime [106]. They demonstrated that in their proposed design, they were able to convert as much as 36.8% of RF power to DC power. In addition, they obtained a V_{OC} of 7.3 V by connecting the SRR structures in parallel. Chandrasekaran *et al* proposed a compact right/left-handed base dual-band receiver antenna for EM energy harvesting at operating frequencies of the Wi-Fi (2.5 GHz) and WiMAX (3.6 GHz) bands [107]. They showed that the proposed MTM-based dual-band rectenna has high efficiency over the wide fractional bandwidths required for RF energy harvesting systems. Unal *et al* presented the design, characterization, theoretical, and

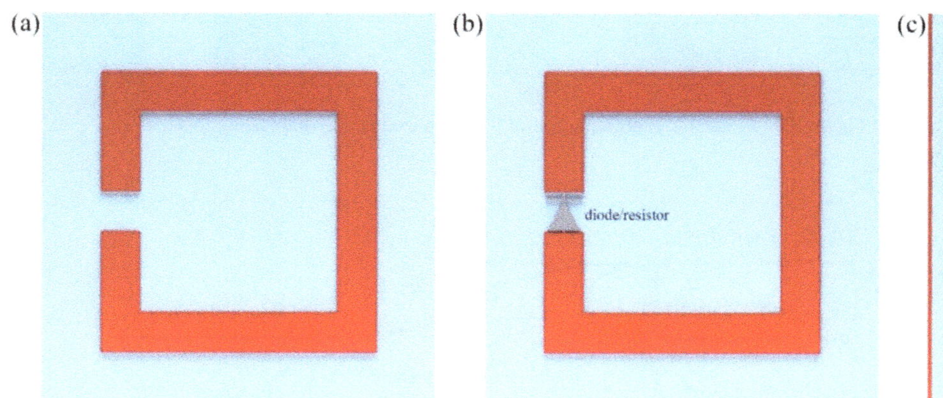

Figure 8.10. (a) The unit cell, (b) diode/resistor integrated unit cell, and (c) left view of a generalized SRR MTM design.

experimental applications of an MTM-based EM energy harvesting and sensor application, which has tunable perfect absorption [4]. They created a design that allows easy fabrication and simple configuration. Besides, they used the golden ratio for the resonator design. In addition, they showed that the proposed MTM structure has 99.9% absorption at 6.78 GHz and can be used for EM energy harvesting. In MTM energy harvesting studies, researchers first examine the real part of the $\varepsilon(\omega)$ and $\mu(\omega)$ parameters to obtain the negative refractive index. However, the loss components of these parameters are also remarkably important and can be used to design structures with perfect absorption [108, 109]. A study proposed by Bakir *et al* presents an EM energy harvesting and density sensor application based on the perfect MTM absorber design in the microwave frequency regime. They obtained the magnitude of the absorption value as 0.9 at the resonance frequency of 4.75 GHz and they preferred chip resistors to collect the EM energy [110]. In addition to EM energy, the energy collection potential of MTM structures for different types of energy such as acoustic energy and solar energy harvesting is investigated by researchers. The working logic of acoustic MTM structures is to compose appropriate resonance defects in a defective acoustic MTM to limit the strain energy resulting from acoustic incidence. The energy collected by this method is obtained as electrical energy by integrating a piezoelectric material to the acoustic MTM structure. Qi *et al* theoretically presented an innovative acoustic energy harvester with a defective acoustic MTM using piezoelectric material [111]. They achieved power density of 0.54 cm^{-3} and maximum output voltage of 1.3 V at the frequency of 2257.5Hz. Bagmanci *et al* proposed an MTM absorber consisting of three layers of different thicknesses that generate plasmonic resonances for solar energy harvesting [112]. The proposed structure's operating frequency is between 430 and 770 THz, which is the visible light regime, and the structure is also analyzed in the ultraviolet region. They demonstrated that an ultra-broadband MTM absorber can be used for solar radiation absorption, solar spectrum imaging, and solar energy harvesting applications.

With the development of wireless communication technologies, the use of devices with low power consumption in electronic textiles has become widespread. Energy harvesting methods are being developed to overcome the battery problem, which is one of the most important limitations of wearable technologies. The MTM-based structures used for energy harvesting have attracted attention as significant candidates in this regard. In order to make MTM-based energy harvesting systems compatible with wearable technologies, there is a need to design these systems with flexible materials. Structural characteristics, energy harvesting potentials, and bending application of MTMs designed with flexible materials can be listed as crucial details for flexible technologies. Bait-Suwailam *et al* proposed a metamaterial-based EM energy harvester for wearable technologies that can operate at the microwave regime, using flexible materials [113]. The proposed structure was constructed using a modified unit cell from a conventional SRR and its resonator is integrated on a grounded very thin flexible substrate. In this way, it is thought to be valuable for many wearable applications due to its cost-effective and sustainability solution in cases where it needs to be integrated into flexible electronic structures. When the proposed structure is integrated on wearables, it will be exposed to bending effects and deformation due to the mobility and flexibility of the wearer. In this respect, the effect of the proposed flexible MTM structure on the total energy conversion potential against bending is also investigated. Kaur *et al* presented a tri-band ultra-thin dual layer flexible MTM absorber for RF energy harvesting [114]. The proposed MTM structure, which gains flexibility by using adhesive copper foil tape, consists of six separate concentric rings with an octagonal inner circumference. They showed that this flexible structure has triple absorption peaks of 96.91%, 96.41%, and 90.12% for the frequency values of 1.75, 2.17, and 2.6 GHz, respectively. In addition, to analyze the bending performance, they obtained an absorption value above 76% for the incidence angle up to 45° in transverse electric (TE) mode and 94% for incidence angle up to 60° in transverse magnetic (TM) mode operation. The calculated and measured results in this study are compatible, showing that the proposed structure has energy harvesting potential. Studies demonstrate that body worn electronic textile technologies attract much attention due to the variety of application branches. In [48], a wearable MTM-based EM energy harvesting design is proposed in the 3–6 GHz frequency range. In this study, the surface current distribution and power characteristics were analyzed by integrating the resistor into the gap region of the proposed structure and the harvesting potential was obtained. Umino *et al* presented a vibration energy harvesting design with high power generation efficiency that can operate in the low frequency band of 21.1 Hz [115]. The proposed design consists of three layers; two layers have a piezoelectric structure, and the middle elastic layer, which is bimorph type, consists of a flexible mechanical MTM structure. This study aimed to improve the energy harvesting capacity of the piezoelectric layer by controlling the amount of flexibility of the middle layer, and thus to produce around 47.4 nW of power at low frequency. The resulting power provides the minimum amount of electrical energy required as a sensor node for systems such as WAN.

8.6 Flexible mechanical MTMs

Mechanical MTM structures can be used to achieve extraordinary values for parameters such as Poisson's ratio, density, and compression/relaxation that cannot be achieved with conventional mechanical structures [16, 116–118]. Mechanical structures, which use mechanical energy according to parameters such as movement, tension, compression, and various deformations, can be expressed as one of the new and important areas for MTMs. These structures, which have rapid development potential, exhibit extraordinary functions such as reprogrammable, shape/pattern morphing, nonlinear responses, and topological protection/guidance according to the applied mechanical forces [119]. When MTMs designed using solid materials are formed with flexible materials, flexible MTM properties are obtained, while mechanical MTM structures are basically expected to have a flexible characteristic. The structural design of flexible mechanical MTMs can be created from periodic geometric arrays, biological structures like cells, crystal structures, and artistic fields such as origami or kirigami. The expansion or contraction behavior of these MTMs, different from the direction of the applied force, can be considered as the early studies include structures formed by regular repeat of simple unit cells. As the scope of studies expands, many different flexible mechanical MTM designs down to the nanometer scale are being developed and increasingly complex unit cells and architectures are being designed. On the other hand, in flexible mechanical MTM designs, important problems such as lack of continuity in adjustability, structural instability, delayed response, insufficient robustness, and plastic deformations exist in the studies.

Linear flexible mechanical MTM structures can be broadly classified as isotropic and anisotropic. Isotropic structures can be described by two elastic coefficients, the Young's modulus and Poisson's ratio. Anisotropic structures have an elasticity tensor that can contain more than twenty independent coefficients, which offer a sufficient opportunity for mechanical MTM design. Auxetic materials with an unusual elasticity tensor are examples of such linear flexible mechanical MTMs. Stretch applied to a material from any axis usually produces a lateral contraction and thus a positive Poisson's ratio is obtained. Depending on the type of material, this parameter may approach zero. In structures called auxetic metamaterials, it is possible to obtain a negative Poisson's ratio by designing patterns like inverted hexagons. Even if the material used in the design of these MTMs has a positive Poisson's ratio, extraordinary behaviors such as expansion or contraction occur due to the design created if a stretching or compressing process is applied. Ebrahimi *et al* created an MTM class with a 3D cellular architecture with a unique node design based on anti-chiral topology to connect planar structures [120]. The node design of the proposed structure is formed by connecting the circular elements of the anti-chiral system. When this planar design is subjected to any load, the circular elements begin to rotate and the bonds are bent, resulting in out-of-plane deformation of the MTM cells. Thus, a negative Poisson's ratio can be obtained. Sorrentino *et al* proposed a bioinspired mechanical MTM design using a rotating square unit cell [121]. The structure with advanced mechanical and topological properties was designed using titanium material and exhibits a negative Poisson's ratio. Chen *et al* presented a lattice MTM structure by designing it in three dimensions [122], inspired by an auxetic MTM structure formed in two dimensions. Theoretical and numerical analysis were performed to analyze the flexible properties and geometric parameters of the proposed MTM design, and it was shown to have a negative Poisson's ratio.

Origami and kirigami studies, defined as the art of paper folding and cutting, inspire geometric designs of mechanical MTM structures to achieve multiple functionalities and reconfigurability. Origami designs, which are preferred for the mechanical basis of devices created in fields such as EM [123], medical [124], and electronics [125], are important options for flexible mechanical MTM designs as they can provide multiple and separate mechanical movements with a continuous degree of freedom. Mechanical MTM structures created using kirigami designs are used to obtain properties such as configuration of elastic characteristics, creating large strains and deformation. Kirigami designs can be used to maintain the electrical conductivity performance of rigid materials, giving them flexibility, creating periodic structures with adjustable Poisson's ratio, and design of complex 3D structures. In addition, flexible mechanical MTMs with extraordinary properties can be created by combining origami and kirigami designs. Silverberg *et al* showed that each unit cell of the Miura-ori tessellation crease pattern is mechanically bistable [126]. In this way, they stated that by switching between states, the compressive modulus of the overall structure can be adjusted rationally and reversibly. Because of such interactions, mechanically stable lattice defects are formed, and these defects produce emergent crystallographic structures such as vacancies, grain boundaries, and dislocations. Such advanced capabilities and scale-free geometries allow the arrangement of reversible folds for programmable mechanical MTMs and can be transferred to systems such as the milli-, micro-, and nano-meter. Walker and Stankovic presented a generalized approach to construct an algorithmic scheme for folding origami mechanism designs with one degree of freedom [127]. This also enables functions to create and analyze kirigami and generic 3D panel-hinge assemblies. The proposed design is created by a mirroring operation that mirrors an origami about a plane defined by three vertices (figure 8.11(a)). By repeating the procedure, structures similar to those shown in figures 8.11(b–e) are obtained. The proposed approach inspires the designs of flexible mechanical MTMs due to their versatility. He *et al* proposed a new mechanical MTM structure using curved crease origami with adjustable stiffness and Poisson's ratio [128]. They have obtained a new generation MTM design by using the structure strategies called the closed, unclosed, and special closed. The proposed cellular MTMs are constructed in figure 8.11(f) and its experimental setup is depicted in figure 8.11(g). They modified the models of the structures they proposed, showing that curved curvaceous origami mechanical MTM structures have adjustable stiffness.

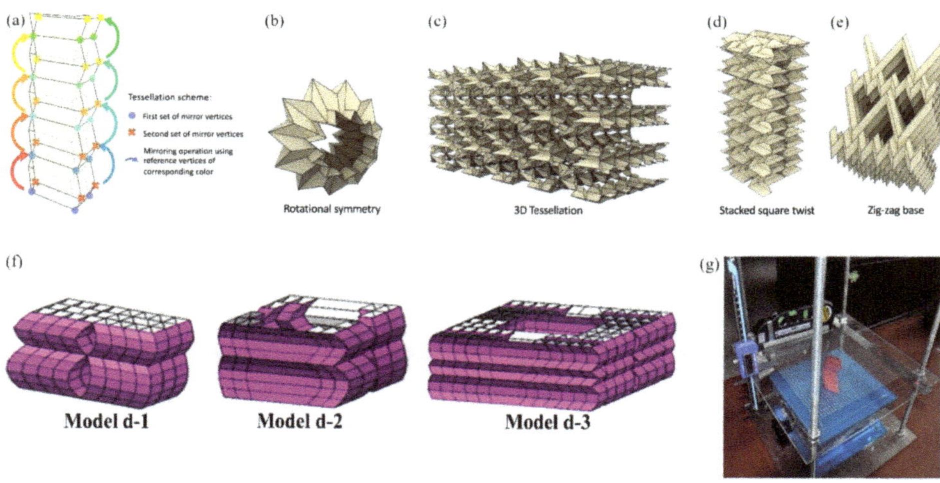

Figure 8.11. (a–e) The generation and analysis of origami/kirigami patterns and three-dimensional structures. (a) Scheme to tessellate patterns using the mirroring operation, (b) structure obtained with a scheme with non-parallel mirror planes, (c) application in two directions, and (d) application in the square twist pattern. (e) Tessellation obtained by application of the translating and mirroring operation. (f) The next-generation 3D cellular MTM and (g) its experimental setup.

One of the most important examples of flexible mechanical MTMs can be expressed as programmable structures. The use of MTMs in the design of flexible devices significantly contributes to the number of degrees of freedom and geometric parameters in the deformation criteria [79]. These programmable structures require a series of integration processes related to the geometric structure. Although such integration processes present several challenges, they provide opportunities for functionality that cannot be achieved with conventional methods in robotic devices. The ability of the system created with these structures to change the mechanical flexible movement in a stable manner is considerably important. An example is that hard disks stably store, read, and write data, considering each reversibly magnetizing bit in a magnetic medium. Similar to this system, Chen *et al* proposed a periodically tileable mechanical MTM structure consisting of a set of physical two elements (m-bits) like digital bits, which considers read and write stages with stable memory on a unit cell basis [129]. The proposed mechanical MTM structure is designed to be flexible for state changes during programming of m-bits, such as the read/write cycle of digital bits. Parts of the m-bit structure consisting of the bistable elastic shell (i), a magnetic cap (ii), four curved columns (iii), two stoppers (iv), and a top lid (v) are shown in figure 8.12(a). Figure 8.12(b) represents programming and testing a 6 × 6 m-bit array. In addition, considering the way non-volatile storage works in the proposed design, write-induced state changes are planned to be stable indefinitely and the programming resolution to be at a single m-bit level without interactions with nearby m-bits. Interchangeability, adjustability, and reconfigurability capabilities are highly desirable features for technologies such as

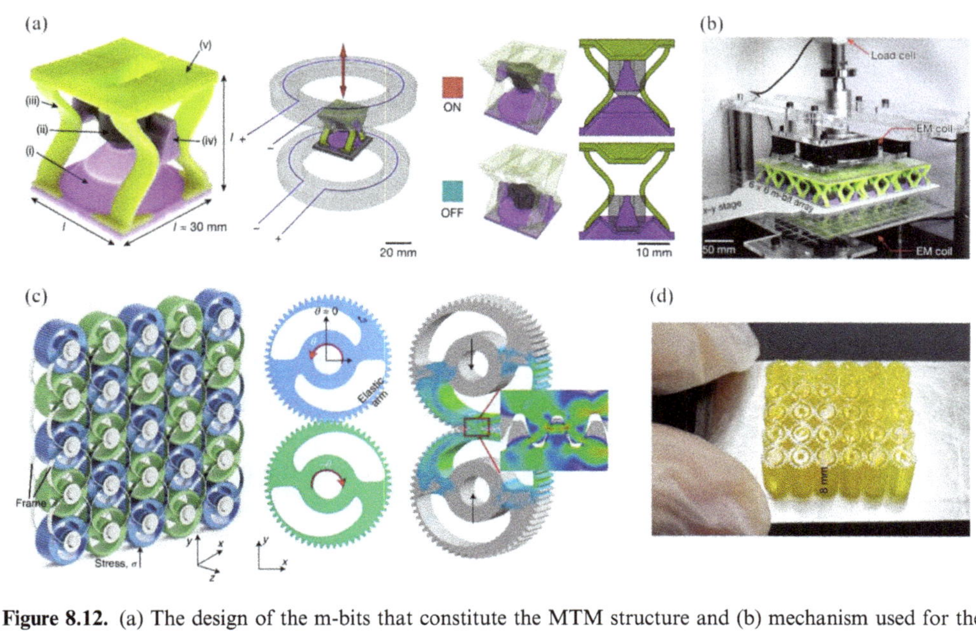

Figure 8.12. (a) The design of the m-bits that constitute the MTM structure and (b) mechanism used for the programming of a tiled array of 6 × 6 m-bits. (c) The architecture of programmable gear based mechanical MTM and (d) printed micro MTM consisting of taiji gears.

robots and smart devices. Based on these capabilities, a programmable mechanical MTM design using gear clusters offering versatile functionality was presented by Fang *et al* [16]. It is seen that the proposed MTM structure maintains robust maneuverability and stability when subjected to a heavy load, as well as offers flexible features that can be adjusted continuously. Figures 8.12(c) and (d) show the programmable gear based mechanical MTM architecture and the printed microstructure consisting of taiji gears, respectively. MTM structures, which are based on a gear mechanism, have useful properties such as Young's modulation, rapid response, and shape morphing between soft and solid states. These features are significant to the design of programmable structures and configurable flexible robots.

8.7 Flexible THz MTMs

Another major research area of MTMs is considered to be the THz spectrum. The THz frequency range, typically spanning from 0.1 to 10 THz, has unique properties and potential applications in various fields such as communications, imaging, sensing, energy harvesting, and spectroscopy [112, 130–132]. THz MTMs refer to a class of artificial materials designed to manipulate and control EM waves in the THz frequency range, which lies between the microwave and infrared regions of the EM spectrum. These structures are constructed by arranging subwavelength-scale structures, often referred to as meta-atoms or resonators, periodically. Conventional materials generally exhibit limited interaction with THz waves, making it challenging to fully exploit the potential of the THz frequency. MTMs offer solutions by providing unprecedented control over THz waves through subwavelength structuring. The periodic structures that make up THz MTMs are designed to exhibit specific EM properties at THz frequencies. These properties arise from the interaction between the incident EM wave and the meta-atoms' geometry, composition, and arrangement. By carefully optimizing the geometric parameters, researchers can adapt the behavior of MTMs and achieve unique manipulation of THz waves. One key characteristic of THz MTMs is their ability to exhibit negative refractive index, as happens in other frequency ranges, which means they can bend light in a direction opposite to that which is observed in naturally occurring materials. This property, known as negative refraction, allows for the creation of super lenses and other devices capable of resolving fine details beyond the diffraction limit of conventional optical systems. Another important feature of THz MTMs is their ability to support surface plasmon polaritons. Surface plasmon polaritons are hybrid EM waves that propagate along the interface between a MTM and a dielectric material [133, 134]. These waves can be tightly confined to the MTM's surface, enabling enhanced light–matter interactions and the concentration of EM energy in nanoscale regions. THz MTMs are typically fabricated using techniques such as electron beam lithography, focused ion beam milling, or nanoimprint lithography. The choice of fabrication method depends on the desired MTM design and the available manufacturing capabilities.

THz MTMs are generally designed using materials that are described as flexible. Flexible THz MTMs refer to a class of artificially designed materials that exhibit unique electromagnetic properties in the terahertz frequency range, typically spanning from a few hundred GHz to a few THz. These materials are similarly designed to manipulate and control the propagation of THz waves, which are EM waves with frequencies between microwaves and infrared light. In the case of THz MTMs, the dimensions of the meta-atoms are typically in the order of micrometers or smaller, allowing for flexibility and conformability. Flexibility in THz MTMs can be achieved through the use of flexible substrates, such as polymers or thin films, on which the MTM structures are fabricated. This flexibility enables the MTMs to conform to various shapes and surfaces, making them suitable for applications where conformability or bending is required, such as wearable electronics, flexible sensors, or conformal coatings. The unique properties of flexible THz MTMs have opened new possibilities for a wide range of applications. Some potential applications include THz imaging and sensing, where MTMs can be used to enhance the resolution and sensitivity of THz devices, as well as THz communication systems, where they can be employed to manipulate and control THz waves for efficient data transmission. Other applications include THz spectroscopy, security screening, non-destructive testing, and THz optics.

With the development of THz MTM structures, the use of flexible materials in MTM designs has started to attract attention. Tao *et al* realized the design, fabrication, and characterization of a highly flexible MTM absorber that experimentally achieves an absorptivity of 0.96 at 1.6 THz [135]. Figure 8.13(a) shows the fabricated photograph and design of their proposed flexible MTM absorber. Their proposed MTM design provides a wide angular range for TE and TM radiation. Such MTM structures can be used for numerous applications ranging from the active element in a thermal detector to THz stealth technology. They similarly fabricated resonant THz MTMs on free-standing PI substrates [18]. Their results appear to provide a path forward for creating multi-layer non-planar MTMs at the THz frequency range. Much work has been done in recent years to develop compact, flexible, and high-precision THz MTM sensors. Yao *et al* proposed a THz MTM refractive index/thickness sensor with flexible substrate, exhibiting low frequency Fano resonance and high frequency electromagnetically induced transparent resonance [44]. An overall photo, partially enlarged photo, photo after being transferred to the substrate, and photo under bending of the THz MTM sensor they designed and produced are shown in figure 8.13(b). The proposed MTM sensor was fabricated using femtosecond laser etching technology. The sensing performance of their proposed MTM structure, confirmed by experimental results, shows that they can distinguish different liquids and detect polyimide film with different thicknesses less than 180 μm. The proposed THz MTM sensor structure has important application areas in the field of biosensing. Most of the studies carried out in recent years aimed to develop compact, flexible, and high-precision THz MTM sensors. In addition, studies on sub-THz communication bands are increasing for the development of wireless communication technologies, which mostly use the GHz frequency. Wu and Lin proposed a flexible THz MTM filter consisting of cross-shaped MTM configurations on a PDMS substrate and the schematic drawing and geometric parameters of their proposed flexible THz MTM structure are

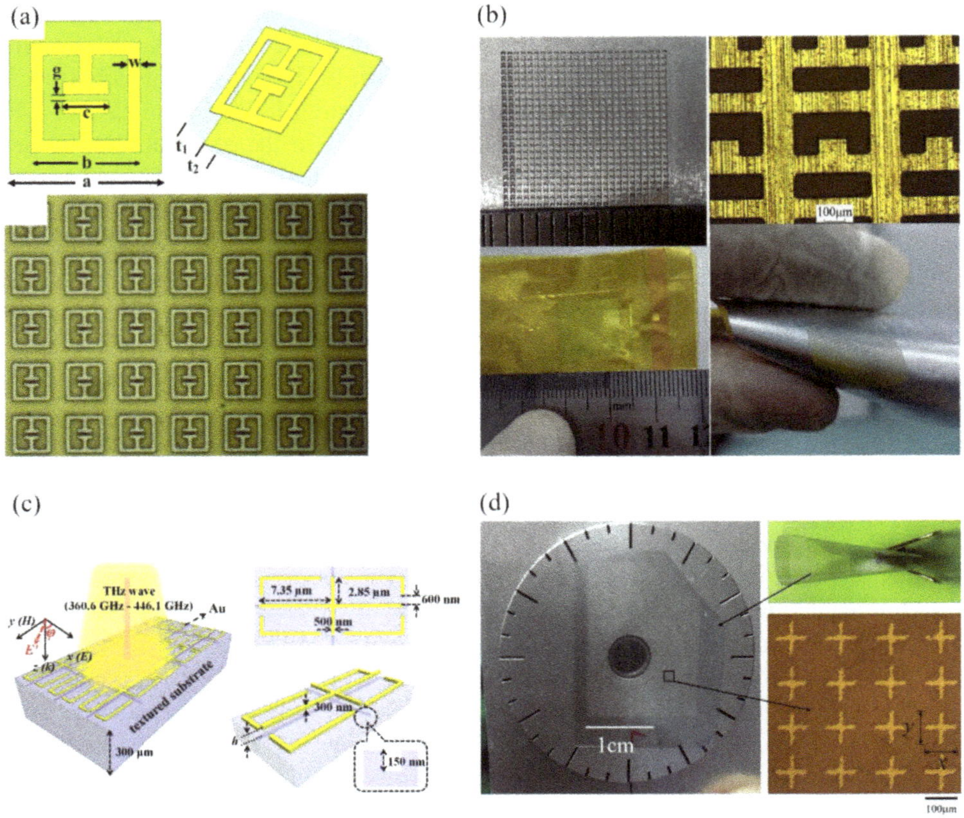

Figure 8.13. Top view, perspective view, and photograph of the THz MTM absorber (a). Photograph of the processed THz MTM sensor (b). Schematic drawing of the flexible THz MTM device (c). Photograph of the fabricated transparent and flexible tunable THz MTM absorber (d).

expressed in figure 8.13(c) [136]. It is seen that the proposed MTM structure exhibits a controllable transmission carrier property for a broad frequency range in the sub-THz window. It is stated that the proposed structure is important for future indoor wireless communication systems, due to its large tuning range and high transmission intensity features. Xiao *et al* presented a tunable THz absorber based on an indium tin oxide MTM [137]. A fabricated sample photograph of the proposed tunable THz MTM absorber is given in figure 8.13(d). The thickness of the middle dielectric layer of the proposed MTM structure is 60 μm, which makes the absorber flexible and very transparent. These and similar MTM structures are expected to be widely used in sensing, THz imaging, and biological detection. It is important to note that while THz MTMs hold great promise, there are still challenges to be overcome in terms of fabrication techniques, scalability, and integration with other technologies. However, ongoing research and advancements in materials science and nano-technology are steadily advancing the field, bringing us closer to the realization of practical and flexible THz MTM-based devices.

8.8 Discussion, challenges, and future perspectives

Over the past decade, advances in wireless communication technologies have greatly increased the interest in wearable devices. Wearable devices are examined in many respects, taking into account their advantages and disadvantages. It can be stated that one of the most attractive advantages is the level of flexibility. Although flexibility provides an advantage in use, it can reveal disadvantages such as performance problems. Flexible features of body-integrated antennas and electronic circuits are crucial in terms of ease of use. Since such structures adapt to the structure of the human body, they can be easily transported and provide significant benefits for wireless communication. On the other hand, these devices can be easily deformed and become passive, which can cause unexpected communication breakdowns. Such problems highlight the durability of the materials used in flexible devices. Durability and continuity are essential for the efficient operation of many electronic devices, from antennas to screens. The fact that flexible materials can be used in wearable materials is of great importance in overcoming this problem. This situation leads researchers to seek new flexible materials. Due to the extraordinary properties of MTMs, they have great potential in the production of flexible devices. Researchers utilizing MTM structures, which have many research areas such as energy collection, sensor, mechanics, and seismic, can expand their effect by using them together with flexible materials. On the other hand, factors such as deformation and cost negatively affect the development of flexible MTM-based devices. Researchers are working to overcome these challenges with methods such as next-generation fabrication techniques, periodic/non-periodic unit cell design, and durable flexible materials. It is clear that flexible MTM structures will make significant contributions to the development of future flexible devices when proper analysis of what MTMs can provide and what flexible devices need is performed.

The integration of MTMs with flexible materials reveals a new and emerging concept of MTMs that covers key areas such as mechanics, electronics, THz, energy harvesting, and antennas. Studies in these areas focus on using the supernatural features of MTMs to improve the functionality of devices required in these areas. For example, mechanical MTM structures offer functional solutions and strategies to address the deformation problems of flexible devices. Mechanical MTMs will have an important place in the development of technologies such as antennas, THz, and electronic devices in the future. Moreover, the creation of mechanical MTMs inspired by origami/kirigami patterns reveals the diversity of unit cell designs. Programmable mechanical MTMs, obtained using elastic properties and functional unit cells, exhibit remarkable properties for building stable intelligent machines in the future. A structural design paradigm for producing flexible robotic devices is also introduced. Another important research area is to produce high-efficiency devices to create automated designs using mechanical MTMs. Thus, design algorithms with mechanical movements that take human design one step ahead can be achieved. These algorithms, together with machine learning, may have the potential to create deep neural networks by producing efficient mechanical devices.

Considering the properties provided by flexible MTMs, it is expected to be widely used for technologies such as communications, sensing, imaging, and spectroscopy in the THz frequency range. Conventional materials often lack the necessary properties to efficiently manipulate THz waves. However, MTMs can overcome these limitations and offer new possibilities. In this respect, the capabilities of the MTMs designed using flexible materials at THz frequency are very interesting for researchers. By carefully designing the properties, size, and arrangement of these unit cells and manipulating the MTM's response to THz waves, researchers can achieve extraordinary effects not found in natural materials. In this way, the production of flexible MTM-based devices operating at the THz frequency can pave the way for new technologies. Although flexible THz MTMs show great promise in terms of their properties, there are still challenges in terms of fabrication techniques, scalability, cost, and integration with other technologies. The fabrication of MTM-based devices at the THz scale is challenging when considering using traditional fabrication techniques, as it requires consideration of materials with precise flexible structures and dimensions. With new production techniques to be developed in the future, researchers can overcome these difficulties. In addition, as with all advanced technologies, fabrication cost and scalability are major problems. Developing cost-effective and scalable fabrication processes for the THz MTMs is critical to their widespread adoption. Another challenge is environmental stability. Flexible MTMs can be sensitive to environmental factors such as humidity, temperature, and mechanical stress in terms of materials and design used. For practical applications, it is very important to ensure environmental stability and robustness in various operating conditions. To achieve practical applications, THz MTMs may need to be integrated with existing devices and systems. Compatibility and seamless integration with other technologies also create various problems. On the other hand, bandwidth limitations may also arise in terms of the operating frequency of THz MTMs. Thus, structures designed for specific THz frequencies may have limited bandwidth, making them difficult to cover a wide frequency range. Extending the operational bandwidth while maintaining the desired functionality is a significant problem to overcome. Despite these challenges, the field of the flexible THz MTMs continues to progress, and researchers are working to overcome these limitations. Ongoing research and integration studies will produce remarkable results in addressing these challenges and enabling the practical use of THz MTMs in the future.

Since wearable technologies are mostly integrated with the human body, a sustainable, light, and durable energy source has become indispensable. It is known that conventional batteries are not very useful for wearable technologies, especially due to their weight. These and similar problems reveal the search for energy sources for wearable devices. Since the amount of energy collected in the energy collection mechanism created for wearable devices is generally low, the conversion system and the obtained efficiency gain importance. Accordingly, instead of conventional structures, the focus is on innovative, adjustable, and highly energy absorbing structures. With the preference of flexible materials in MTM designs, the EM energy harvesting potential of structures designed using the unusual properties of MTMs has shown that they are a remarkable alternative for wearable technologies. On the other hand, considering the absorber properties of MTMs, their energy harvesting abilities are important, but they are not considered sufficient for many flexible electronic devices. In recent years, research on low power consumption of electronic circuits has made significant progress. These developments support the notion that flexible MTMs are still important candidates for wearable electronic circuits, where a continuous energy source is always at the forefront, due to their EM energy harvesting capability. Considering that the MTMs designed using flexible materials will become an important component for wearable/portable electronic circuits in the future, it is imperative to increase the number of studies in this research area. Similar developments apply to wearable antenna technologies. MTM structures, which have a multi-functional working mechanism, can be used as an EM energy source for antenna technologies, as well as to prevent the damage of radiation generated by antennas. Multi-functional MTM research, such as both energy source and harmful radiation prevention for electronic circuits, will guide the technologies of the future.

8.9 Conclusion

In this study, a rapidly developing research area that combines flexible materials and MTMs used for different purposes such as energy harvesting, antennas, mechanical and electronic fields is summarized. The developments in mechanical structures, the energy needed in portable devices, the increase in the usage area of wireless communication devices, and their integration into the human body, greatly increases the interest in flexible materials. These developments imply various problems such as functionality, sustainability, and ease of use. Such problems have led researchers to look for new materials with different characteristics. For areas where flexible materials are used, the extraordinary properties of MTMs are quite remarkable. MTM designs, developed using flexible conductor/insulator materials, provide significant advantages in areas such as wireless communication devices, EM wave manipulation, and outstanding mechanical properties. Flexible MTMs have become a new and important interdisciplinary field developed by considering these advantages and disadvantages. Research is carried out considering the features provided by MTMs and the needs of flexible devices form the basis of this field. The studies already presented are the beginning for flexible MTMs and this emerging field has great potential. Many of the devices presented regarding flexible MTMs are in the prototype process and have theoretical performance in the laboratory environment, but the results demonstrate the strength of the practical products of this field. Flexibility draws attention as a very important trend in human life. The continuation of the current growth trend in flexible MTMs' research and the spread of industrial products are seen as inevitable. This review firstly provides a brief overview of flexible materials and MTMs. Then, features such as flexible materials, usage areas, and design areas are examined. Considering the important application areas of MTMs, structures designed with flexible materials are discussed. Additionally, the advantages and disadvantages encountered in flexible MTM design to advance this remarkable field are summarized.

References

[1] Di Falco A, Ploschner M and Krauss T F 2010 Flexible metamaterials at visible wavelengths *New J. Phys.* **12**

[2] Iwaszczuk K, Strikwerda A C, Fan K, Zhang X, Averitt R D and Jepsen P U 2012 Flexible metamaterial absorbers for stealth applications at terahertz frequencies *Opt. Express* **20** 635–43

[3] Zhang H, Guo X, Wu J, Fang D and Zhang Y 2018 Soft mechanical metamaterials with unusual swelling behavior and tunable stress-strain curves *Sci. Adv.* **4** eaar8535

[4] Unal E, Dincer F, Tetik E, Karaaslan M, Bakir M and Sabah C 2015 Tunable perfect metamaterial absorber design using the golden ratio and energy harvesting and sensor applications *J. Mater. Sci., Mater. Electron.* **26** 9735–40

[5] Alaukally M N N, Elwi T A and Atilla D C 2022 Miniaturized flexible metamaterial antenna of circularly polarized high gain-bandwidth product for radio frequency energy harvesting *Int. J. Commun. Syst.* **35** e5024

[6] Ercan M, Tetik G and Karaaslan M 2021 Design and simulation analysis of a textile based metamaterial perfect absorber for energy harvesting applications at WIFI band *Usak Univ. J. Eng. Sci.* **4** 124–31

[7] Tetik E 2020 Flexible perfect metamaterial absorber and sensor application at terahertz frequencies *J. Optoelectron. Adv. Mater.* **22** 213–8

[8] Bakır M, Karaaslan M, Dincer F, Delihacioglu K and Sabah C 2016 Tunable perfect metamaterial absorber and sensor applications *J. Mater. Sci., Mater. Electron.* **27** 12091–9

[9] Dincer F, Karaaslan M, Colak S, Tetik E, Akgol O, Altıntas O and Sabah C 2016 Multi-band polarization independent cylindrical metamaterial absorber and sensor application *Mod. Phys. Lett.* B **30** 1650095

[10] Pendry J B 2000 Negative refraction makes a perfect lens *Phys. Rev. Lett.* **85** 3966–9

[11] Iba A, Ikeda M, Mag-usara V K, Agulto V C and Nakajima M 2022 Sub-diffraction focusing using metamaterial-based terahertz super-oscillatory lens *Appl. Sci. (Switz.)* **12** 12770

[12] Padilla W J and Averitt R D 2022 Imaging with metamaterials *Nat. Rev. Phys.* **4** 85–100

[13] Alu A and Engheta N 2008 Plasmonic and metamaterial cloaking: physical mechanisms and potentials *J. Opt.* A **10** 093002

[14] Guven K, Saenz E, Gonzalo R, Ozbay E and Tretyakov S 2010 Metamaterial based cloaking with sparse distribution of spiral resonators *Metamaterials* V **7711** 771111

[15] Zhou X, Ren L, Song Z, Li G, Zhang J, Li B, Wu Q, Li W, Ren L and Liu Q 2023 Advances in 3D/4D printing of mechanical metamaterials: from manufacturing to applications *Compos. Part B: Eng.* **254** 110585

[16] Fang X, Wen J, Cheng L, Yu D, Zhang H and Gumbsch P 2022 Programmable gear-based mechanical metamaterials *Nat. Mater.* **21** 869–76

[17] MacDonald W A 2004 Engineered films for display technologies *J. Mater. Chem.* **14** 4–10

[18] Tao H, Strikwerda A C, Fan K, Bingham C M, Padilla W J, Zhang X and Averitt R D 2008 Terahertz metamaterials on free-standing highly-flexible polyimide substrates *J. Phys. D: Appl. Phys.* **41** 232004

[19] Choi M C, Kim Y and Ha C S 2008 Polymers for flexible displays: from material selection to device applications *Prog. Polym. Sci. (Oxford)* **33** 581–630

[20] Costanzo S, Venneri I, Di Massa G and Borgia A 2010 Benzocyclobutene as substrate material for planar millimeter-wave structures: dielectric characterization and application *J. Infrared Millim. Terahertz Waves* **31** 66–77

[21] Kane C F and Krchnavek R R 1995 Benzocyclobutene optical waveguides *IEEE Photonics Technol. Lett.* **7**

[22] Si L M, Yong Y, Sun H J and Xin L 2008 Characterization and application of planar terahertz narrow bandpass filter with metamaterial resonators *META08—Proc. of the 2008 Int. Workshop on Metamaterials* pp. 351–4

[23] Bothra S, Kellam M and Garrou P 1994 Feasibility of BCB as an interlevel dielectric in integrated circuits *J. Electron. Mater.* **23** 819–25

[24] Khodasevych I E, Shah C M, Sriram S, Bhaskaran M, Withayachumnankul W, Ung B S Y, Lin H, Rowe W S T, Abbott D and Mitchell A 2012 Elastomeric silicone substrates for terahertz fishnet metamaterials *Appl. Phys. Lett.* **100** 061101

[25] Hayes G J, So J H, Qusba A, Dickey M D and Lazzi G 2012 Flexible liquid metal alloy (EGaIn) microstrip patch antenna *IEEE Trans. Antennas Propag.* **60** 2151–56

[26] Rubino E and Ioppolo T 2016 Young's modulus and loss tangent measurement of polydimethylsiloxane using an optical lever *J. Polym. Sci. B Polym. Phys.* **26** 747–51

[27] Birch J R 1992 The far-infrared optical constants of polypropylene, PTFE and polystyrene *Infrared Phys.* **33** 33–8

[28] Ling K, Yoo M, Su W, Kim K, Cook B, Tentzeris M M and Lim S 2015 Microfluidic tunable inkjet-printed metamaterial absorber on paper *Opt. Express* **23** 110–20

[29] Nateghi A, Sangiuliano L, Claeys C, Deckers E, Pluymers B and Desmet W 2019 Design and experimental validation of a metamaterial solution for improved noise and vibration behavior of pipes *J. Sound. Vib.* **455** 96–117

[30] Shen Y, Zhang J, Pang Y, Wang J, Ma H and Qu S 2018 Transparent broadband metamaterial absorber enhanced by water-substrate incorporation *Opt. Express* **26**

[31] Alsulami Q A, Wageh S, Al-Ghamdi A A, Bilal R M H and Saeed M A 2022 A tunable and wearable dual-band metamaterial absorber based on polyethylene terephthalate (PET) substrate for sensing applications *Polymers (Basel)* **14** 4503

[32] Zechmeister J and Lacik J 2019 Complex relative permittivity measurement of selected 3D-printed materials up to 10 GHz *2019 Conference on Microwave Techniques (COMITE)* pp. 1–4

[33] Shabbir T, Islam M T, Al Bawri S S, Aldhaheri R W, Alharbi K H, Aljohani A J and Saleem R 2020 16-Port non-planar mimo antenna system with near-zero-index (NZI) metamaterial decoupling structure for 5g applications *IEEE Access* **8** 157946–58

[34] Shabbir T, Saleem R, Al-Bawri S S, Shafique M F and Islam M T 2020 Eight-port metamaterial loaded UWB-MIMO antenna system for 3D system-in-package applications *IEEE Access* **8** 106982–92

[35] Del-Rio-Ruiz R, Lopez-Garde J M, Legarda J, Lemey S, Caytan O and Rogier H 2020 Reliable lab-scale construction process for electromagnetically coupled textile microstrip patch antennas for the 2.45 GHz ISM band *IEEE Antennas Wirel. Propag. Lett.* **19** 153–7

[36] Loss C, Gouveia C, Salvado R, Pinho P and Vieira J 2021 Textile antenna for bio-radar embedded in a car seat *Materials* **14** 213

[37] Mantash M, Tarot A C, Collardey S and Mahdjoubi K 2012 Investigation of flexible textile antennas and AMC reflectors *Int. J. Antennas Propag* **2012** 1–10

[38] Hossain K, Sabapathy T, Jusoh M, Abdelghany M A, Soh P J, Osman M N, Yasin M N M, Rahim H A and Al-Bawri S S 2021 A negative index nonagonal CSRR metamaterial-based compact flexible planar monopole antenna for ultrawideband applications using viscose-wool felt *Polymers (Basel)* **13** 2819

[39] Hossain K *et al* 2021 Electrically tunable left-handed textile metamaterial for microwave applications *Materials* **14**
[40] Ali U, Ullah S, Shafi M, Shah S A A, Shah I A and Flint J A 2019 Design and comparative analysis of conventional and metamaterial-based textile antennas for wearable applications *Int. J. Numer. Model.: Electron. Netw., Devices Fields* **32**
[41] Jais M I, Malek M F, Jamlos M F and Jusoh M 2012 Conductive E-textile analysis of 1.575 GHz rectangular antenna with H-slot for GPS application *2012 Loughborough Antennas & Propagation Conference (LAPC), IEEE* pp. 1–4
[42] Afridi A, Ullah S, Khan S, Ahmed A, Khalil A H and Tarar M A 2013 Design of dual band wearable antenna using metamaterials *J. Microw. Power Electromagn. Energy* **47** 126–37
[43] Jiang C, Rist F, Wang H, Wallner J and Pottmann H 2022 Shape-morphing mechanical metamaterials *CAD Computer Aided Design* **143** 103146
[44] Yao H, Mei H, Zhang W, Zhong S and Wang X 2022 Theoretical and experimental research on terahertz metamaterial sensor with flexible substrate *IEEE Photonics J.* **14** 1–9
[45] Xu Z and Lin Y S 2019 A stretchable terahertz parabolic-shaped metamaterial *Adv. Opt. Mater.* **7** 1900379
[46] Anjali M, Rengaswamy K, Ukey A, Stephen L, Krishnamurthy C V and Subramanian V 2023 Flexible metamaterial based microwave absorber with epoxy/graphene nanoplatelets composite as substrate *J. Appl. Phys.* **133** 063105
[47] Yan S and Vandenbosch G A E 2016 Radiation pattern-reconfigurable wearable antenna based on metamaterial structure *IEEE Antennas Wirel. Propag. Lett.* **15** 1715–8
[48] Tetik E 2022 Flexible metamaterial design based on wearable material for microwave energy harvesting *Bull. Mater. Sci.* **45** 1–7
[49] Tak J and Choi J 2017 A wearable metamaterial microwave absorber *IEEE Antennas Wirel. Propag. Lett.* **16** 784–7
[50] Jalil M E, Rahim M K A, Samsuri N A, Dewan R and Kamardin K 2017 Flexible ultra-wideband antenna incorporated with metamaterial structures: multiple notches for chipless RFID application *Appl. Phys. A Mater. Sci. Process* **123** 1–5
[51] De Araújo M, Fangueiro R and Hong H September): 2003 Modelling and simulation of the mechanical behaviour of weft-knitted fabrics for technical applications: Part I: General considerations and experimental analyses *AUTEX Res. J.* **3** 111–23
[52] Tetik E and Tetik G D 93–97, March 2018 The effect of a metamaterial based wearable monopole antenna on the human body *Celal Bayar Üniv. Fen Bilimleri Dergisi* **14** 93–7
[53] Zhang F, Liu Z, Qiu K, Zhang W, Wu C and Feng S 2015 Conductive rubber based flexible metamaterial *Appl. Phys. Lett.* **106** 061906
[54] Zhang F, Feng S, Qiu K, Liu Z, Fan Y, Zhang W, Zhao Q and Zhou J 2015 Mechanically stretchable and tunable metamaterial absorber *Appl. Phys. Lett.* **106** 091907
[55] Ling K, Kim K and Lim S 2015 Flexible liquid metal-filled metamaterial absorber on polydimethylsiloxane (PDMS) *Opt. Express* **23** 21375–83
[56] Zhang X, Wang Y, Cui Z, Zhang X, Chen S, Zhang K and Wang X 2021 Carbon nanotube-based flexible metamaterials for THz sensing *Opt. Mater. Express* **11** 1470–83
[57] Nikolaenko A E, Papasimakis N, Chipouline A, Angelis F D, Fabrizio E D and Zheludev N I 2012 THz bandwidth optical switching with carbon nanotube metamaterial *Opt. Express* **20** 6068–79

[58] Zhang Y, Xu Y, Liu H and Sun B 2022 Ultrahigh sensitivity nitrogen-doping carbon nanotubes-based metamaterial-free flexible terahertz sensing platform for insecticides detection *Food Chem.* **394** 133467

[59] Moroz A 2009 Some negative refractive index material headlines long before Veselago work and going back as far as to 1905 (http://wave-scattering.com/negative.html)

[60] Veselago V G 1968 The electrodynamics of substances with simultaneously negative values of\epsilon and μ *Sov. Phys. Usp.* **10** 509–14

[61] Pendry J 1996 Extremely low frequency plasmons in metallic microstructures *Phys. Rev. Lett.* **76** 4773

[62] Pendry J B, Holden A, Robbins D and Stewart W 1998 Low frequency plasmons in thin wire structures *J. Phys. Condens. Matter* **10** 4785–809

[63] Pendry J B, Holden A J, Robbins D J and Stewart W J 1999 Magnetism from conductors and enhanced nonlinear phenomena *IEEE Trans. Microw. Theory Tech.* **47** 2075–84

[64] Smith D R, Padilla W J, Vier D C, Nemat-Nasser S C and Schultz S 2000 Composite medium with simultaneously negative permeability and permittivity *Phys. Rev. Lett.* **84** 4184–7

[65] Shelby R a, Smith D R and Schultz S 2001 Experimental verification of a negative index of refraction *Science* **292** 77–9

[66] Joannopoulos J D, Johnson S G, Winn J N and Meade R D 2011 Photonic crystals: molding the flow of light 2nd edn. (Princeton, NJ: Princeton Univ. Press)

[67] Foteinopoulou S and Soukoulis C M 2005 Electromagnetic wave propagation in two-dimensional photonic crystals: a study of anomalous refractive effects *Phys. Rev. B Condens. Matter Mater. Phys.* **72** 165112

[68] Segal N, Keren-Zur S, Hendler N and Ellenbogen T 2015 Controlling light with metamaterial-based nonlinear photonic crystals *Nat. Photonics* **9** 180–4

[69] Boltasseva A and Shalaev V M 2008 Fabrication of optical negative-index metamaterials: recent advances and outlook *Metamaterials* **2** 1–17

[70] Ni X, Wong Z J, Mrejen M, Wang Y and Zhang X 2015 An ultrathin invisibility skin cloak for visible light *Science* **349** 1310–4

[71] Jacob Z, Alekseyev L V and Narimanov E 2006 Optical hyperlens: far-field imaging beyond the diffraction limit *Opt. Express* **14** 8247–56

[72] Walia S, Shah C M, Gutruf P, Nili H, Chowdhury D R, Withayachumnankul W, Bhaskaran M and Sriram S 2015 Flexible metasurfaces and metamaterials: a review of materials and fabrication processes at micro- and nano-scales *Appl. Phys. Rev.* **2** 011303

[73] Liu J, Guo H and Wang T 2020 A review of acoustic metamaterials and phononic crystals *Crystals* **10** 305

[74] Chen S, Fan Y, Fu Q, Wu H, Jin Y, Zheng J and Zhang F 2018 A review of tunable acoustic metamaterials *Appl. Sci. (Switz.)* **8** 1480

[75] Peng W and Wu H 2019 Flexible and stretchable photonic sensors based on modulation of light transmission *Adv. Opt. Mater.* **7** 1900329

[76] Huang X *et al* 2022 Flexible mechanical metamaterials enabled electronic skin for real-time detection of unstable grasping in robotic manipulation *Adv. Funct. Mater.* **32** 2109109

[77] Yang S, Liu P, Yang M, Wang Q, Song J and Dong L February 2016 From flexible and stretchable meta-atom to metamaterial: a wearable microwave meta-skin with tunable frequency selective and cloaking effects *Sci. Rep.* **6** 21921

[78] Peng W, Liu Z, Xiao H, Zhang H and Song W 2023 High-efficiency and flexible inkjet-printed quasi metasurface harvester for microwave power transfer application *IEEE Antennas Wirel. Propag. Lett.* **22** 1706–10

[79] Rafsanjani A, Bertoldi K and Studart A R 2019 Programming soft robots with flexible mechanical metamaterials *Sci Robot* **4** eaav7874

[80] Tetik E and Tetik G D December 2017 The effect of a metamaterial-based wearable microstrip patch antenna on the human body *Can. J. Phys.* **96** 796–800

[81] Jia Q, Xu H, Xiong M F, Zhang B and Duan J 2019 Omnidirectional solid angle beam-switching flexible array antenna in millimeter wave for 5G micro base station applications *IEEE Access* **7** 157027–36

[82] Kiourti A and Volakis J L 2014 Stretchable and flexible e-fiber wire antennas embedded in polymer *IEEE Antennas Wirel. Propag. Lett.* **13** 1381–84

[83] Simorangkir R B V B, Kiourti A and Esselle K P 2018 UWB wearable antenna with a full ground plane based on PDMS-embedded conductive fabric *IEEE Antennas Wirel. Propag. Lett.* **17** 493–6

[84] Muduli A, Kanneboina M K, Mudumunthala K V and Valluri S 2021 A reconfigurable wearable antenna for mid band 5G applications *J. Phys. Conf. Ser.* **1921** 012051

[85] Srilatha K, Madhav B T P, Anil Babu B, Rishikesh Raj M, Somala T, Nimmaraju V and Rao M C 2021 Design and analysis of jeans based wearable monopole antenna with enhanced gain using AMC backing *J. Phys. Conf. Ser.* **1804** 012189

[86] Ahmed M I and Ahmed M F 2020 A wearable flexible antenna integrated on a smart watch for 5G Applications *J. Phys. Conf. Ser.* **1447** 012005

[87] Salayma M, Al-Dubai A, Romdhani I and Nasser Y 2017 Wireless Body Area Network (WBAN): a survey on reliability, fault tolerance, and technologies coexistence *ACM Comput. Surv.* **50** 1–38

[88] Yan S, Soh P J and Vandenbosch G A E 2018 Wearable ultrawideband technology—a review of ultrawideband antennas, propagation channels, and applications in wireless body area networks *IEEE Access* **6** 42177–85

[89] Bharadwaj R and Koul S K 2022 Wearable UWB technology for daily physical activity tracking, detection, and classification *IEEE Sens. J.* **22** 20684–94

[90] Stoppa M and Chiolerio A 2014 Wearable electronics and smart textiles: a critical review *Sensors (Switzerland)* **14** 11957–92

[91] Mendes C and Peixeiro C 2018 On-body transmission performance of a novel dual-mode wearable microstrip antenna *IEEE Trans. Antennas Propag.* **66** 4872–7

[92] Al-Sehemi A, Al-Ghamdi A, Dishovsky N, Atanasov N and Atanasova G 2023 A flexible miniature antenna for body-worn devices: design and transmission performance *Micromachines (Basel)* **14** 514

[93] Enoch S, Tayeb G, Sabouroux P, Guérin N and Vincent P 2002 A metamaterial for directive emission *Phys. Rev. Lett.* **89** 213902

[94] Salonen P 2001 A low-cost 2.45 GHz photonic band-gap patch antenna for wearable systems *11th Int. Conf. on Antennas and Propagation (ICAP 2001)* **2001** 719–23

[95] Venkateswara Rao M, Madhav B T P, Anilkumar T and Prudhvinadh B 2020 Circularly polarized flexible antenna on liquid crystal polymer substrate material with metamaterial loading *Microw. Opt. Technol. Lett.* **62** 866–74

[96] Yan S, Soh P J and Vandenbosch G A E 2015 Compact all-textile dual-band antenna loaded with metamaterial-inspired structure *IEEE Antennas Wirel. Propag. Lett.* **14** 1486–9

[97] Gao G P, Hu B, Wang S F and Yang C 2018 Wearable circular ring slot antenna with EBG structure for wireless body area network *IEEE Antennas Wirel. Propag. Lett.* **17** 434–7
[98] TETİK G and TETİK E 2020 Metamaterial based flexible coplanar antenna design and simulation for human body applications *J. Inst. Sci. Technol.* **10** 2541–50
[99] Di Y H, Liu X Y and Tentzeris M M 2014 A conformable dual-band antenna equipped with AMC for WBAN applications Proc. of 3rd Asia-Pacific Conf. on Antennas and Propagation, APCAP 2014
[100] Abirami B S and Sundarsingh E F 2017 EBG-backed flexible printed yagi-uda antenna for on-body communication *IEEE Trans. Antennas Propag.* **65** 3762–65
[101] Chen Z, Guo B, Yang Y and Cheng C 2014 Metamaterials-based enhanced energy harvesting: a review *Phys. B Condens. Matter.* **438** 1–8
[102] Bağmancı M, Karaaslan M, Altıntaş O, Karadağ F, Tetik E and Bakır M 2018 Wideband metamaterial absorber based on CRRs with lumped elements for microwave energy harvesting *J. Microwave Power Electromagn. Energy* **52** 45–59
[103] Bakır M, Karaaslan M, Karadağ F, Ünal E, Akgöl O, Alkurt F Ö and Sabah C 2018 Metamaterial-based energy harvesting for GSM and satellite communication frequency bands *Opt. Eng.* **57** 087110
[104] Lee G, Lee D, Park J, Jang Y, Kim M and Rho J 2022 Piezoelectric energy harvesting using mechanical metamaterials and phononic crystals *Commun. Phys.* **5** 94
[105] Ramahi O M, Almoneef T S, Alshareef M and Boybay M S 2012 Metamaterial particles for electromagnetic energy harvesting *Appl. Phys. Lett.* **101** 173903
[106] Hawkes A M, Katko A R and Cummer S A 2013 A microwave metamaterial with integrated power harvesting functionality *Appl. Phys. Lett.* **103** 163901
[107] K T *et al* 2020 Compact dual-band metamaterial-based high-efficiency rectenna: an application for ambient electromagnetic energy harvesting *IEEE Antennas Propag. Mag.* **62** 18–29
[108] Landy N I, Sajuyigbe S, Mock J J, Smith D R and Padilla W J 2008 Perfect metamaterial absorber *Phys. Rev. Lett.* **100** 207402
[109] Tetik E 2017 An overview of metamaterial researches *Researches on Science and Art in 21St Century Turkey* (Gece Publishing) pp 1817–25
[110] Bakir M, Karaaslan M, Dincer F, Akgol O and Sabah C 2016 Electromagnetic energy harvesting and density sensor application based on perfect metamaterial absorber *Int. J. Mod. Phys.* B **30** 1650133
[111] Qi S, Oudich M, Li Y and Assouar B 2016 Acoustic energy harvesting based on a planar acoustic metamaterial *Appl. Phys. Lett.* **108** 263501
[112] Bagmanci M, Karaaslan M, Unal E, Akgol O, Baklr M and Sabah C 2019 Solar energy harvesting with ultra-broadband metamaterial absorber *Int. J. Mod. Phys.* B **33** 1950056
[113] Bait-Suwailam M M, Almoneef T S and Saeed S M 2020 Flexible metamaterial electromagnetic harvester using modified split-ring resonator *Prog. Electromagn. Res.* M **95** 135–44
[114] Kaur K P, Upadhyaya T, Palandoken M and Gocen C 2019 Ultrathin dual-layer triple-band flexible microwave metamaterial absorber for energy harvesting applications *Int. J. RF Microwave Comput. Aided Eng.* **29** e21646
[115] Umino Y, Tsukamoto T, Shiomi S, Yamada K and Suzuki T 2018 Development of vibration energy harvester with 2D mechanical metamaterial structure *J. Phys. Conf. Ser.* **1052** 012103
[116] Lee J B *et al* 2012 A mechanical metamaterial made from a DNA hydrogel *Nat. Nanotechnol.* **7** 816–20

[117] Frenzel T, Kadic M and Wegener M 2017 Three-dimensional mechanical metamaterials with a twist *Science* **358** 1072–4

[118] Zheng X *et al* 2014 Ultralight, ultrastiff mechanical metamaterials *Science (1979)* **344** 1373–7

[119] Christensen J, Kadic M, Kraft O and Wegener M 2015 Vibrant times for mechanical metamaterials *MRS Commun.* **5** 453–62

[120] Ebrahimi H, Mousanezhad D, Nayeb-Hashemi H, Norato J and Vaziri A 2018 3D cellular metamaterials with planar anti-chiral topology *Mater. Des.* **145** 226–31

[121] Sorrentino A, Castagnetti D, Mizzi L and Spaggiari A 2022 Bio-inspired auxetic mechanical metamaterials evolved from rotating squares unit *Mech. Mater.* **173** 104421

[122] Chen Y and Fu M H 2017 A novel three-dimensional auxetic lattice meta-material with enhanced stiffness *Smart Mater. Struct.* **26** 105029

[123] Kuribayashi K, Tsuchiya K, You Z, Tomus D, Umemoto M, Ito T and Sasaki M 2006 Self-deployable origami stent grafts as a biomedical application of Ni-rich TiNi shape memory alloy foil *Mater. Sci. Eng.,* A **419** 131–7

[124] Wang Z, Jing L, Yao K, Yang Y, Zheng B, Soukoulis C M, Chen H and Liu Y 2017 Origami-based reconfigurable metamaterials for tunable chirality *Adv. Mater.* **29** 1700412

[125] Le D H and Lim S 2019 Four-mode programmable metamaterial using ternary foldable origami *ACS Appl. Mater. Interfaces* **11** 28554–61

[126] Silverberg J L, Evans A A, McLeod L, Hayward R C, Hull T, Santangelo C D and Cohen I 2014 Using origami design principles to fold reprogrammable mechanical metamaterials *Science (1979)* **345** 647–50

[127] Walker A and Stankovic T 2022 Algorithmic design of origami mechanisms and tessellations *Commun. Mater.* **3** 4

[128] He Y L, Zhang P W, You Z, Li Z Q, Wang Z H and Shu X F 2020 Programming mechanical metamaterials using origami tessellations *Compos. Sci. Technol.* **189** 108015

[129] Chen T, Pauly M and Reis P M 2021 A reprogrammable mechanical metamaterial with stable memory *Nature* **589** 386–90

[130] Zhao X, Lin Z, Wang Y, Yang X, Yang K, Zhang Y, Peng J, Lamy de la Chapelle M, Zhang L and Fu W 2019 Label-free self-referenced sensing of living cells by terahertz metamaterial-based reflection spectroscopy *Biomed. Opt. Express* **10** 1196–206

[131] Roh Y, Lee S H, Kwak J, Song H S, Shin S, Kim Y K, Wu J W, Ju B K, Kang B and Seo M 2022 Terahertz imaging with metamaterials for biological applications *Sensors Actuators* B **352** 130993

[132] Park S J, Hong J T, Choi S J, Kim H S, Park W K, Han S T, Park J Y, Lee S, Kim D S and Ahn Y H 2014 Detection of microorganisms using terahertz metamaterials *Sci. Rep.* **4** 4988

[133] Ioannidis T, Gric T and Rafailov E 2020 Surface plasmon polariton waves propagation at the boundary of graphene based metamaterial and corrugated metal in THz range *Opt. Quantum Electron.* **52** 1–12

[134] Cui Z, Wang Y, Yue L, Zhao X, Zhang D, Yao Z, Zhang X, Hou L and Zhang X 2021 Absorption-mode splitting of terahertz metamaterial mediated by coupling of spoof surface plasmon polariton *IEEE Trans. Terahertz Sci. Technol.* **11** 626–34

[135] Tao H, Bingham C M, Strikwerda A C, Pilon D, Shrekenhamer D, Landy N I, Fan K, Zhang X, Padilla W J and Averitt R D 2008 Highly flexible wide angle of incidence terahertz metamaterial absorber: design, fabrication, and characterization *Phys. Rev. B Condens. Matter Mater. Phys.* **78** 241103

[136] Wu L and Lin Y S 2023 Flexible terahertz metamaterial filter with high transmission intensity and large tuning range for optical communication application *Phys. E Low Dimens. Syst. Nanostruct.* **146** 115563

[137] Xiao J, Xiao R, Zhang R, Shen Z, Hu W, Wang L and Lu Y 2020 Tunable terahertz absorber based on transparent and flexible metamaterial *Chin. Opt. Lett.* **18** 092403

Chapter 9

Acoustic metamaterials

Zafer Özer

Symbols

β	Bulk modulus
ε	Permittivity
κ	Dielectric constant
λ	Wavelength (m)
ρ	Density (kg/m^3)
k	Wave vector
f	Frequency (Hz)
n	Refractive index
θ	Angle of incidence
a	Lattice constant
p	Pressure (Pa)

9.1 Introduction

Acoustic metamaterial, also known as a sonic crystal or phononic crystal (PnC), is a type of material specifically designed to control, direct, and manipulate sound waves or phonons across gases, liquids, and solids, including crystal lattices. The manipulation of sound waves is achieved by controlling and manipulating various parameters such as bulk modulus (β), density (ρ), and chirality. Depending on their engineering utilization, acoustic metamaterials can be designed to either transmit or trap and amplify sound waves at specific frequencies. In cases where sound waves are trapped and amplified, the material functions are denoted as an acoustic resonator.

Acoustic metamaterials represent a fascinating group of materials capable of manipulating sound waves in ways that were once considered implausible. Through precise engineering of these materials' properties and structures, scientists have been able to construct architectures that can bend, amplify, or even completely obstruct sound waves. These materials have the potential to revolutionize a wide range of

doi:10.1088/978-0-7503-5754-8ch9

applications, from noise reduction and soundproofing to medical imaging and communication systems.

In this chapter, we will explore the basic principles behind acoustic metamaterials, their unique properties and applications, and the challenges and opportunities that lie ahead in this exciting field of research. Acoustic metamaterials find applications in the research and modeling of acoustic phenomena, ranging from extremely large-scale events like earthquakes and seismic waves to extremely small-scale events like atoms. One key feature that enables this is the concept of band gap engineering, where acoustic metamaterials can be specifically designed to exhibit band gaps for phonons, similar to how electrons have band gaps in solids or electron orbitals in atoms. This has led to increased research into PnCs, which are a type of acoustic metamaterial that have band gaps for phonons. The design and development of acoustic metamaterials have also contributed to advances in quantum technologies and experiments that probe quantum mechanics. The fields of negative refractive index material research and optomechanics, including quantum optomechanics, also strongly rely on acoustic metamaterials.

For acoustic materials and acoustic metamaterials, two important component parameters are the bulk modulus and density, which together define their refractive index. The concept of refractive index is similar to what is used in optics, but in this case, it is used to describe the behavior of pressure or shear waves, rather than electromagnetic waves. The acoustic refractive index plays an important role in determining the behavior of sound waves as they propagate through different materials, and it can be modified through the use of acoustic metamaterials to achieve specific desired effects.

Acoustic metamaterials or PnCs can be thought of as the acoustic equivalent of photonic crystals. In photonic crystals, electromagnetic waves or photons propagate through a material with a periodically modified optical refractive index, which results in a modification of the speed of light. Similarly, in acoustic metamaterials or PnCs, pressure waves or phonons also propagate through a material with a periodically modified acoustic refractive index, resulting in a modification of the speed of sound. By controlling the periodicity and geometry of the structure, PnCs can be designed to exhibit specific band gaps for phonons, which can be used to manipulate sound waves in various ways.

Structures with electrical permeability (ε) and magnetic permeability (μ) properties that are not naturally found in nature are called PnCs [1]. PnCs are periodic structures that consist of two or more materials with different mechanical properties and can be designed as acoustic metamaterials, such as negative refractive index, sound focusing, wave confinement, and waveguide. The main feature of PnCs is that they do not allow the propagation of mechanical or acoustic waves in a certain frequency range. Frequencies that are not allowed to pass through the periodic structure are called phononic bands [2, 16].

In the literature, the band structure of periodic, semi-periodic, irregular, one-dimensional (1D), two-dimensional (2D), and three-dimensional (3D) PnCs, transmission reflection coefficients, and negative refractive index properties have been studied, along with acoustic lens design, sound barriers at low frequencies, acoustic filter design, and seismic structures [1–22]. Due to their characteristic structures, they have potential application in micro/nano-mechanical structures, MHz, GHz frequencies, wireless communication, sensing, shielding, etc [23, 24].

The formation of forbidden bands in PnCs is similar to electronic band structures in solid-state physics [3]. The first study on wave propagation in periodic structures was done by Leon Brillouin [4]. Considering that the electron behaves like a wave, diffraction occurs when the electron wavelength satisfies the Bragg conditions demonstrated in equations (9.1) and (9.2).

$$2d.\ \sin\theta = n.\ \lambda \tag{9.1}$$

$$k.\ \sin\theta = \frac{n.\ \lambda}{d} \tag{9.2}$$

Here, θ is the angle of incidence, λ is the wavelength, and n is an integer.

$$k.\ \sin\theta = k_x \tag{9.3}$$

If the wave vector component of k in the Brillouin zone is $k_x = \frac{n.\lambda}{d}$, Bragg scattering takes place and reflection occurs from the lattice points [2]. A standing wave is formed as a result of the interaction of incident and reflected waves.

If the periodic structure consists of two different materials, there may be maximum points of the envelope function in $a/2$ wide material with high sound velocity, and materials with low sound velocity. In this case, two frequencies may occur for the same wave number, depending on the k-dependent frequency. This phenomenon leads to the opening of bands and the creation of gaps. The width of the band gap depends on the magnitude of the difference between the wave velocities. Generally, a large difference in the properties of the two periodic materials results in a larger band gap.

In PnCs, the mechanical wave primarily propagates through the fluid due to the rigid structure of the cylinders. Assuming that the cylinders extend in the z direction and the lattice is infinitely periodic in the x–y plane, the mechanical properties of the PnC change periodically in the x–y plane. The motion of the acoustic Bloch wave in a PnC, which consists of a periodic solid–fluid material, is described by the acoustic wave equation in equation 9.4;

$$\nabla\left(\frac{1}{\rho}\nabla p\right) = -\frac{1}{\rho c^2}\left(\omega\left(\tilde{k}\right)\right)^2 .\ p \tag{9.4}$$

In this equation, $\rho = \rho(\tilde{r})$ is the density, $c = c(\tilde{r})$ is the longitudinal velocity of the acoustic wave in the PnC, and $p = p_k(\tilde{r})$ in the phononic crystal with a pressure of $p = p_k(\tilde{r})$ with $\tilde{k}$ wave vector acoustic Bloch which is the positional portion of the wave.

In a two-dimensional PnC, the wave equation becomes:

$$\frac{\partial}{\partial x}\left(\frac{1}{\rho}\frac{\partial p}{\partial x}\right) + \frac{\partial}{\partial y}\left(\frac{1}{\rho}\frac{\partial p}{\partial y}\right) = -\frac{1}{\rho c^2}\left(\omega\left(\tilde{k}\right)\right)^2 . p \tag{9.5}$$

Here, $\rho = \rho(x, y)$ and $c = c(x, y)$ are the mechanical properties of the 2D phononic crystal and p is spatial pressure ($p_k(x, y)$).

The presence of a band gap in a periodic structure is a result of multiple reflections of the mechanical wave at the interface between materials with different mechanical properties. When a periodic structure possesses a band gap, mechanical waves within this range are completely reflected back, as they cannot propagate through the structure. Phononic crystals confine and restrict the propagation of mechanical waves within their structure. Depending on the lattice constant, PnCs prevent the passage of mechanical waves within the frequency range of 20 to 20 000 Hz, whether the lattice is measured in meters or centimeters.

The behavior of acoustic or mechanical waves in periodic media is associated with acoustic band gaps, similar to electronic band gaps in solid-state physics. The finite element method (FEM) is the most commonly used technique for determining the band structure. The calculation of the band structure involves considering waves polarized perpendicular to the propagation plane.

This chapter discusses the combination of theoretical analyzes and numerical simulations to explore various properties and applications of acoustic metamaterials. Methodologies include the use of mathematical models and simulations to study the behavior and potential functions of these unique structures in manipulating acoustic waves and obtaining new acoustic effects.

The methodology of the article can be summarized as follows:

Phononic crystals and acoustic lens property:

In this section, a unique analysis of the phonon band structure of 2D phononic crystals consisting of $LiNbO_3$ cylinders immersed in water was performed. The finite element method (FEM) is used to calculate and demonstrate the formation of negative refraction in acoustic waves at the interfaces between a phononic crystal and water. To further investigate the effects of negative refraction and acoustic lensing, the study involves the use of 8 series $LiNbO_3$ inclusions in a phononic crystal configuration.

Fractal phononic crystals and band structures:

The section focuses on investigating the band structure of phononic crystals with fractal structures, with particular emphasis on utilizing the Hausdorff fractal structure. By applying Floquet boundary conditions to the edges of the square lattice, the first 15 bands are obtained and analyzed. The fractal phononic crystal exhibits a wide band gap, showcasing its unique properties. In addition, the research proposes an innovative adjustable compound waveguide (TCW) built upon 2D photonic band gap (PBC) structures. This novel waveguide incorporates piezoelectric rods and external inductor circuits in each unit cell, enabling independent control over the unit cell properties. Notably, the study demonstrates the capability to adjust the working frequency gap of the waveguide through the application of electric current and piezoelectric coupling effects, offering a promising platform for tunable waveguide technologies.

Second harmonic production in acoustic metamaterials:

The section focuses on the discovery and analysis of acoustic second harmonic generation (SHG), a fascinating nonlinear phenomenon that generates a new wave at twice the frequency when two waves with different frequencies interact. The investigation employs numerical simulations utilizing the FEM and the Murnaghan model to study SHG on aluminum plates. The results reveal the presence of both static and second harmonic components in the frequency domain, highlighting the nonlinearity of the acoustic system and providing valuable insights into this intriguing phenomenon.

Acoustic subwavelength structures:

The section explores the investigation of subwavelength structures in acoustics, where the focus is on objects or structures smaller than the wavelength of the sound wave. The research involves the design of sonic crystal barriers using concentric symmetric complementary split rectangular resonators (CSRRs) to achieve wide attenuation bands in the low-frequency range. Numerical methods, especially the FEM, are used to comprehensively analyze and calculate the band structures and transmission properties of proposed systems. This approach aims to explore the potential of subwavelength structures in acoustics to create efficient sound barriers with wide attenuation bands in the desired frequency range.

Acoustic Weyl point materials methodology:

The section proposes a two-band tight coupling model that adheres to the symmetries of space group 89 to obtain type-III Weyl points in a 3D sonic crystal. Parameters such as jump parameters (t_1, t_2, t_3, t_4) can be set to control the type and chirality of Weyl points. The designed acoustic Weyl sonic crystal with Type-III Weyl points is simulated using the finite element method to analyze the band structure and determine the presence of type-III Weyl points. The simulations show the linear distribution of type-III Weyl points along the k_z direction and the quadratic distribution in the k_x–k_y plane.

9.1.1 Negative refractive index of phononic crystals and acoustic lens property

In the present study, we conduct a unique analysis on the phonon band structure of two-dimensional (2D) phononic crystals. These crystals are composed of periodically arranged $LiNbO_3$ cylinders immersed in water. Employing the FEM, the presence of negative refraction in the acoustic waves has been calculated and examined at the interfaces between the phononic crystal slab and the water. The findings provide compelling evidence for the existence of negative refraction and reveal an intriguing phenomenon known as the 'acoustic lens effect,' which manifests itself due to the negative refractions occurring within the slab.

The calculations were performed along the path of Γ-X-M-Γ by applying Floquet boundary conditions at the edges of the square lattice (see figure 9.1). The lattice parameter was set to $a = 25$ mm, and the cylinder radius was $r = 9.5$ mm.

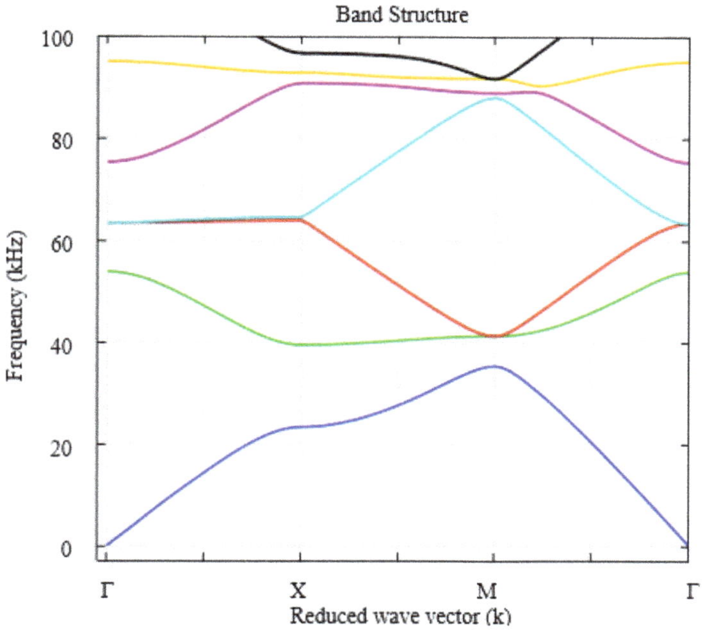

Figure 9.1. Band structure and transmission of water/$LiNbO_3$ phononic crystal.

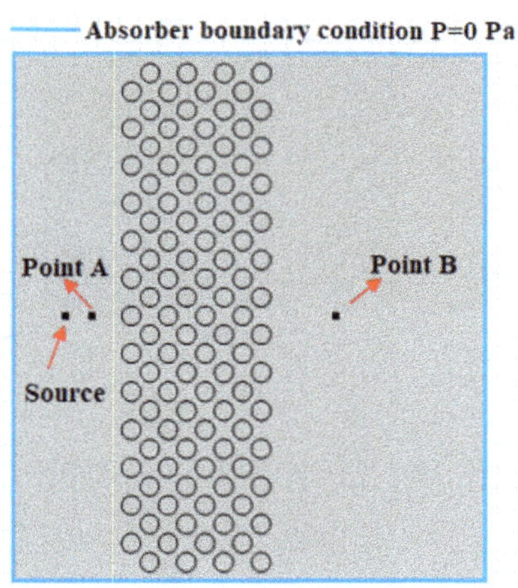

Figure 9.2. Diagram of a phononic crystal plate with negative refractive index.

Regarding the issue of negative refraction, it has been found that the concept of super lensing is not solely dependent on a negative refractive index. Furthermore, when the sound speed is low, the first band is positioned at a lower energy level. This is because, at each k-point along the circular path, the gradient $d\omega/dk$ is directed towards the M center. Examining the band structure of the PnC in figure 9.1, a peak has been observed at the M point in the first band, which leads to negative refraction.

Figure 9.2 illustrates the utilization of an 8-array of $LiNbO_3$ inclusions to showcase the negative refraction and lens effect using a 2D PnC. Absorber boundary

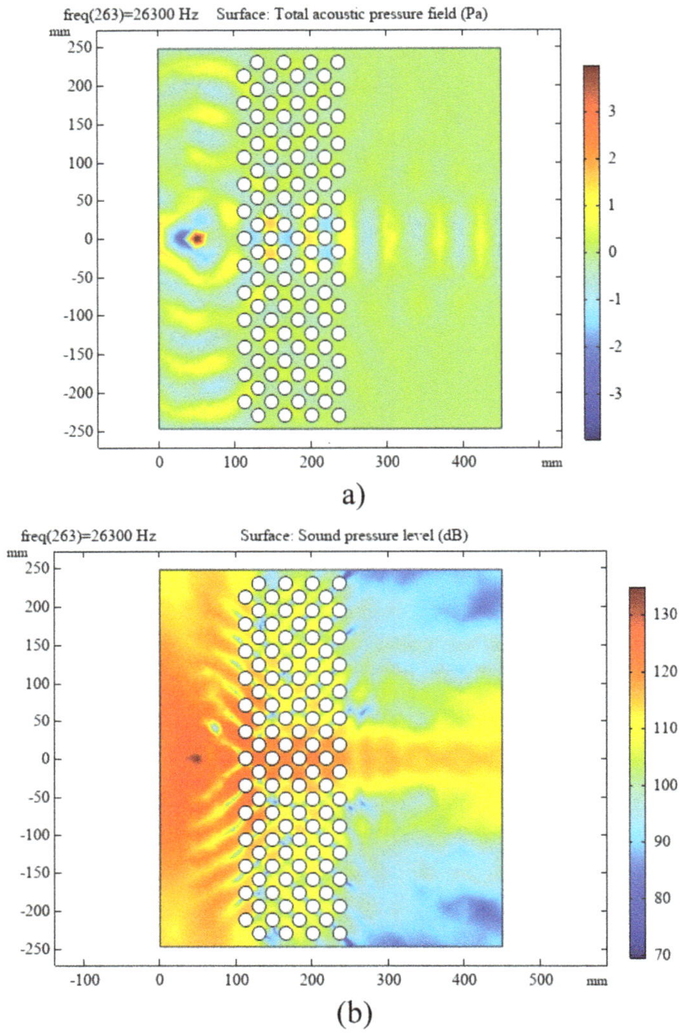

Figure 9.3. Lens effect of an acoustic wave. (a) Normalized total pressure field (Pa), (b) sound pressure level (dB).

conditions were applied to the edges, and a point source was positioned on the left side of the PnC. Observing the right side, it is evident that the waves converge and focus. This outcome clearly demonstrates the redirection and concentration of the wave that propagates from the point source towards the PnC located on the left side of the array, ultimately converging on the right side of the PnC.

Subsequently, FEM simulations were conducted to demonstrate the acoustic lens effect and negative refraction using a phononic crystal slab (figure 9.3).

Using this property of phononic crystals, new acoustic metamaterials such as acoustic and ultrasonic cloaks [21] and devices with negative refractive index can be developed [22].

9.1.2 Fractal phononic crystals and their band structure

Fractal phononic crystals, which utilize fractal geometry to create complex and irregular structures, can be classified as a type of acoustic metamaterial. They are engineered to manipulate the propagation of sound waves, similar to other acoustic metamaterials, with a specific focus on achieving a wide band gap. Phononic crystals, a subtype of acoustic metamaterials, employ periodic structures to create band gaps in sound wave propagation. In contrast, fractal PnCs utilize fractal geometry to generate more intricate structures that can exhibit multiple band gaps and broader frequency ranges for sound wave blocking. This property makes them potentially more effective for applications such as noise control. Hence, a fractal PnC has been designed to manipulate sound wave propagation and, featuring a broad band gap, can be regarded as a form of acoustic metamaterial.

In this section, the band structure of PnCs with fractal structures has been investigated. Several studies in the literature have explored the band structures of PnCs with fractal configurations [25, 26]. One example of such a fractal structure is the Hausdorff fractal structure, illustrated in figure 9.4.

In this fractal structure, the lattice parameter is set to $a = 5$ mm, divided into 6×6 equal squares. The blue regions in figure 9.4 represent air, while the remaining parts correspond to steel. By applying Floquet boundary conditions to the edges of the square lattice, the first 15 bands have been obtained. As shown in figure 9.5, the fractal phononic crystal exhibits a wide band gap.

Solid material phononic crystals facilitate the propagation of elastic waves comprising both longitudinal and transverse wave components. Conversely, sonic crystals only account for the longitudinal wave component. The band structure has been obtained by applying the same fractal structure to the phononic crystal composed of rubber/titanium solid materials.

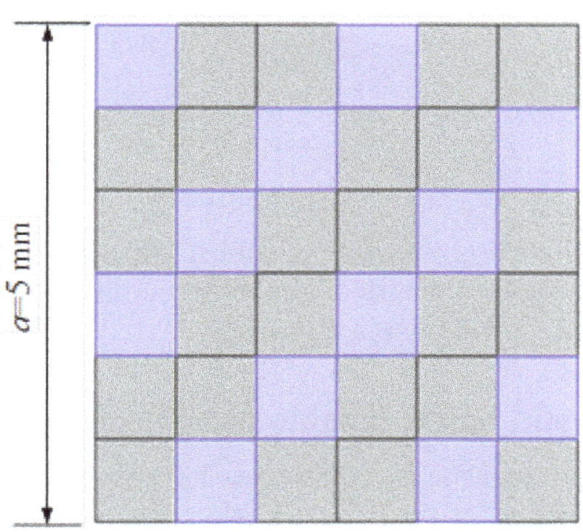

Figure 9.4. Hausdorff fractal structure.

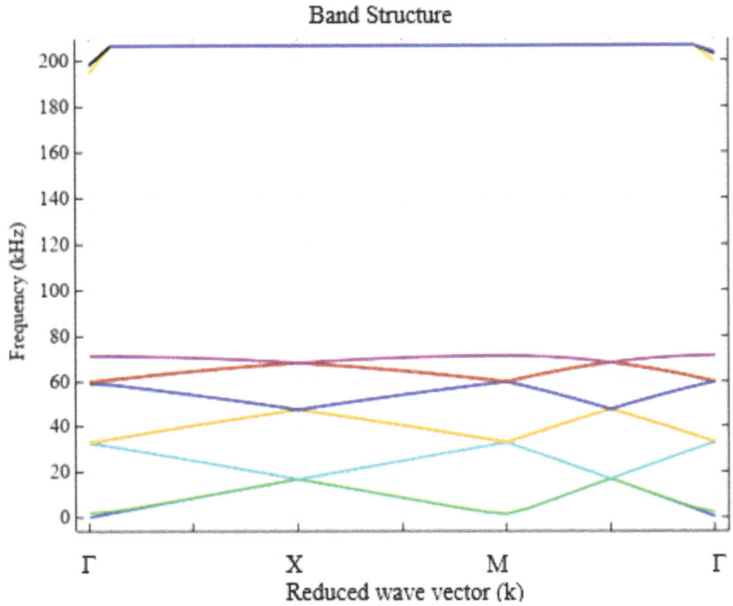

Figure 9.5. Band structure of a Hausdorff fractal PnC.

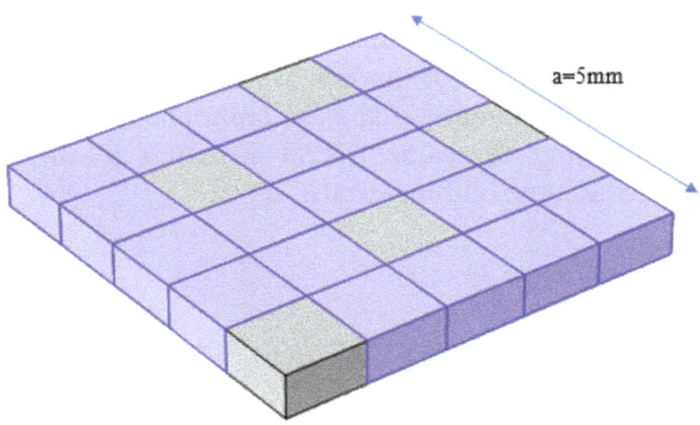

Figure 9.6. 3D Hausdorff fractal unit cell.

In figure 9.6, the blue parts are made of rubber [26] and the other parts are made of titanium [27]. The band structure of rubber/titanium solid PnC and conduction losses of finite structure are shown in figure 9.7.

9.2 Phononic crystal based tunable piezoelectric waveguide

Phononic crystals and acoustic metamaterials have garnered significant attention over the past few decades for their remarkable properties, making them the most prevalent types of acoustic artificial structures. These structures play a crucial role in controlling the propagation of acoustic waves and elastic waves, exhibiting diverse

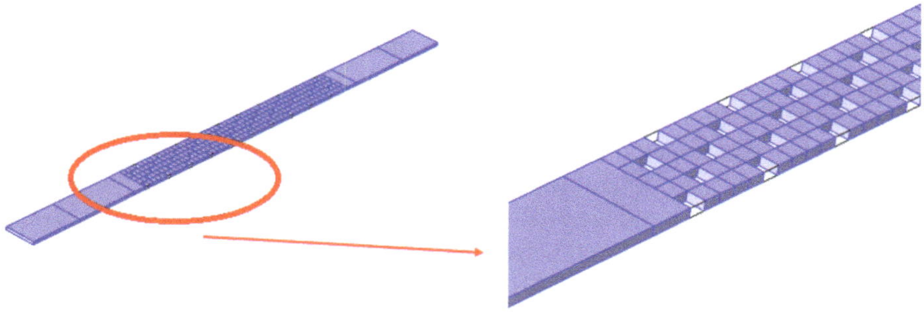

(a)

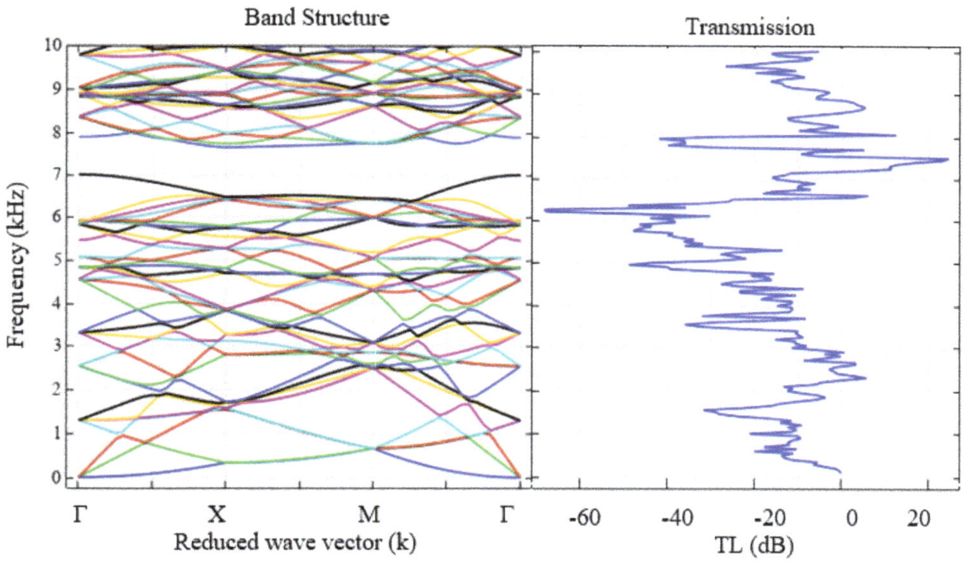

(b)

Figure 9.7. (a) Finite structure consisting of 10-unit cells, (b) band structure and transmission loss of the PnC.

dispersion characteristics such as forbidden propagation within band gaps (BGs) and negative refraction within pass bands (PBs).

In phononic crystals, the occurrence of band gaps is closely tied to the wavelength, typically approaching the size of the unit cell, resulting in what is known as Bragg-scattering BGs. Additionally, in acoustic metamaterials, a distinct type of BGs, referred to as local-resonant BGs, can be achieved through the local resonance of the unit cell, where the wavelength is significantly larger than the size of the unit cell.

Band gaps possess the unique ability to impede wave propagation, making them valuable for applications such as noise control and wave filters. Phononic crystal waveguides enable the transmission of acoustic energy and information within specific frequency ranges, consequently attracting considerable research interest. Typically, phononic crystal waveguides are constructed by modifying the geometry parameters or material components of the phononic crystal scatters. As a result, their properties, including shapes and operational frequency ranges, are fixed once the waveguides are designed.

In this section, we propose a tunable composite waveguide (TCW) based on two-dimensional photonic band gap (PPC) structures. By incorporating piezoelectric rods and external inductor circuits into each unit cell, independent control over the unit cell properties is achieved. This allows for the emergence of a local-resonant pass band within the complete band gap, enabling the TCW to have arbitrary shapes and an adjustable operational frequency range.

Several studies have explored waveguides using defective structures in phononic crystals (PnCs) with forbidden bands [28–32]. Alternatively, a tunable waveguide can be achieved by employing active materials like piezoelectric rods, allowing control of each rod with a separate circuit to shape the desired waveguide. While millimeter-scale studies have been ongoing since the 1990s, recent advancements have led to PnC applications at the micrometer scale in radio frequency (RF) communication and microelectromechanical systems (MEMS). The structural equations for piezoelectric materials are defined in equations (9.6) and (9.7).

$$T = cS - e^{T}.\, E \tag{9.6}$$

$$D = eS + \varepsilon.\, E \tag{9.7}$$

In these equations, T, S, E, and D represent stress, strain, electric field, and electric density, respectively. The parameters c, e, and ε correspond to elastic stiffness, piezoelectric coupling, and electrical permittivity values. When the piezoelectric rods are energized, the coupling effect e in equations (9.6) and (9.7) becomes negligible. In this scenario, the dispersion effect of the PnC varies depending on whether the circuit connected to the piezoelectric rods is open or closed. If the piezoelectric rods are periodically arranged in the x–y plane with polarization in the z-axis direction, the displacement in the out-of-plane (z-axis) direction is influenced by the piezoelectric coupling. When the circuit connected to the piezoelectric rods is open, forbidden bands appear, preventing the wave from propagating. Conversely, when the circuit is closed, the bands disappear, allowing the wave to progress.

The PnC structure consists of a 10×9 array of $LiTaO_3$ piezoelectric square rods embedded in a three-dimensional (3D) epoxy matrix (figure 9.8). The lattice constant is $a = 3$ mm, with square bars having a side length of 2 mm and a height of $0.25 \times a$. The 3D unit cell depicted in figure 9.9 is utilized to obtain the PnC's band structure. By applying Floquet periodic boundary conditions to the unit cell's sides along the x and y axes, the band structure along the Γ-X-M-Γ path in the first Brillouin zone is obtained.

To compare the band structure and transmission losses of the PnC, a finite structure as shown in figure 9.9(b) was created. Perfectly matched layer (PML) boundary conditions were applied to the right and left edges of the model, while symmetry boundary conditions were applied to the lower and upper edges. To induce an elastic

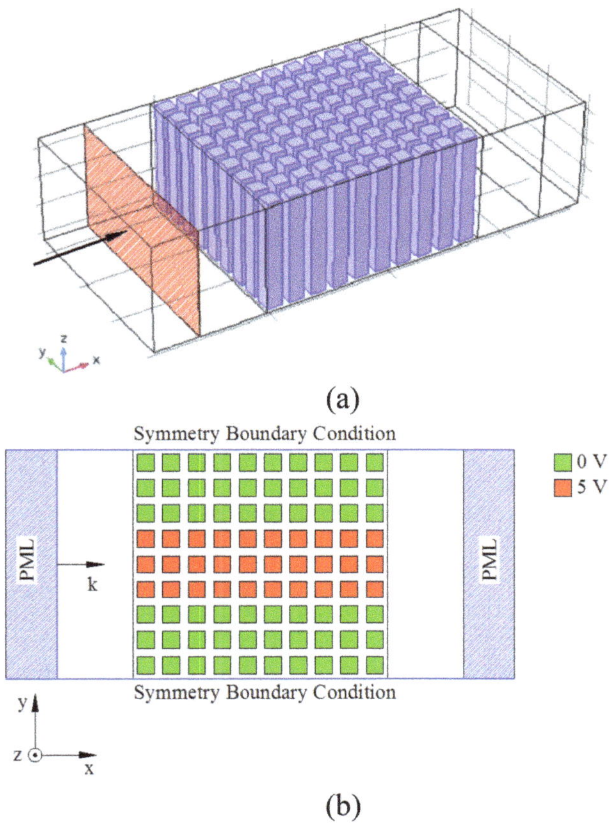

Figure 9.8. (a) 3D phononic crystal in finite structure. (b) View in the x–y plane.

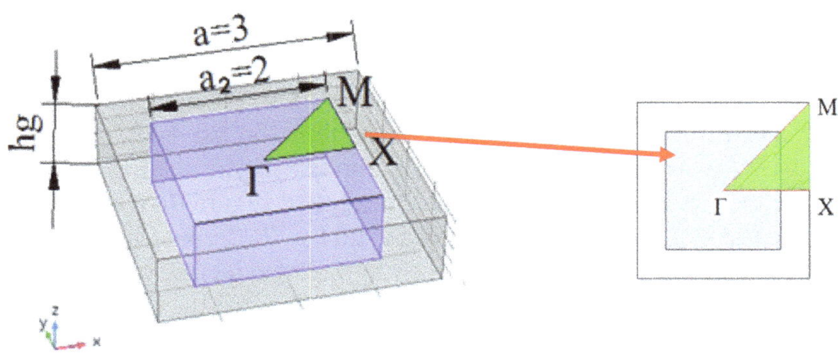

Figure 9.9. 3D unit cell and first Brillouin zone.

wave, parametric scanning was performed across the frequency range of 10 kHz to 400 kHz. A force of 1 N mm^{-2} was applied in the x direction and -1 N mm^{-2} in the z direction to the shaded area (figure 9.9(a)). The base area of all bars in figure 9.9 was set as the reference ground (0 V). The upper area of the green-colored bars was also set

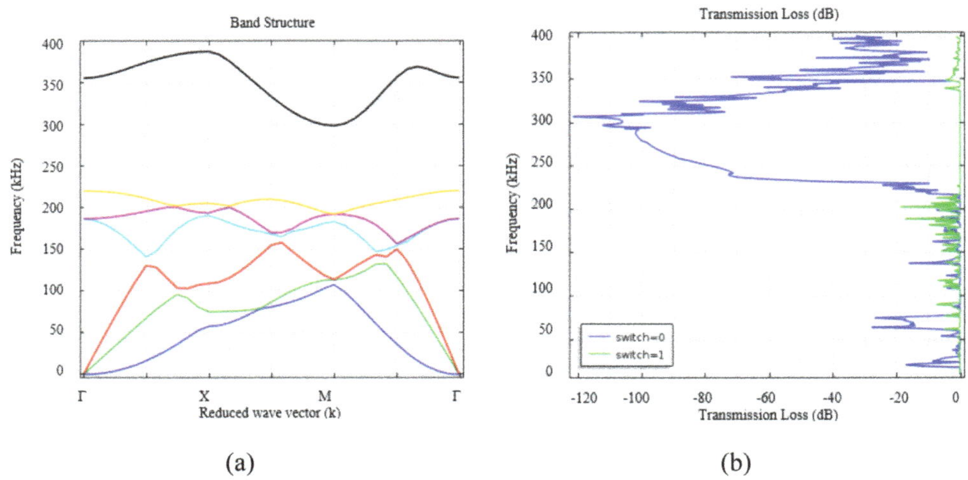

Figure 9.10. (a) Band structure and (b) transmission loss.

to 0 V, while the upper area of the bars (highlighted in red) along the desired waveguide path was initially set to 0 V and then increased to 5 V.

Figure 9.10 depicts the band structure (a) and the conduction losses (b) of the finite structure PnC. In the band structure graph, it is evident that a complete band exists within the frequency range of 225–300 kHz, enabling the PnC waveguide to operate effectively. The transmission losses of the adjustable waveguide are presented in figure 9.10(b). The blue graph represents the conduction loss when 0 V is applied to the bars, while the green graph corresponds to a voltage of 5 V applied to the same bars. The figure clearly demonstrates that the structure prohibits wave propagation when a voltage of 0 V is applied, whereas the PnC functions as a waveguide and allows wave transmission when a voltage of 5 V is applied to the same bars. In figure 9.11(a) and (b), the propagation of a wave with a frequency of 279 kHz within the band gap is observed in the finite PnC structure. It is evident from the figure that the piezoelectric rods within the waveguide are unable to transmit waves when a voltage of 0 V is applied (figure 9.11(a), whereas wave propagation occurs when a voltage of 5 V is applied (figure 9.11(b)).

The outcomes of this study demonstrate that a PnC constructed using piezoelectric rods has the capability to generate waveguides of various shapes. To transform the PnC into an adjustable waveguide, electrical current is supplied to the circuits connected to the piezoelectric rods along the desired wave path. By closing these circuits, the PnC effectively operates as a waveguide, taking advantage of its forbidden band within the specific frequency range. The periodic arrangement of piezoelectric rods connected to individual circuits allows for easy independent control of each rod, enabling precise wave manipulation. This guiding approach opens up possibilities for the development of integrated elastic waveguide devices with enhanced functionality.

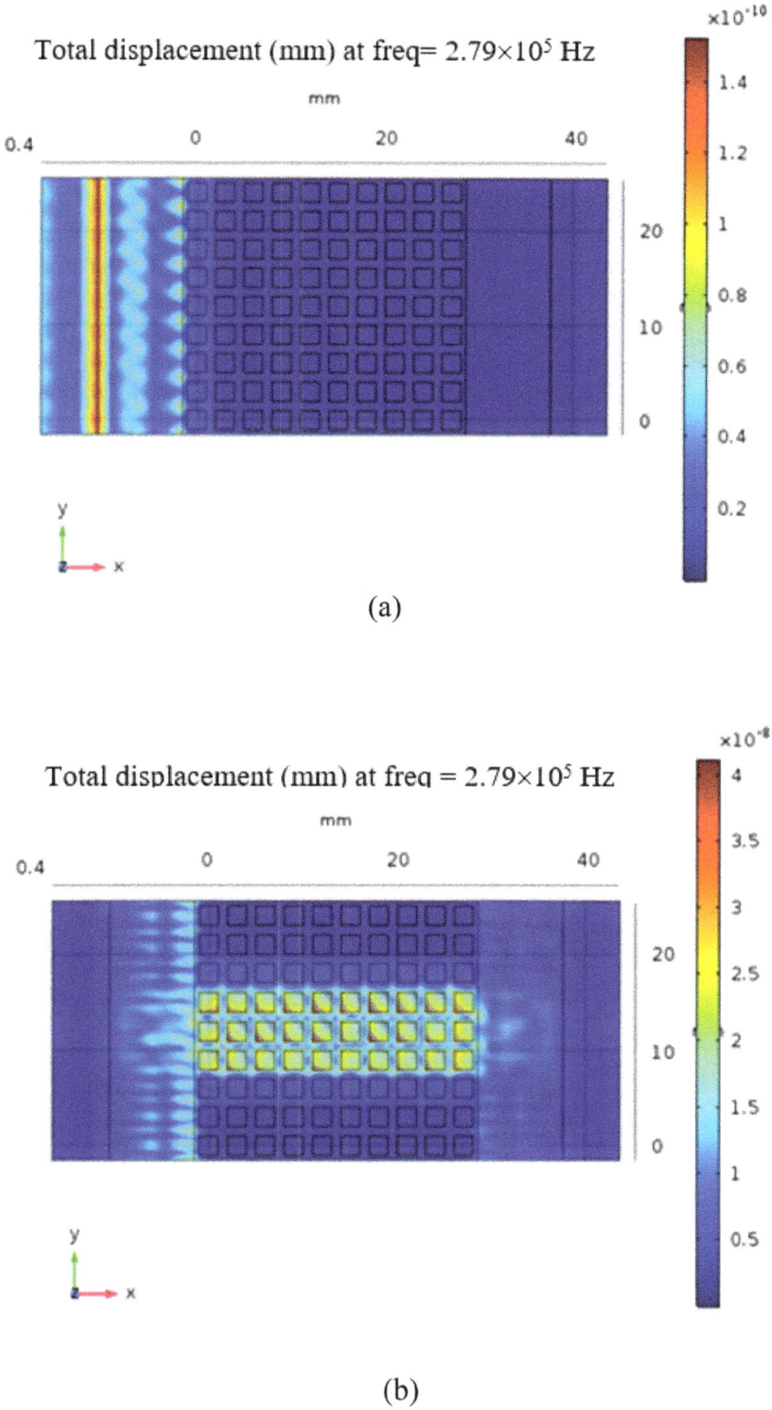

Figure 9.11. Progress of the 279 kHz wave when (a) 0 V and (b) 5 V are applied.

9.3 Second harmonic generation in acoustic metamaterials

Acoustic waves, like other types of waves, have the ability to interact with each other, and result in the production of new waves with frequencies different from the original waves. This phenomenon is known as nonlinear acoustic wave propagation. Specifically, when two acoustic waves with different frequencies interact within a material, they can generate a new wave with twice the frequency of the original waves. This process is referred to as acoustic second harmonic generation (SHG). Acoustic SHG serves as a valuable tool for examining the nonlinear properties of materials. By measuring the strength of the generated second harmonic signal and observing how it varies with the amplitude and frequency of the original waves, researchers can gain valuable insights into the material's nonlinear response. Additionally, acoustic SHG can be employed to create high-resolution images of materials since the second harmonic signal is only generated in the region where the two original waves overlap.

Acoustic SHG finds numerous potential applications in various fields such as nondestructive testing, material characterization, and biomedical imaging. It is important to note, however, that compared to other nonlinear phenomena like optical SHG, acoustic SHG is relatively weak. Consequently, accurately measuring the generated second harmonic signal typically necessitates the use of high-power acoustic sources and sensitive detection systems.

In this section, we present results obtained from numerical simulations carried out using the FEM. All simulations are performed using the Murnaghan model (equation 9.8) for aluminum, with elastic constants listed in table 9.1. Figure 9.12 depicts the 2D model used in the simulation. The plate's thickness is set to 1 mm, and the plate's length is assumed to be 100 mm unless otherwise noted.

Table 9.1. Murnaghan model parameters of the model.

λ	μ	l	m	n
51	26	−250	−333	−350

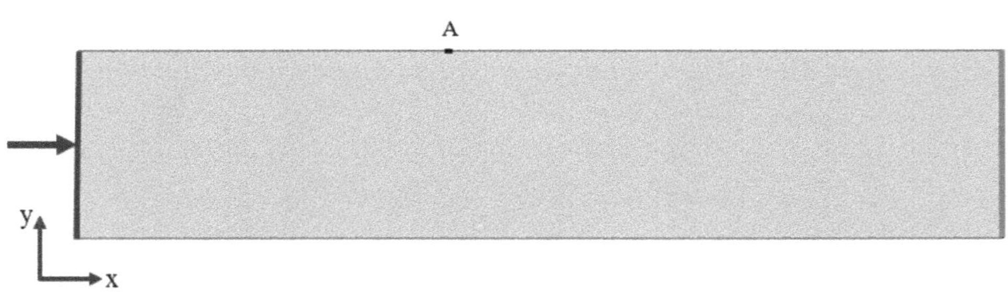

Figure 9.12. The model used for the simulation.

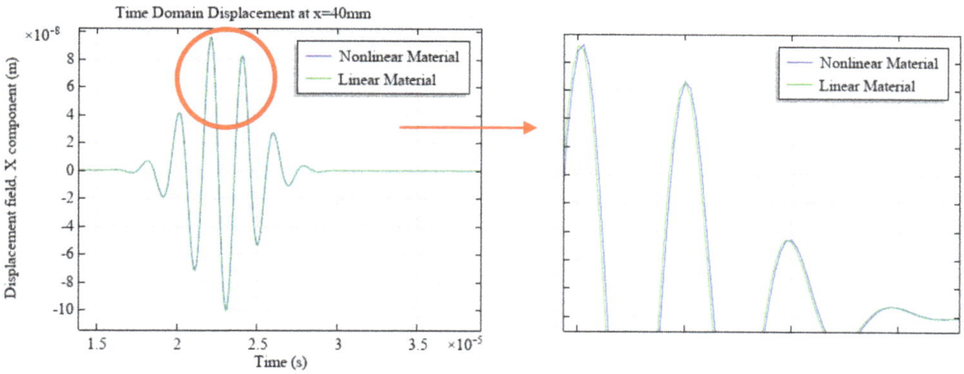

Figure 9.13. Excitation signal.

$$W(E) = \frac{1}{2}\lambda(\mathrm{tr}(E))^2 + \mu\mathrm{tr}(E^2) + \frac{1}{3}(l + 2m)(tr(E))^3 - \mathrm{mtr}(E) \times (\mathrm{tr}(E))^2 - \mathrm{tr}(E^2) + \mathrm{ndet}(E) \tag{9.8}$$

Here, l, m, and n are Murnaghan model parameters, det($\boldsymbol{E}$) and tr($\boldsymbol{E}$) show the determinant and the trace of the tensor, respectively.

The diagram of the 2D model used in the simulation is depicted in figure 9.23. The plate's thickness is selected to be 1 mm, and the plate's length is presumed to be 100 mm unless stated otherwise. The prescribed displacement excitation signal with a maximum amplitude of 1×10^{-7} m in figure 9.13 is applied to the left edge to create an S_0 wave in the aluminum sheet. Three control observation points are added to x = 40 mm (point A) from the left edge.

The material is firstly defined as nonlinear and then linear, and time-dependent analysis is performed with increments of 0.1 µs. The displacement of the point at x = 40 mm distance in the x direction in the time domain, depending on whether the material is nonlinear or linear, is shown in figure 9.13.

At observation point A, as seen in figure 9.13, there is not a significant difference between the displacements in the time domain in the nonlinear material and the linear material. Figure 9.14 shows the displacements in the time domain converted to the frequency domain by fast Fourier transform (FFT).

However, in the frequency domain, nonlinearity becomes apparent because of the existence of static and second harmonic components in the displacement, as demonstrated in the FFT depicted in figure 9.14. It is important to mention that a brief tone burst results in a fairly broad frequency range. While the linear scenario does not display any second harmonic (or static) component, as expected, the other two scenarios exhibit significant second harmonic (and static) components. The signals are received at the left end of the plate, and their respective FFTs, presented on a logarithmic scale, are illustrated in figure 9.14.

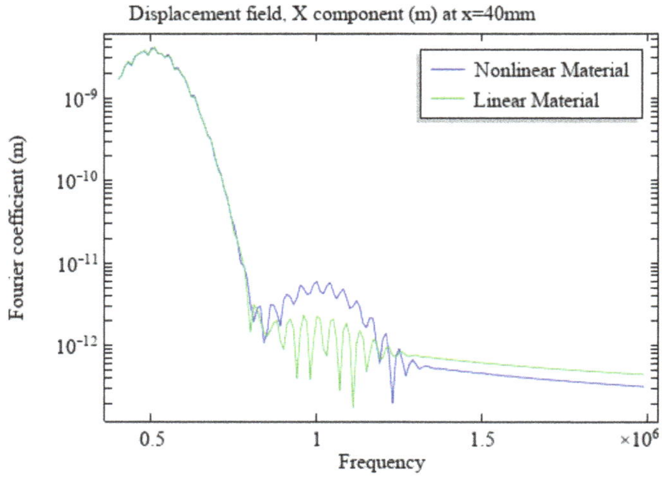

Figure 9.14. Frequency domain transformation of displacements in nonlinear and linear materials at $S_0 = 0.5$ MHz.

The FFT analysis in figure 9.14 reveals the following results:

- In the frequency domain, nonlinearity becomes evident, showing the presence of static and second harmonic components in the displacement.
- In the linear scenario, no second harmonic (or static) component is observed, confirming expectations.
- However, in the nonlinear scenarios, significant second harmonic (and static) components are present.

The obtained results demonstrate the effectiveness and potential of acoustic SHG imaging:

- The nonlinear response observed in the frequency domain provides valuable insights into the material's behavior.
- Acoustic SHG holds promise for a variety of applications, such as non-destructive testing, material characterization, and biomedical imaging.
- The ability to generate high-resolution images based on the overlapping of original waves highlights the usefulness of acoustic SHG.

The findings validate its value as a tool for investigating material behavior and imaging techniques.

9.4 Acoustic subwavelength structures

The term "subwavelength" refers to sizes or distances that are smaller than the wavelength of a given wave. In physics, wavelength represents the distance between successive crests or troughs of a wave. For example, in electromagnetic waves like light, wavelength is typically measured in nanometers (nm) or micrometers (μm). In various scientific fields, subwavelength structures are utilized to interact with waves on a scale smaller than their wavelength. In nanophotonics, for instance, these structures manipulate light, impacting its intensity, polarization, and propagation direction. Subwavelength structures also find applications in nanoelectronics, metamaterials, and plasmonics.

In acoustics, wavelength refers to the distance between consecutive points of maximum displacement in a sound wave. Acoustic subwavelength structures are objects or structures that are smaller than the wavelength of the sound wave propagating through a medium. When sound waves encounter a subwavelength-sized obstacle, they scatter and diffract around the object, resulting in the formation of an acoustic shadow. This principle is applied in noise barriers designed to effectively reduce sound transmission by utilizing subwavelength structures.

Sonic crystals, typically composed of periodic arrays of cylindrical rods made of high mechanical impedance materials, exhibit high attenuation at specific but narrow frequency bands due to multiple scattering phenomena. Sonic crystal barriers allow the passage of light and do not obstruct the flow of air, making them distinct from traditional solid sound barriers.

However, for reducing low-frequency noise like traffic noise, it is important to consider the relationship between the operating frequency and lattice parameter. Extremely large barriers would be required, making locally resonant sonic materials (LRSMs) more suitable as they can form decoupled band gaps that cover a narrow range of attenuation. Despite this limitation, LRSM is not suitable to use as a noise barrier. The proposed sonic crystal structure is specifically designed to generate broad attenuation bands in the lower frequency range. It comprises a concentric symmetric complementary split rectangular resonator (CSRR).

In a previous study, a sonic crystal lens was developed using an array of two-dimensional Helmholtz resonators [33]. This lens demonstrated a high relative acoustic refractive index (n) and a small acoustic impedance (Z) mismatch with the surrounding air for airborne sound in the long-wave regime. The study also investigated wave propagation in a sonic crystal featuring a Helmholtz resonator defect, leading to localized resonance phenomena [34]. Acoustic metamaterials with similar characteristics have displayed sound confinement and focusing capabilities [35].

In this study, our objective is to examine an arrangement of resonant elements that exhibit wide resonances below the Bragg band gaps. Specifically, we focus on the inclusion of complementary split rectangular resonators (CSRRs) with various configurations as the chosen elements. The interaction of these resonances leads to the formation of band gaps and acoustic attenuation. To investigate the proposed systems, we employ numerical methods, particularly the FEM. Analytical techniques for calculating the band structure and transmission become impractical for phononic crystals with complex geometries. Expressing intricate geometries using elementary functions poses challenges, whereas FEM allows for easy representation of unusual geometries and discretization of the continuous problem [36, 37].

9.4.1 FEM model of resonant arrays for numerical analysis

To initiate our analysis, we begin by examining the band structures of infinite arrays of solid cylinders through the utilization of FEM. For our calculations, we adopt the

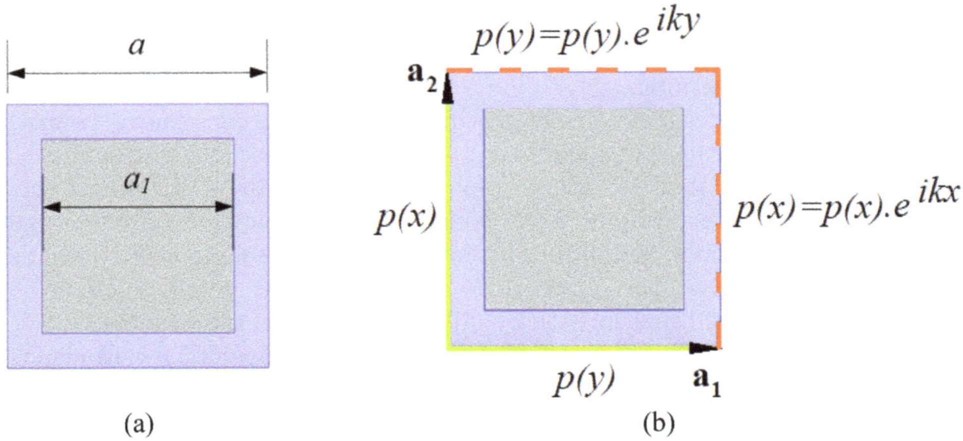

Figure 9.15. Size of a unit cell of the sonic crystal and applied boundary conditions.

unit cell of a two-dimensional square array (depicted in figure 9.15) as the fundamental basis. The structure is considered to be infinite and exhibits periodicity in the x direction with a period of a_1 and in the y direction with a period of a_2. The structure is described by two basis vectors: $(a_1, 0)$ and $(0, a_2)$.

According to the Floquet–Bloch theorem, the pressure distribution p for nodes located on the boundary of the unit cell can be expressed as:

$$p(\vec{x} + \vec{a_1} + \vec{a_2}) = p(x)\exp\left[i\left(k_x a_1 + k_y a_2\right)\right] \tag{9.9}$$

In this equation, x represents the position vector within the unit cell, and $\vec{k} = k_x + k_y$ represents the Bloch wave vector. Initially, we apply Neumann-type boundary conditions, which are necessary on boundaries where the pressure p is governed by a periodic boundary condition. Subsequently, a phase relation is imposed at the unit cell boundary to establish boundary conditions between adjacent units. This phase relation is associated with the wave number of the incident wave in the periodic structure.

Periodic boundary conditions are applied in both the x and y directions to ensure the continuity of pressure components at the edges of the computational domain. This means that pressure values at the edges are shifted to the corresponding edges on the opposite side of the domain. For example, a wave penetrating from the top edge of the domain will reappear outside the domain through the bottom periodic boundary condition.

The same applies to waves approaching the left, right, and bottom edges of the domain. Within the unit cell, the square is assumed to be rigid, and therefore a Neumann boundary condition is imposed on its surface. The Bloch wave vector is defined within the first Brillouin zone, where for the Γ-X direction, k_x ranges from 0 to π while k_y is fixed at 0; for the Γ-M direction, k_y ranges from 0 to π while k_x is fixed at π; and for the X-M direction, both k_x and k_y vary from 0 to π.

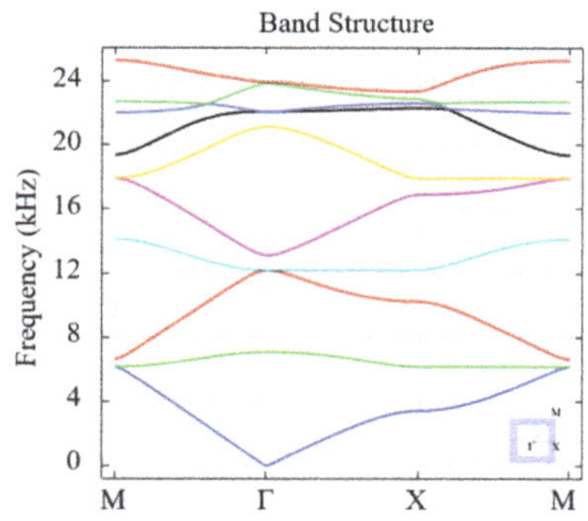

Figure 9.16. Band structure for a sonic crystal consisting of steel scatterers embedded in the air (a_1 = 22 mm, a = 30 mm).

In our computations, we analyze the first 10 eigenfrequencies and their corresponding eigenvectors, which represent the pressure distribution of each mode. The infinite sonic crystal is constructed using square inclusions in air, with a lattice parameter of a = 30 mm and a steel scatterer edge length of a_1 = 22 mm. The characteristic band structure of this system is depicted in figure 9.16, showing the plotted results in the three principal symmetry directions.

At lower frequencies, the system displays isotropic dispersion characterized by a linear relationship between phase velocity (c) and wave number (k), given by $c = \omega/k$. In this regime, the propagating wave cannot distinguish the intricate details of the cylinders due to the longer wavelengths involved. In the Γ-X direction, a sonic band gap appears between the first two bands. As the Brillouin zone approaches the boundaries, the dispersion relation deviates from linearity, resulting in a curved band with zero group velocity observed at the band edges.

9.4.2 Transmission analysis

The transmission analysis involves the finite element method to calculate the pressure field behind a sonic crystal and generate a pressure map of the system at specific frequencies. The equation used to analyze acoustic wave problems is expressed as equation (9.10), which can be simplified to the Helmholtz equation for a time-harmonic pressure wave excitation, $p = p_0.e^{i\omega t}$, where $\omega = 2\pi f$ is the angular frequency. By solving equation (9.10), we can obtain the pressure field.

$$\frac{1}{\rho_0 \partial t^2}\frac{\partial^2 t}{\partial t^2} + \nabla.\left(-\frac{1}{\rho_0}\partial p\right) \tag{9.10}$$

In the simulations, we consider a two-dimensional sonic crystal system with a 10 × 10 square lattice described using the finite element method. The system has a lattice parameter of a = 30 mm and square inclusions with an edge length of a_1 = 22 mm. For the rigid squares in the sonic crystal, we apply sound-hard boundary conditions, which set the normal component of the air particle velocity to zero at the walls of the square inclusions.

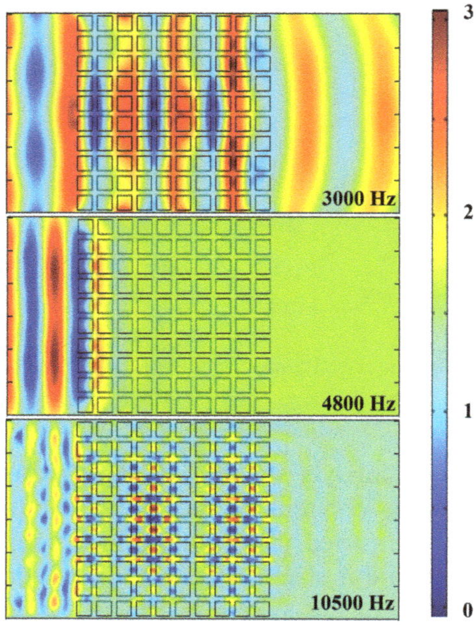

Figure 9.17. The obtained pressure maps of solid steel squares in the air at three frequencies by FEM (3000, 4800, and 10 500 Hz).

The pressure maps shown in figure 9.17 illustrate the distinct frequency ranges and their effects on the behavior of the sonic crystal system. These maps are captured at three specific frequencies: 4000 Hz (prior to the formation of the Bragg band gap), 8000 Hz (at the center of the band gap), and 12 000 Hz (following the formation of the Bragg band gap).

At 3000 Hz, the pressure map demonstrates that the sonic crystal system behaves as a homogeneous material for low frequencies. The periodic structure has minimal impact on the propagation of acoustic waves since the lattice parameter is significantly smaller than the corresponding wavelength. Moving to the pressure map at 4800 Hz, we observe the emergence of a band gap. At this frequency, the wavelength of the incident acoustic wave aligns with the lattice parameter, resulting in significant attenuation caused by multiple scattering effects. This leads to the formation of a shadow zone behind the sonic crystal.

At 10 500 Hz, after the band gap formation, the wavelength of the acoustic wave becomes smaller than the lattice parameter of the sonic crystal system. As a result, the wave can freely propagate through the sonic crystal since the plane wave is unable to resolve the individual scatterers.

These findings illustrate how frequency effects the behavior of the sonic crystal system, emphasizing the formation and effects of band gaps within the periodic structure. By conducting a parametric sweep of frequency and solving for various values, a frequency spectrum can be generated to reveal the attenuation properties of the sonic crystal. Figure 9.18 provides a comparison between the transmission

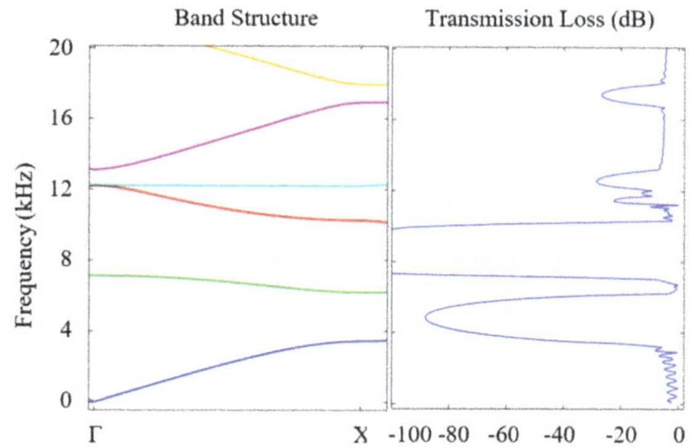

Figure 9.18. An evaluation of the FEM band structure in the Γ-X direction is compared to the finite element computed frequency spectra for sound pressure level in the case of a conventional sonic crystal.

spectrum spanning from 0 to 20.1 kHz and the computed band structure. It is important to note that the study focuses on the Γ-X direction.

The finite element transmission calculations show wider band gaps than those predicted by the band structure analysis. This difference can be attributed to the finite number of scatterers used in the transmission simulations and the diffraction effects occurring around the edges of the sonic crystal structure. However, it is worth highlighting that the overall positions of the band gaps align well with those obtained from the band structure calculations. This comparison enhances our understanding of the transmission properties of the sonic crystal and demonstrates the agreement between the calculated band structure and the experimental transmission spectrum.

9.4.3 Complementary split rectangular resonator (CSRR) locally resonant sonic crystal

The conventional sonic crystal, which incorporates solid scattering inclusions, produces band gaps solely based on periodicity, in line with theory and our experimental results [38]. To operate below this Bragg gap, we now explore the design of a locally resonant sonic crystal (LRSC) utilizing an array of complementary split rectangular resonators (CSRRs). One advantage of using the FEM to compute the acoustic band structure is its capability to model more intricate scatterer geometries. Similar to the previous modeling of the conventional sonic crystal system, periodic boundary conditions have been implemented, as shown in figure 9.18.

In figure 9.19(a), the lattice periodicity is specified as follows: a = 30 mm, a_1 = 22 mm, a_2 = 18 mm, a_3 = 14 mm, a_4 = 10 mm, and w = 2 mm. By varying the wave vector within the first Brillouin zone for the first 10 eigenvalues, the band structure can be constructed, as depicted in figure 9.19(b). This figure illustrates the computed band structure of a two-dimensional sonic crystal composed of slotted squares, as shown in figure 9.20. These squares are arranged in a square lattice within air, with each resonating inclusion referred to as a CSRR resonator.

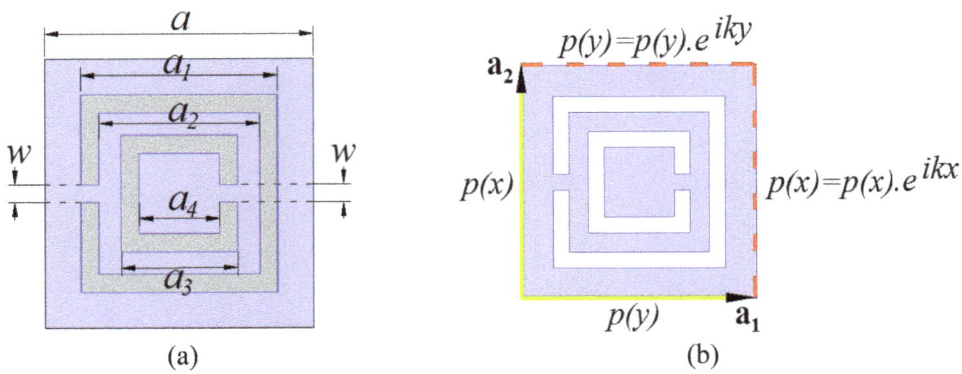

Figure 9.19. (a) Size of the unit cell for a CSRR-shaped locally resonant (b) applied boundary conditions.

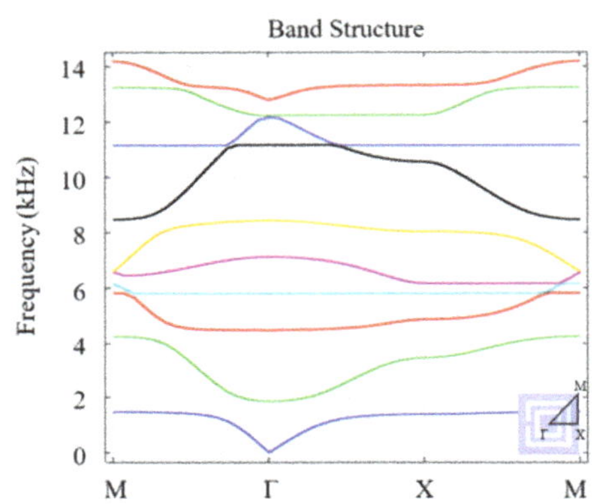

Figure 9.20. The computed band structure of a two-dimensional sonic crystal consisting of a CSRR.

We can observe the presence of a flat band in the band structure, as depicted in figure 9.20. Modes associated with a flat band exhibit a group velocity of zero and display strong spatial localization. In practical applications, such localized modes are often achieved by introducing defects or cavities within the periodic structures. In the case of a sonic crystal, the acoustic resonance originating from the CSRR inclusions gives rise to the emergence of this flat band. This acoustic resonance contributes significantly to the formation of a complete acoustic band gap, primarily induced by the local acoustic resonance of each individual scatterer.

The combined effect of these resonators leads to the formation of a degenerate state, resulting in a symmetrical band gap centered at 1400 Hz, covering the frequency range of 1308–1492 Hz.

The CSRR inclusions within the structure still give rise to Bragg band gaps, particularly in the Γ-X direction. The first Bragg band gap covers the frequency range of 3345–4416 Hz. Additionally, six more Bragg band gaps occur as the Bragg condition is met, located at 4816–5616 Hz, 5616–6176 Hz, 7075–8078 Hz, 8502–10 468 Hz, 7075–8078 Hz, and 12 256–12 5880 Hz. The presence of an additional flat resonance band offers the potential for constructing effective low-frequency acoustic barriers, providing sound attenuation across all crystal lattice planes. This flat band, resulting from the localized acoustic resonance observed in the band structure, forms a significant anti-crossing gap known as a hybridization gap in the context of sonic crystals.

To analyze the transmission properties of the CSRR inclusion LRSC, finite element (FE) simulations are conducted. The same boundary conditions used in the previous investigation of the conventional sonic crystal are applied to this new system. Essentially, this system replicates the conventional sonic crystal described earlier but includes slots to create resonant cavities.

Pressure maps are computed at four specific frequencies to illustrate the propagation of an acoustic plane wave through the CSRR inclusion LRSC. Similar to the conventional sonic crystal, below the active frequency (870 Hz), the incident wave propagates as if the system is a homogeneous medium. At 1410 Hz, the pressure map demonstrates the attenuation of the wave within this frequency range by the CSRR inclusion LRSC. Figure 9.21 highlights localized regions of maximum pressure corresponding to the resonance frequency of the inclusions. Over the resonance frequency, around 3200 Hz, the acoustic wave can freely propagate

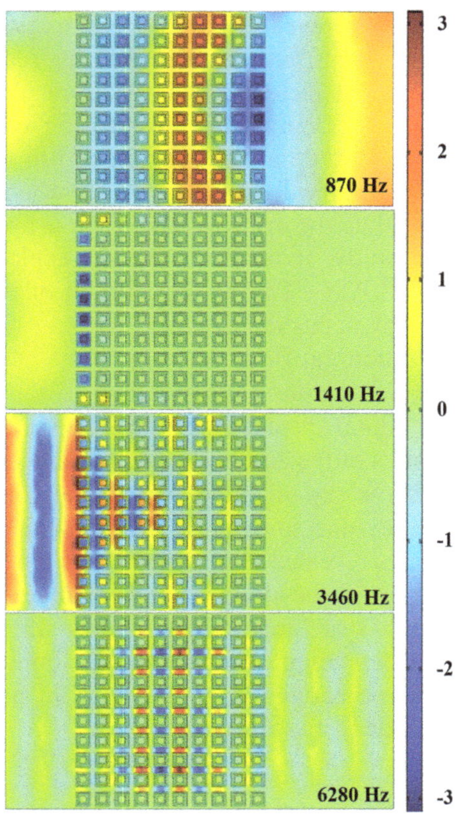

Figure 9.21. Computed pressure level maps for a CSRR locally resonant sonic crystal at 870, 1410, 3460, and 6280 Hz.

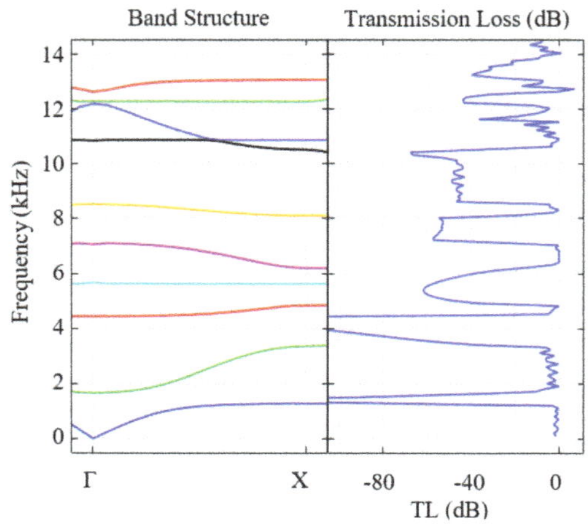

Figure 9.22. Computed band structure and transmission loss of the CSRR locally resonant sonic crystal.

through the system. As we approach the frequency of the Bragg band gap, approximately 6280 Hz, we observe the onset of attenuation within the band gap. Examining the pressure maps reveals that the two regions of attenuation are governed by different mechanisms: resonance and Bragg scattering.

The transmission spectrum, depicted in figure 9.22, displays additional peaks of attenuation around 12.4 kHz. Comparing the location of these peaks with the computed band structure suggests that they correspond to regions of anti-crossing present in the band structure. Typically, such gaps arise when two bands of the same symmetry avoid crossing. While the investigation of these anti-crossing regions falls outside the scope of this study, it warrants further exploration to enhance the performance of the CSRR LRSC. The physical origin of these anti-crossing gaps differs from those induced by acoustic resonance.

The flat band (hybridization gap) originates from the acoustic resonance of the CSRR scatterer, whereas the regions corresponding to the anti-crossing gaps are formed due to the longitudinal displacement field in the homogeneous effective medium [39]. The narrowness of the anti-crossing gaps suggests their weaker nature.

The proposed CSRR sonic crystal offers a practical solution to overcome the inherent spacing dependence encountered in conventional sonic crystal designs. Extensive investigations have revealed that such systems have the ability to generate multiple resonance band gaps in the lower frequency range, below the formation of Bragg gaps. Subwavelength structures are also being utilized to create acoustic metamaterials, which possess unique properties not found in natural materials. These metamaterials hold potential for applications in acoustic imaging, noise control, and other areas of acoustics research.

9.5 Acoustic Weyl point materials

Topological semimetals are materials that exhibit band structures with topologically nontrivial touching points. These materials can support quasiparticle excitations resembling Dirac or Weyl fermions [40–44]. The presence of such quasiparticles, such as Weyl/Dirac fermions, in Weyl/Dirac semimetals has attracted significant attention [45–47]. In Weyl semimetals, the conduction and valence bands intersect linearly, resulting in twofold-degenerate Weyl points characterized by divergent Berry curvature and quantized topological charges. These unique properties give rise to a variety of extraordinary phenomena, including the chiral magnetic effect [48], quantized circulation of anomalous shift [49], and other notable optical and magnetic responses [50, 51].

These Weyl points, or similar features, are not limited to electronic systems but can also be engineered in artificially created periodic structures using classical waves, such as electromagnetic waves in photonic crystals [52–55] and acoustic waves in phononic crystals [56, 57]. However, achieving topological acoustic materials poses challenges due to the absence of spin and the difficulty in breaking time-reversal symmetry for sound, unlike in electronic or optical systems.

An acoustic Weyl point metamaterial refers to an engineered material that enables topologically protected sound propagation in three dimensions. This type of material exhibits symmetry-breaking structures that confine sound waves to propagate along specific directions and frequencies. The Weyl points in the metamaterial are associated with topologically protected surface states, allowing sound waves to propagate around obstacles and defects without scattering.

One potential application of acoustic Weyl point devices in underwater communication is the development of novel underwater acoustic transducers. These transducers could be designed to emit sound waves with specific frequencies and directions, facilitating long-distance information transmission in the ocean.

Another potential application of acoustic Weyl point devices is underwater sensing, specifically in the development of new types of underwater acoustic sensors. These sensors could be designed to detect sound waves with specific frequencies and directions, enabling the monitoring of underwater environments for changes in temperature, pressure, and other physical parameters. By harnessing the topological protection provided by Weyl points, these sensors could operate reliably even in the presence of oceanic noise and other environmental disturbances.

The development of acoustic Weyl point devices for underwater applications would generally involve a combination of theoretical modeling, material design, and experimental testing. The devices would need to be optimized for the specific underwater environment in which they will be deployed, considering factors such as water temperature, pressure, salinity, and turbulence. Successful development of acoustic Weyl point devices for underwater applications could lead to significant advancements in underwater communication and sensing, with potential applications in fields like oceanography, marine biology, and defense.

In solids, rich crystal symmetries give rise to Weyl points with nonlinear dispersion and larger topological charges, such as charge-2 Weyl points ($C = \pm 2$), charge-3 Weyl points ($C = \pm 3$), and charge-4 Weyl points ($C = \pm 4$) [58, 59]. These unconventional Weyl points can exhibit even more novel phenomena due to their higher topological charge [60, 61]. Moreover, the nonlinear energy dispersions associated with these points can result in remarkable non-Fermi-liquid behaviors

[62, 63]. Aside from classifying Weyl points based on topological charge, they can also be categorized based on the geometry of their Fermi surfaces, leading to the distinction between type-I and type-II Weyl points [64].

Type-I Weyl points are characterized by point-like Fermi surfaces, whereas type-II Weyl points exhibit electron-hole pocket Fermi surfaces due to linearly tilted bands. However, recent research has revealed the possibility of a Fermi surface consisting of two intersecting electron (or hole) pockets for unconventional Weyl points with nonlinear dispersion, known as type-III Weyl points [65]. Type-III Weyl points can generate this Fermi surface configuration due to quadratically tilted bands. It's important to note that the type-III Fermi surface is not observed in regular charge-1 Weyl points but is only found in Weyl points with higher topological charge.

In the literature, it has been shown that for type-I Weyl points, the slopes of the two bands have opposite signs along any k direction, whereas for type-II Weyl points, the slopes can have the same sign along certain k directions. Type-III Weyl points can be observed in charge-2 Weyl systems, which exhibit a parabolic band dispersion with the same opening direction along a specific k direction. The low-energy behaviors of solid-state systems are typically governed by the geometry of the Fermi surface; thus type-II and type-III Weyl points can exhibit various distinctive physical properties, including magnetic breakdown [66], Landau level collapse [67–69], and other typical magneto responses [68].

Type-I and type-II conventional charge-1 Weyl points, as well as type-I charge-2 Weyl points, have been reported in electronic systems [70–74] and artificially created periodic structures such as photonic crystals [75–77], sonic crystals [78–83], and electrical circuits [84, 85].

The existence of type-III Weyl points has only been confirmed in a single electronic material to date, but it was anticipated that these points would exhibit a gap due to the wave effect of charge density at low temperatures [65]. Therefore, it is crucial to identify ideal platforms that can host type-III Weyl points. In this context, the term "ideal" refers to a system that contains only a single type-III Weyl point in the Brillouin zone (BZ) and where the two bands forming the Weyl points are isolated and well-separated from other bands. This ensures that the signal solely originates from the type-III Weyl points without interference from other band transitions or insignificant bands.

In our study, we present a comprehensive account of our experimental findings on the state of ideal type-III Weyl points in a 3D sonic crystal. Specifically, we have observed a single pair of type-III Weyl points formed by the lowest two bands, and these two bands are distinctly separated from the other bands.

To illustrate the origins of this concept, we have developed a two-band tight-binding (TB) model in a tetragonal lattice.

9.5.1 Design of a phononic crystal with type-III Weyl points

To achieve type-III Weyl points in sonic systems, it is essential to create a spinless crystal with unconventional charge-n ($n > 1$) Weyl points, where the band dispersion of these points is quadratically tilted. In a previous study [59], it was demonstrated that a charge-2 Weyl point can emerge in space group (SG) No. 89 as a crucial spinless excitation located at one of the four high symmetry points (Γ, Z, M, A). Consequently, we construct a lattice model to explore the potential existence of type-III Weyl points in SG 89.

We present a two-band tight-binding model that adheres to the symmetries of space group 89:

$$H = h_x\sigma_x + h_y\sigma_y + h_z\sigma_z + h_0\sigma_0 \tag{9.11}$$

Here, σ_0 represents the 2 × 2 identity matrix, and σ_i (with $i = x, y, z$) represents the Pauli matrices.

$$h_x = 4\cos\frac{k_x}{2}\cos\frac{k_y}{2}(t_1 + t_2\cos k_z),$$

$$h_y = 4t_2\sin\frac{k_x}{2}.\ \sin\frac{k_y}{2}\sin\quad k_z),$$

$$h_z = 2(t_3 - t_4)(\cos k_x - \cos k_y),$$

$$h_0 = 2(t_3 + t_4)\left(\cos\quad k_x + \cos\quad k_y\right)$$

Using real parameters for t in this context, it becomes evident that the two bands will exhibit degeneracy at the M (π; π; 0) and A (π; π; π) points, resulting in the formation of two Weyl points. To investigate the low-energy characteristics of these Weyl points, we expand the tight-binding Hamiltonian around the M and A points. The corresponding effective Hamiltonians, to the leading order, can be expressed as follows:

$$H_M = (t_1 + t_2)\, k_xk_y\sigma_x + 4\, t_2k_z\sigma_y + (t_3 - t_4)\left(k_x^2 - k_y^2\right)\sigma_z + (t_3 + t_4)(k_x^2 + k_y^2)\sigma_0 \tag{9.12}$$

for M point and

$$H_A = (t_1 - t_2)\, k_xk_y\sigma_x + 4\, t_2k_z\sigma_y + (t_3 - t_4)\left(k_x^2 - k_y^2\right)\sigma_z + (t_3 + t_4)(k_x^2 + k_y^2)\sigma_0 \tag{9.13}$$

for the A point. From equations (9.12) and (9.13), it can be observed that the Weyl points at the M and A points exhibit characteristics of charge-2 Weyl points. These points demonstrate linear dispersion along the k_z direction and quadratic dispersion in the k_x–k_y plane, dictated by the two-dimensional irreducible representation of the D4 point group [59].

Importantly, the presence of the σ_0 matrix in the last term of equations (9.12–13) indicates the spectrum's tilt in the k_x–k_y plane, which plays a crucial role in achieving type-III Weyl points [65]. The Weyl points at both the M and A points transform into type-III when the condition $|t_3 + t_4| > |t_3 – t_4|$ is satisfied. In such cases, both bands (parabolas) along the k_x and k_y axes open in the same direction, determined solely by the sign of $t_3 + t_4$.

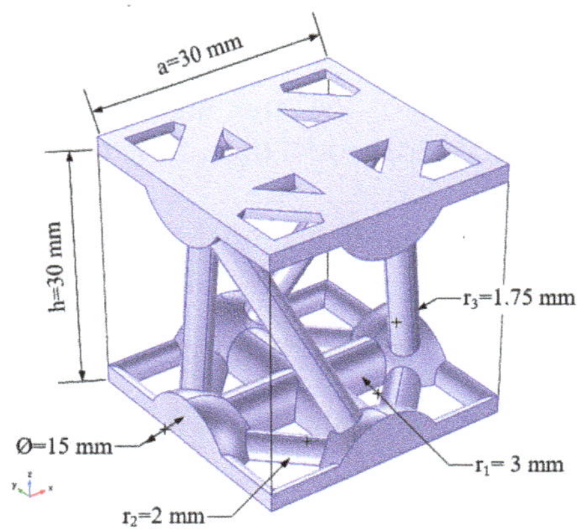

Figure 9.23. Unit cell of the Weyl point phononic crystal.

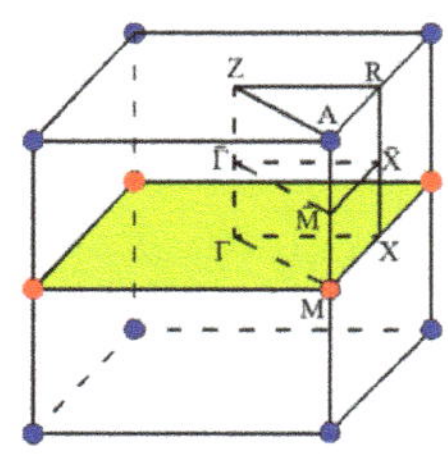

Figure 9.24. First Brillouin zone.

When $|t_3 + t_4| < |t_3 - t_4|$, the two points represent conventional charge-2 Weyl points with type-I dispersion. When $|t_3 + t_4| = |t_3 - t_4|$, the system reaches a critical state that is generally unstable. It is worth noting that the hopping parameters t_i (where $i = 1, 2, 3, 4$) can be easily adjusted in a sonic crystal, making it an ideal acoustic platform for experimentally achieving the type-III Weyl phase.

Moreover, the chirality of the two Weyl points can be independently controlled by manipulating the model parameters t_1 and t_2. Specifically, when $|t_1| > |t_2|$, the two Weyl points exhibit opposite chirality, while $|t_1| < |t_2|$ results in the same chirality. Consequently, in the case of $|t_1| > |t_2|$, the Weyl points located at the M and A points can be the only band crossings in the system, leading to a single pair of Weyl points in nonmagnetic crystals. The design of the sonic crystal sample was accomplished using the FEM, as depicted in figure 9.23. The unit cell, characterized by a lattice constant of $a = 30$ mm and $h = 30$ mm, with $r_1 = 3$ mm, $r_2 = 2$ mm, $r_3 = 1.75$ mm, and $D = 15$ mm, is illustrated in figure 9.23.

Figure 9.24 illustrates the corresponding bulk Brillouin zone (BZ) containing the two type-III Weyl points. The structure is acoustically rigid, featuring air-filled cavities, and serves as a simulation platform for studying novel phenomena in spinless systems with the same crystalline symmetry.

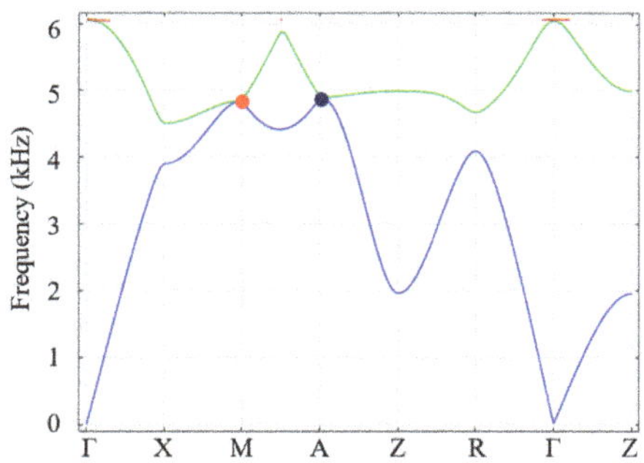

Figure 9.25. The simulated band structure of the type-III Weyl point phononic crystal.

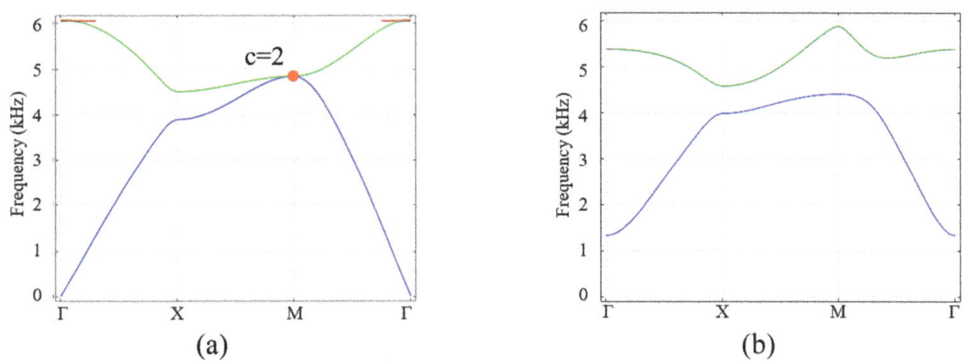

Figure 9.26. The bulk bands of the two lowest modes. (a) In planes $k_z = 0$, (b) $k_z = 0.5\pi/\text{H}$.

To provide evidence for the presence of type-III Weyl points, we conducted a simulation of the sample's band structure. The results, shown in figure 9.25, reveal two distinct Weyl points labeled *WM* and *WA*, located at the *M* and *A* points, respectively. Within the frequency range of 0 to 6.04 kHz, *WA* and *WM* appear as two topological bands without interference from trivial bands. Additionally, it is evident that *WA* and *WM* exhibit linear dispersion along the k_z direction (*M*-*A* path) and quadratic dispersion in the k_x–k_y plane. Thus, *WA* and *WM* correspond to charge-2 Weyl points, in agreement with the findings from symmetry analysis and the discussed tight-binding (TB) model.

Figure 9.26(a) depicts the conduction band and valence band with parabolic dispersion opening in the same orientation along the k_x (k_y) direction. This characteristic confirms that *WA* and *WM* represent type-III Weyl points. Furthermore, figure 9.26(b) clearly demonstrates the distinct dispersion characteristics of type-III Weyl points, providing strong verification of the existence of a single pair of type-III Weyl points in the sample.

Furthermore, in figure 9.26(b), we performed band structure measurements in the Γ-*X*-*M* plane (with $k_z = 0.5\pi/\text{H}$), which is considerably far from the type-III Weyl points. The results indicate the formation of a band gap at the band crossing point, indicating that the slice at $k_z = 0.5\pi/H$ exhibits the properties of a Chern insulator with $|C| = 1$.

In summary, we have successfully designed an acoustic Weyl sonic crystal of type-III, featuring a single pair of type-III Weyl points with a topological charge of ±2 at the high-symmetry points within the BZ. Through simulations, we have discovered two remarkable long Fermi arc states associated with the type-III Weyl points, as well as captured the chiral edge state of the two-dimensional Chern insulator at a fixed k_z. It is important to note that the type-III Weyl points discussed in our acoustic semimetal can also be realized in photonic, mechanical, and cold atom systems.

In an acoustic Weyl point metamaterial, the symmetry-breaking structure is typically achieved by patterning the material on a subwavelength scale, such as using a periodic array of acoustic resonators. This structure leads to the formation of Weyl points in the frequency domain, which can be adjusted by manipulating the parameters of the metamaterial. Acoustic Weyl point metamaterials have numerous potential applications, including sound manipulation, acoustic imaging, and sensing. For instance, they could be utilized to create novel types of acoustic filters, amplifiers, and waveguides with improved performance and functionality. Additionally, they could enhance the sensitivity and resolution of acoustic sensing and imaging techniques, such as medical ultrasound or underwater sonar [86].

9.6 Challenges and future works

One of the main challenges for the study is the experimental validation of the findings. While numerical simulations using the FEM provide valuable insights, experimental verification is essential to confirm the existence and behavior of negative refraction and the acoustic lens effect in the phononic crystal. Constructing and testing the proposed 2D phononic crystal structures in the laboratory can help validate the theoretical results and enhance the credibility of the research. For the proposed sonic crystal lens using complementary split rectangular resonators, experimental validation is necessary to confirm its performance in focusing acoustic waves and forming band gaps.

Further research can focus on optimizing the parameters of the phononic crystals, such as the lattice parameter and cylinder radius, to enhance the negative refraction and acoustic lens properties. Understanding how different geometrical configurations impact the performance of the phononic crystals will be crucial in designing more efficient acoustic metamaterials and devices. Further exploration of different fractal geometries and their impact on band structures and wave manipulation can provide valuable insights into designing more efficient noise control and wave blocking devices.

Future studies may examine in more depth the design and optimization of such waveguides for specific applications, such as integrated elastic waveguide devices and advanced wave control systems. Research on acoustic second harmonic generation can be extended to explore the nonlinear properties of other materials and investigate different scenarios where the generated second harmonic signal is more pronounced. Improving the sensitivity of detection systems can help in practical applications of acoustic SHG, such as material characterization and biomedical imaging. With further studies, subwavelength acoustic structures can be developed with optimized geometries to achieve more efficient sound reduction and noise reduction. It will be very important to explore the scalability and adaptability of such structures for real-world applications. The integration of phononic crystals and acoustic metamaterials with other technologies, such as electronics or optics, may lead to the development of multifunctional devices with wider applications.

Although the theoretical basis and the results of simulation studies are promising, there are many challenges to overcome and exciting future research directions to explore in order to realize and exploit the unique properties of these metamaterials. Taking advantage of the unique properties of acoustic Weyl point metamaterials to manipulate sound waves could lead to advances in acoustic technologies. For example, exploring the possibility of acoustic cloaking or designing unconventional acoustic waveguides that take advantage of the properties of Weyl points can be fruitful areas of exploration.

Exploring the behavior of acoustic Weyl point metamaterials under different conditions, such as temperature, pressure or external stimuli, can reveal interesting topological phase transitions. Understanding how these transitions affect the topological properties and acoustic response of metamaterials may have practical implications for device design and performance. Exploring the integration of acoustic Weyl point metamaterials with other types of metamaterials or functional materials may lead to hybrid systems with improved or novel functions. Combining acoustic and photonic or acoustic and mechanical metamaterials could open up new avenues for acoustic signal processing and manipulation. The development of acoustic Weyl point devices can lead to transformative developments in acoustic technologies and improve our understanding of topological acoustics.

9.7 Conclusion

In summary, acoustic metamaterials, such as phononic crystals and fractal phononic crystals, nonlinear acoustic materials, locally resonant sonic crystals, and Weyl point phononic crystals offer unique properties for controlling the propagation of sound waves. These materials can exhibit negative refraction, lensing effects, broad band gaps, and tunable waveguiding capabilities, making them suitable for various applications in noise control, wave filtering, underwater communication and sensing, and the development of integrated elastic waveguiding devices.

Acoustic second harmonic generation (SHG) is a phenomenon where two acoustic waves with different frequencies interact within a material and generate a new wave with twice the frequency. It can be used as a tool for examining the nonlinear properties of materials and for creating high-resolution images of materials.

The nonlinear response of materials can be observed in the frequency domain through the presence of static and second harmonic components in the displacement. This provides valuable insights into the material's behavior.

Locally resonant sonic crystals (LRSCs) utilize resonators, such as complementary split rectangular resonators (CSRRs), to create band gaps and achieve acoustic attenuation. The design of LRSCs allows for operation below the Bragg gap of conventional sonic crystals.

The realization of acoustic Weyl point metamaterials typically involves patterning the material on a subwavelength scale, such as using a periodic array of acoustic resonators. This structure leads to the formation of Weyl points in the frequency domain, and their properties can be adjusted by manipulating the parameters of the metamaterial.

Author contributions

Conceptualization; methodology; investigation; writing—original draft preparation; writing—review and editing, and visualization, Z. O.

Data availability statement

The datasets generated during and/or analyzed during the current study are available from the corresponding author on reasonable request.

Contact person: zaferozer@mersin.edu.tr.

Acknowledgments

The author would like to thank Amirullah M. Mamedov and Ekmel Özbay, for making Comsol Multiphysics available.

Conflicts of Interest

The authors declare no conflict of interest.

References

[1] Maetani N, Kurose T and Tsuruta K 2010 Numerical simulation of acoustic waves in two-dimesional phononic crystal: negative refraction *Mem. Fac. Eng.* **44** 7–12

[2] Elford D P 2010 Band gap formation in acoustically resonant phononic crystals *Doctoral Thesis* (Loughborough University)

[3] Glusker J P 1985 *Crystal Structure Analysis* 2nd (Oxford: Oxford University Press)

[4] Brillouin L 1931 *Wave Propagation in Periodic Structures* (Berlin: Springer)

[5] Su M F, Olsson R H, Leseman Z C and El-Kady I 2010 Realization of a phononic crystal operating at gigahertz frequencies *Appl. Phys. Lett.* **96** 053111

[6] Wang G, Yu D, Wen J, Liu Y and Wen X 2004 One-dimensional phononic crystals with locally resonant structures *Phys. Letters* A **327** 512–21

[7] Sheng P, Zhang X X, Liu Z and Chan C T 2003 Locally resonant sonic materials *Phys.* B **338** 201–5

[8] Chen. L A and Wang. Y S 2006 Study on band gaps of elastic waves propagating in one-dimensional disordered phononic crystals *Physica* **B 392** 369–78

[9] Kurose T, Tsuruta K, Totsuji C and Totsuji H 2009 Negative refraction of acoustic waves in a two-dimensional phononic crystal via FDTD simulation *Mater. Trans.* **50** 1004–7

[10] Alaie S, Su M F, Goettler D F, El-Kady I and Leseman Z 2013 Effects of flexural and extensional excitation modes on the transmission spectrum of phononic crystals operating at gigahertz frequencies *J. Appl. Phys.* **113** 103513

[11] Feng Z X, Sheng-Chun L, TaoXu T, Tie-Hai W and Jian-Chun C 2010 Investigation of a silicon-based one-dimensional phononic crystal plate via the super-cell plane wave expansion method *Chin. Phys.* B **19** 044301

[12] Meidani M, Kim E, Li F, Yang J and Ngo D 2015 Tunable evolutions of wave modes and band gaps in quasi-1D cylindrical phononic crystals *J. Sound Vib.* **334** 270–81

[13] Elford D P *et al* 2011 Matryoshka locally resonant sonic crystal *J. Acoust. Soc. Am.* **130**

[14] Chung G and Phan D T 2010 Finite element modeling of surface acoustic waves in piezoelectric thin films *J. Korean Phys. Soc.* **57** 446–50

[15] Lucklum R, Zubtsov M, Oseev A, Schmidt M P, Hirsch S and Hagemann F 2013 Phononic crystals and applications *AMA Conf.* pp 95–9

[16] Liu X J and Y H F 2013 Band structure characteristics of T-square fractal phononic crystals *Chin. Phys.* B **22** 036101

[17] Pachiu C, Ion M J, Izbicki L and Moagar V 2011 Congrès Français de Mécanique *Besançon*

[18] Brûlé S, Javelaudv E H and Enoch S 2013 Guenneau. Seismic metamaterial: how to shake friends and influence waves? (http://arxiv.org/abs/1301.7642) (Accessed 31 December 2015)

[19] Chiyan L, Johnson S G and Joannopoulos J D 2002 All-angle negative refraction without negative effective index *Phys. Rev.* B **65** 201104–7

[20] http://nature.com/articles/srep03351 (Accessed 30 December 2015)

[21] Stenger N, Wilhelm M and Wegener M 2012 Experiments on elastic cloaking in thin plates *Phys. Rev. Lett.* **108** 14301

[22] Lu M *et al* 2007 Negative birefraction of acoustic waves in a sonic crystal *Nat. Mater.* **6** 744–8

[23] Mohammadi S 2010 Phononic band gap micro/nano-mechanical structures for communications and sensing applications *Partial Fulfillment of the Requirements for the Degree Doctor of Philosophy in Electrical and Computer Engineering* (Georgia Institute of Technology)

[24] Thiago P M A, Safavi-Naeini A, Winger M and Painter O 2011 *Quasi-Two-Dimensional Optomechanical Crystals with a Complete Phononic Bandgap* (Optical Society of America)

[25] Ozer Z, Palaz S, Mamedov A M and Ozbay E 2020 Band structure and transmission spectra in multiferroic based Sierpinski-carpet phononic crystal *EPJ Appl. Phys.* **90** 2019035

[26] Ozer Z, Mamedov A M and Ozbay E 2021 Phononic crystals with archimedean-like tiling: band structure and the transformation of sound *Integr. Ferroelectr.* **220** 132–9

[27] Majeed F, Alsakarneh A, Momani L, Abuharaz M and Olayanju L 2018 Comparative study on 1D phononic crystal vibration transmission—a case study on surgical tool structure 2018 Advances in Science and Engineering Technology Int. Conferences (ASET) 5 April 2018

[28] Khelif A, Choujaa A, Benchabane S, Djafari-Rouhani B and Laude V 2004 Guiding and bending of acoustic waves in highly confined phononic crystal waveguides *Appl. Phys. Lett.* **84** 4400

[29] Sun J H and Wu T T 2005 Analyses of mode coupling in joined parallel phononic crystal waveguides *Phys. Rev.* B **71** 174303

[30] Wu T C, Wu T T and Hsu J C 2009 Waveguiding and frequency selection of Lamb waves in a plate with a periodic stubbed surface *Phys. Rev.* B **79** 104306

[31] Pennec Y, Djafari Rouhani B, Larabi H, Akjouj A, Gillet J N, Vasseur J O and Thabet G 2009 Phonon transport and waveguiding in a phononic crystal made up of cylindrical dots on a thin homogeneous plate *Phys. Rev.* B **80** 144302

[32] Oh J H, Lee I K, Ma P S and Kim Y Y 2011 Active wave-guiding of piezoelectric phononic crystals *Appl. Phys. Lett.* **99** 083505

[33] Hu X, Chan C T and Zi J 2005 Two-dimensional sonic crystals with Helmholtz resonators *Phys. Rev.* E **71** 055601

[34] Wu L Y and Chen L W 2010 Wave propagation in a 2D sonic crystal with a Helmholtz resonant defect *J. Phys.* D **43** 055401

[35] Guenneau S, Movchan A, Petursson G and Ramakrishna S A 2007 Acoustic metamaterials for sound focussing and confinement *New J. Phys.* **9** 399

[36] Wu L Y and Chen L W 2007 The dispersion characteristics of sonic crystals consisting of elliptic cylinders *J. Phys.* D **40** 7579

[37] Romero-Garcia V and Fuster-Garcia E 2007 Interferometric method of determining the refractive index of two-dimensional sonic crystals *Phys. Rev.* B **75** 224305

[38] Chalmers L, Elford D, Kusmartsev F and Swallowe G 2009 Acoustic band gap formation in two-dimensional locally resonant sonic crystals comprised of Helmholtz resonators *Int. J. Mod. Phys.* B **23** 4234

[39] Sainidou R and Stefanou A M N 2002 Formation of absolute frequency gaps in three-dimensional solid sonic crystals *Phys. Rev.* B **66** 212301

[40] Xu S *et al* 2015 Discovery of a Weyl fermion state with Fermi arcs in niobium arsenide *Nat. Phys.* **11** 748–54

[41] Xu S Y *et al* 2015 Discovery of a Weyl fermion semimetal and topological Fermi arcs *Science* **349** 613–7

[42] Shekhar C *et al* 2015 Extremely large magnetoresistance and ultrahigh mobility in the topological Weyl semimetal candidate NbP *Nat. Phys.* **11** 645–9

[43] Lv B Q *et al* 2015 Experimental discovery of Weyl semimetal TaAs *Phys. Rev.* X **5** 031013

[44] Lv B Q *et al* 2015 Observation of Weyl nodes in TaAs *Nat. Phys.* **11** 724–7

[45] Murakami S 2007 Phase transition between the quantum spin hall and insulator phases in 3D: emergence of a topological gapless phase *New J. Phys.* **9** 356

[46] Wan X, Turner A M, Vishwanath A and Savrasov S Y 2011 Topological semimetal and fermi-arc surface states in the electronic structure of pyrochlore iridates *Phys. Rev.* B **83** 205101

[47] Weng H, Fang C, Fang Z, Bernevig B A and Dai X 2015 Weyl semimetal phase in noncentrosymmetric transitionmetal monophosphides *Phys. Rev.* **5** 011029

[48] Burkov A 2015 Chiral anomaly and transport in Weyl metals *J. Phys. Condens. Matter* **27** 113201

[49] Liu Y, Yu Z-M, Xiao C and Yang S A 2020 Quantized circulation of anomalous shift in interface reflection *Phys. Rev. Lett.* **125** 076801

[50] Zhou J, Chang H-R and Xiao D 2015 Plasmon mode as a detection of the chiral anomaly in Weyl semimetals *Phys. Rev.* B **91** 035114

[51] de Juan F, Grushin A G, Morimoto T and Moore J E 2017 Quantized circular photogalvanic effect in Weyl semimetals *Nat. Commun.* **8** 1

[52] Noh J *et al* 2017 Experimental observation of optical Weyl points and Fermi arc-like surface states *Nat. Phys.* **13** 611–7

[53] Chen W J, Xiao M and Chan C T 2016 Photonic crystals possessing multiple Weyl points and the experimental observation of robust surface states *Nat. Commun.* **7** 13038

[54] Lu L, Fu L, Joannopoulos J D and Soljacic M 2013 Weyl points and line nodes in gyroid photonic crystals *Nat. Photon.* **7** 294–9

[55] Lu L *et al* 2015 Experimental observation of Weyl points *Science* **349** 622–4

[56] Yang Z and Zhang B 2016 Acoustic type-II Weyl nodes from stacking dimerized chains *Phys. Rev. Lett.* **117** 224301

[57] Xiao M *et al* 2015 Synthetic gauge flux and Weyl points in acoustic systems *Nat. Phys.* **11** 920–4

[58] Zhang T, Takahashi R, Fang C and Murakami S 2020 Twofold quadruple Weyl nodes in chiral cubic crystals *Phys. Rev.* B **102** 125148

[59] Yu Z-M, Zhang Z, Liu G-B, Wu W, Li X-P, Zhang R-W, Yang S A and Yao Y 2022 Encyclopedia of emergent particles in three-dimensional crystals *Sci. Bull.* **67** 375

[60] Nagaosa N, Morimoto T and Tokura Y 2020 Transport, magnetic and optical properties of Weyl materials *Nat. Rev. Mater.* **5** 621

[61] Cui C, Li X-P, Ma D-S, Yu Z-M and Yao Y 2021 Charge-four Weyl point: minimum lattice model and chirality-dependent properties *Phys. Rev.* B **104** 075115

[62] Han S, Lee C, Moon E-G and Min H 2019 Emergent anisotropic non-fermi liquid at a topological phase transition in three dimensions *Phys. Rev. Lett.* **122** 187601

[63] Wang J-R, Liu G-Z and Zhang C-J 2019 Topological quantum critical point in a triple-Weyl semimetal: nonfermi-liquid behavior and instabilities *Phys. Rev.* B **99** 195119

[64] Soluyanov A A, Gresch D, Wang Z, Wu Q, Troyer M, Dai X and Bernevig B A 2015 Type-II Weyl semimetals *Nature* **527** 495

[65] Li X-P, Deng K, Fu B, Li Y, Ma D-S, Han J, Zhou J, Zhou S and Yao Y 2021 Type-III Weyl semimetals:$(TaSe_4)_2I$ *Phys. Rev.* B **103** L081402

[66] O'Brien T, Diez M and Beenakker C 2016 Magnetic breakdown and Klein tunneling in a type-II Weyl semimetal *Phys. Rev. Lett.* **116** 236401

[67] Udagawa M and Bergholtz E J 2016 Field-selective anomaly and chiral mode reversal in type-II Weyl materials *Phys. Rev. Lett.* **117** 086401

[68] Yu Z-M, Yao Y and Yang S A 2016 Predicted unusual magnetoresponse in type-II Weyl semimetals *Phys. Rev. Lett.* **117** 077202

[69] Tchoumakov S, Civelli M and Goerbig M O 2016 Magnetic-field-induced relativistic properties in type-I and type-II Weyl semimetals *Phys. Rev. Lett.* **117** 086402

[70] Huang S-M, Xu S-Y, Belopolski I, Lee C-C, Chang G, Chang T-R, Wang B, Alidoust N, Bian G and Neupane M *et al* 2016 New type of Weyl semimetal with quadratic double weyl fermions *Proc. Natl. Acad. Sci.* **113** 1180

[71] Chang G, Xu S-Y, Sanchez D S, Huang S-M, Lee C-C, Chang T-R, Bian G, Zheng H, Belopolski I and Alidoust N *et al* 2016 A strongly robust type II Weyl fermion semimetal state in Ta_3S_2 *Sci. Adv.* **2** e1600295

[72] Yao M-Y, Xu N, Wu Q, Aut`es G, Kumar N, Strocov V N, Plumb N C, Radovic M, Yazyev O V and Felser C *et al* 2019 Observation of Weyl nodes in robust type-II Weyl semimetal WP_2 *Phys. Rev. Lett.* **122** 176402

[73] Xu S-Y, Alidoust N, Chang G, Lu H, Singh B, Belopolski I, Sanchez D S, Zhang X, Bian G and Zheng H *et al* 2017 Discovery of Lorentz-violating type II Weyl fermions in laalge *Sci. Adv.* **3** e1603266

[74] Belopolski I, Yu P, Sanchez D S, Ishida Y, Chang T-R, Zhang S S, Xu S-Y, Zheng H, Chang G and Bian G *et al* 2017 Signatures of a time-reversal symmetric Weyl semimetal with only four Weyl points *Nat. Commun.* **8** 1

[75] Saba M, Hamm J M, Baumberg J J and Hess O 2017 Group theoretical route to deterministic Weyl points in chiral photonic lattices *Phys. Rev. Lett.* **119** 227401

[76] Shastri K, Yang Z and Zhang B 2017 Realizing type-II Weyl points in an optical lattice *Phys. Rev.* B **95** 014306
[77] Yang B, Guo Q, Tremain B, Liu R, Barr L E, Yan Q, Gao W, Liu H, Xiang Y and Chen J *et al* 2018 Ideal Weyl points and helicoid surface states in artificial photonic crystal structures *Science* **359** 1013
[78] He H, Qiu C, Ye L, Cai X, Fan X, Ke M, Zhang F and Liu Z 2018 Topological negative refraction of surface acoustic waves in a Weyl phononic crystal *Nature* **560** 61
[79] Peri V, Serra-Garcia M, Ilan R and Huber S D 2019 Axial-field-induced chiral channels in an acoustic Weyl system *Nat. Phys.* **15** 357
[80] Xie B, Liu H, Cheng H, Liu Z, Chen S and Tian J 2019 Experimental realization of type-II Weyl points and Fermi arcs in phononic crystal *Phys. Rev. Lett.* **122** 104302
[81] Yang Y, Sun H-x, Xia J-p, Xue H, Gao Z, Ge Y, Jia D, Yuan S-q, Chong Y and Zhang B 2019 Topological triply degenerate point with double Fermi arcs *Nat. Phys.* **15** 645
[82] Huang X, Deng W, Li F, Lu J and Liu Z 2020 Ideal type-II Weyl phase and topological transition in phononic crystals *Phys. Rev. Lett.* **124** 206802
[83] He H, Qiu C, Cai X, Xiao M, Ke M, Zhang F and Liu Z 2020 Observation of quadratic Weyl points and doublehelicoid arcs *Nat. Commun.* **11** 1
[84] Lee C H, Imhof S, Berger C, Bayer F, Brehm J, Molenkamp L W, Kiessling T and Thomale R 2018 Topolectrical circuits *Commun. Phys.* **1** 1
[85] Li R, Lv B, Tao H, Shi J, Chong Y, Zhang B and Chen H 2021 Ideal type-II Weyl points in topological circuits *Natl. Sci. Rev.* **8** nwaa192
[86] Fitzpatrick A, Singhvi A and Arbabian A 2020 An airborne sonar system for underwater remote sensing and imaging *IEEE Access* **8** 189945–59

IOP Publishing

Advanced Metamaterials for Engineers

Lulu Wang and Muharrem Karaaslan

Chapter 10

Data-driven modeling of microstrip reflectarray unit element design

Peyman Mahouti, Aysu Belen, Mehmet A Belen and Özlem Tari

Symbols

ε Permittivity
κ Dielectric constant
λ Wavelength (m)
f Frequency

10.1 Introduction

Reflectarray antennas (RAs) are a class of antenna systems that combine the advantages of parabolic reflector antennas and phased array antennas [1]. They offer a cost-effective and high-performance substitute for various communication applications. RAs are constituted by a planar, two-dimensional array of radiating elements. The aforementioned components are stimulated by a primary feed antenna, which is commonly a waveguide feed or a horn antenna. The feed antenna emits electromagnetic waves that undergo reflection upon reaching the radiating elements [2]. These elements are designed to induce accurate phase modifications in the reflected waves. By altering the phase shifts between radiating elements, it is feasible to attain focused reflected waves, ultimately leading to the formation of a preferred radiation pattern. The radiation manipulating techniques that can be employed in beam modification are a pencil beam, a shaped beam, or multiple beams, for example. The basic components of an RA design consist of a feed antenna and a planar array of radiating elements. The production of electromagnetic waves that illuminate the array of radiating elements is attributed to the feed antenna. The configuration of radiating components on a planar surface, typically composed of a dielectric substance, can be accomplished by employing a variety of designs such as patches, dipoles, or slots [3–5].

doi:10.1088/978-0-7503-5754-8ch10

RAs have become a compelling substitute for conventional parabolic reflector antennas and phased arrays in diverse communication applications. This section provides a comprehensive summary of the primary benefits associated with RAs, which include their capacity for versatile design, economical viability, and appropriateness for advanced communication systems [6]. RAs offer a significant benefit in terms of their design flexibility, as they can be customized to meet specific application requirements by adjusting their radiation patterns. RAs possess the capability to manipulate the radiating elements on an individual basis, thereby facilitating the creation of specific radiation patterns such as pencil beams, shaped beams, and multiple beams [7, 8]. The attribute of flexibility holds significant advantages for applications that necessitate accurate beam pointing, beam shaping, or beam switching, such as radar systems and satellite communications. RAs are a suitable option for applications that are restricted by space and weight limitations due to their lightweight structure and low profile. In contrast to parabolic reflector antennas, which necessitate substantial and weighty frameworks to uphold the reflector surface, RAs are composed of a two-dimensional array of radiating elements. The adoption of a planar configuration leads to a decrease in overall size and weight, thereby simplifying the integration process into communication systems and lowering the total system cost. RAs present a more economical alternative in comparison to phased array antennas. The management of radiation patterns in phased arrays generally necessitates intricate and costly feeding networks and phase shifters. RAs differ from other types of antennas in that they employ a less complex feed system, whereby a single feed antenna is used to illuminate the array. This approach leads to decreased manufacturing and operational expenses. Moreover, RAs can be produced using inexpensive materials and printed circuit board (PCB) techniques, thereby augmenting their cost-efficiency. RAs exhibit various superior performance attributes that render them appropriate for challenging communication scenarios. Achieving high gain, low side lobe levels, and narrow beam widths is crucial for long-range communication and interference mitigation. Furthermore, it has been observed that RAs exhibit a notable capability to sustain a high aperture efficiency, thereby resulting in enhanced system performance [9–12].

The RA operates on the fundamental principles of electromagnetic wave reflection and phase manipulation. The waves originating from the feed antenna encounter the radiating elements, which result in the generation of reflected waves due to the induced currents. The phase of the reflected waves is contingent upon the geometrical and material characteristics of the radiating elements, in conjunction with the angle of incidence. Through meticulous design of the radiating elements to introduce precise phase shifts, the resultant reflected waves can be effectively amalgamated in the far-field region, thereby generating the intended radiation pattern. The performance of the RA is significantly influenced by the design of its radiating elements. The RA can attain a significant level of regulation over the resultant radiation pattern by suitably choosing the geometrical parameters and phase-shifting mechanisms of the radiating elements [13, 14].

The intricate nature of RA design and optimization arises from the multitude of variables and interactions that are implicated, encompassing the radiating elements' geometry and material properties, phase distribution, and feed antenna characteristics. The utilization of surrogate models, specifically artificial intelligence (AI) and machine learning (ML) techniques, has been implemented to address the aforementioned obstacles, providing effective and precise design resolutions. Now the present section delves into the utilization of AI and ML methodologies in the development and enhancement of RAs [15–17].

Surrogate modeling is an approach that utilizes artificial intelligence to construct a computationally efficient approximation of a complex system, such as a RA. Surrogate models are typically trained with a restricted set of data samples, which are commonly acquired from either experimental measurements or full-wave electromagnetic simulations. Upon completion of the training process, the surrogate model has the capability to make predictions regarding the RA's performance across various design parameters. This facilitates efficient exploration and optimization of the design space [18, 19].

Various machine learning (ML) methodologies have been employed to develop substitute models for the design of RAs. These include models based on regression, artificial neural networks (ANNs) [20], support vector machines (SVMs) [21], and Gaussian process regression (GPR) [22]. The techniques exhibit variability in terms of their complexity, precision, and computational efficiency, rendering them appropriate for diverse design scenarios. The utilization of AI and ML-based surrogate modeling presents various benefits in the development and enhancement of RA. Surrogate models have the potential to considerably decrease the computational load that is typically associated with complete electromagnetic simulations [23]. This can result in expedited design iterations and optimization, thereby enhancing computational efficiency. The efficient exploration of design space is facilitated by the capability to swiftly forecast the performance of RAs across a broad spectrum of design parameters, ultimately resulting in the determination of optimal configurations. This process is commonly referred to as design space exploration [23].

The use of AI and ML techniques in adaptive and data-driven design can facilitate the refinement and adaptation of the surrogate model in response to new data. This process can enhance the accuracy and generalizability of the model. The integration of surrogate models within multi-objective optimization frameworks facilitates the concurrent optimization of diverse performance metrics. To sum up, the utilization of AI and ML methodologies, specifically surrogate modeling, presents notable benefits in the development and enhancement of RA. The utilization of AI and ML-driven approaches can potentially improve the performance and applicability of RA in contemporary communication systems through the reduction of computational complexity, facilitation of efficient design space exploration, and enabling adaptive and data-driven design.

10.2 Methods

In this work, firstly, a full-wave 3D model of a RA unit element is created by using the CST microwave studio for analyzing the electromagnetic (EM) response of the unit element with respect to the design variables. Next, a brief explanation of the sampling strategy of unit element design variables for generating the training and test (holdout) data sets will be discussed. At the next stage, a series of artificial intelligence based algorithms will be used to create a surrogate model representation of the studied RA unit element. A series of AI algorithms are deployed to study their potential for creating a mapping between the reflection phase characterization response with respect to the variables of the unit element within the given operation band.

10.3 Modeling of the RA unit element

In this section, the design and characteristics of the selected RA unit element presented in figure 10.1 are studied. For this work, a square-shaped microstrip patch design with a rotatable rectangular slot in its center is taken into consideration, and the geometrical details of the design are presented in figure 10.1 and table 10.1. A detailed analysis of the effect of the geometrical design variables on the reflection phase of the incoming EM waves and their capabilities to manipulate the reflected waves in the desired direction is investigated in this section to determine the surrogate model definition space for gathering training and test data sets.

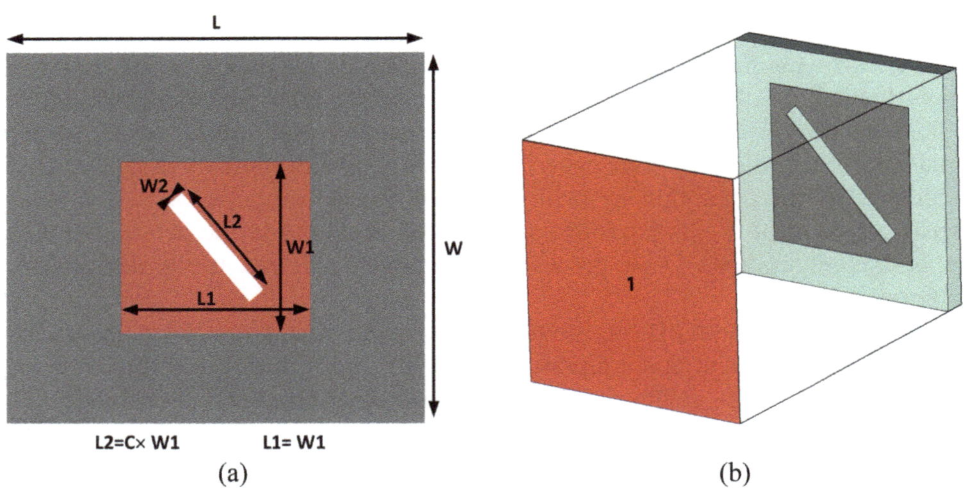

Figure 10.1. (a) Schematic, (b) 3D EM model view of the studied RA unit element.

Table 10.1. Variables of the studied RA unit element and their range of variation.

Variable	Lower bound	Upper Bound
W1 [mm]	4	14
W2 [mm]	0	4
Rot	0	180
C	0	1

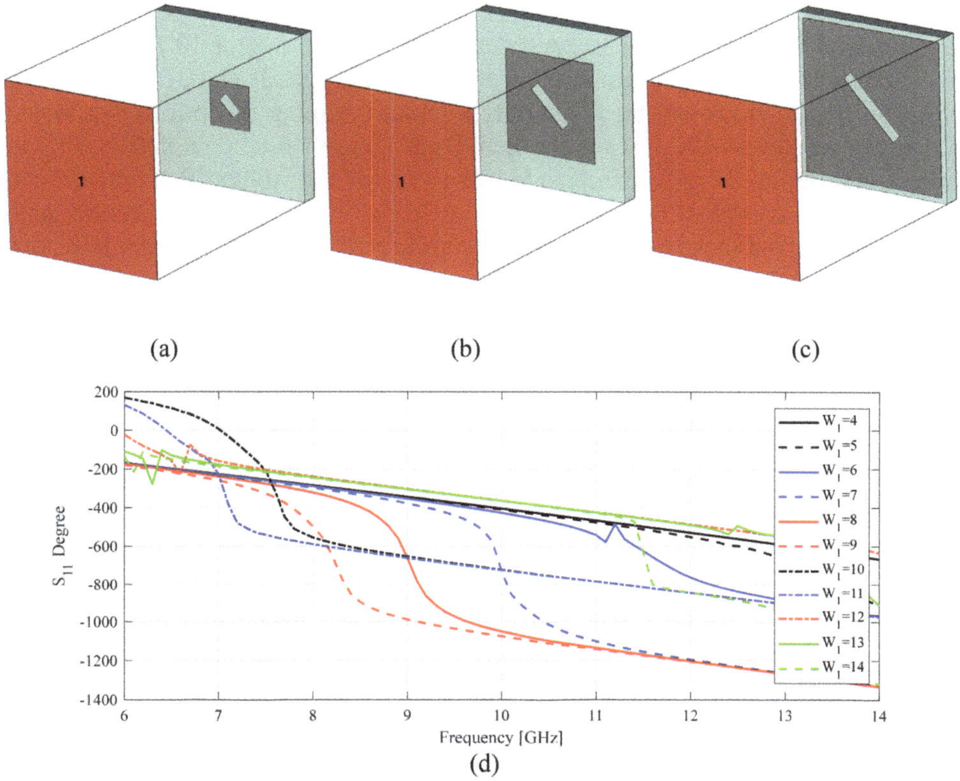

Figure 10.2. Different design geometries with respect to the variation of W1, where (a) W1 = 4 [mm], (b) W1 = 9 [mm], (c) W1 = 14 [mm]; (d) parametric analyses of the reflection phase characteristics with respect to W1. Other design parameters are taken as constants for this analysis as follows: W2 = 1 [mm], Rot = 45°, and C = 0.5.

As can be seen from figures 10.2–10.5, each of the design variables has a unique effect on the EM response characteristics of the unit element. Here it should be mentioned that the reflection phase values are not wrapped between −180° and 180° in order to clearly see the effective range of variables, thus the total range of phase variation is beyond −360° and 360°. While W1 seems to be the dominant variable that can both affect the resonance frequency and the reflection phase value, other parameters have an effect on the magnitude value of the reflection phase. As an example, while W1 is taken to be equal to 9 [mm], the variation in W2 affects the unit element reflection at around 8.5 GHz operation band, so that W2 can only change the reflection phase between −250° and −950°. However, for the same variation of W2, if the analyses were done while W1 is taken equal to 6 [mm], the operation frequency will be shifted to around 12 GHz with a variation range of −660° and −1150°, or if W1 is taken equal to 10 [mm], the operation frequency will be shifted to around 7.5 GHz with a variation range of −50° and −400°.

In figures 10.4–10.5, the variation characteristics of the C and Rot variables are also studied. As it can be observed in these variables, similar to W2, the variable itself does not affect the operating frequency but has a certain level of impact on the magnitude of the reflection phase. As it can be seen from figures 10.2–10.5, each of the variables has a certain amount of effect on the overall characteristics of the reflection phase. However, there are only a limited number of options available in the given scenarios. As it can be seen from figure 10.6, completely different geometrical design samples can also be obtained for the studied RA unit element. Each of these combinations can have a different effect on the reflection phase characteristics, and these variations must be accurately modeled. In order to have an accurate model for such a complex design over a wide range of operation frequencies (6–14 GHz), not only is a modern and powerful surrogate model required, but also a wise strategy is a must for gathering the sample points for this multi-dimensional problem in a cost-efficient manner. In the next section, the computational cost and series of sampling strategies for gathering training and test samples for creating the data-driven surrogate model are studied.

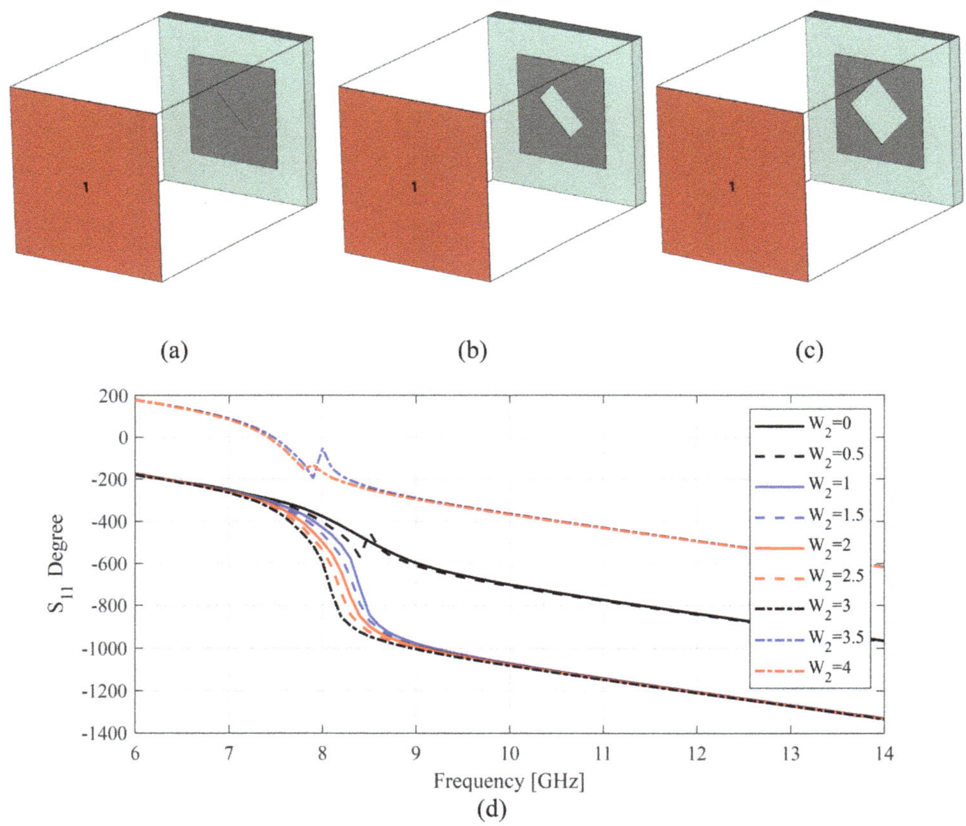

Figure 10.3. Different design geometries with respect to the variation of W2, where (a) W2 = 0 [mm], (b) W2 = 2 [mm], (c) W2 = 4 [mm]; (d) parametric analyses of the reflection phase characteristics with respect to W2. Other design parameters are taken as constants for this analysis as follows: W1 = 9 [mm], Rot = 45°, C = 0.5; the reflection phase characteristics with respect to W2 are changed to (e) W1 = 6 [mm], (f) W1 = 10 [mm], and the others are taken as Rot = 45°, C = 0.5.

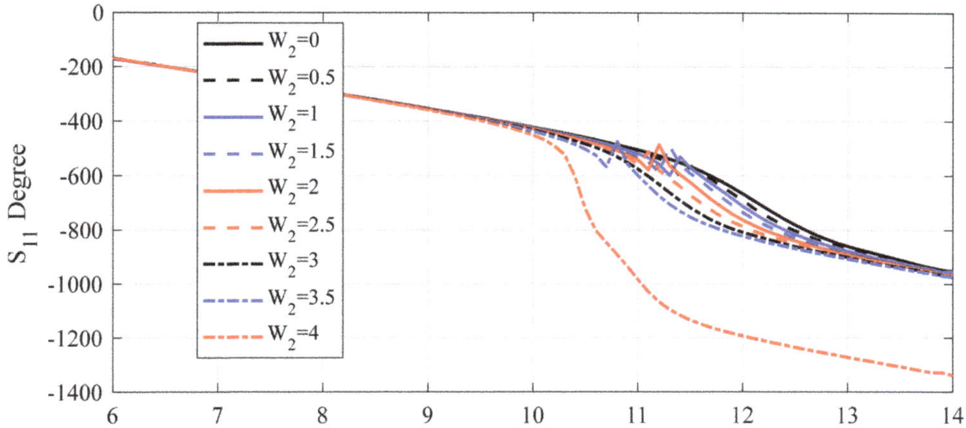

(e)

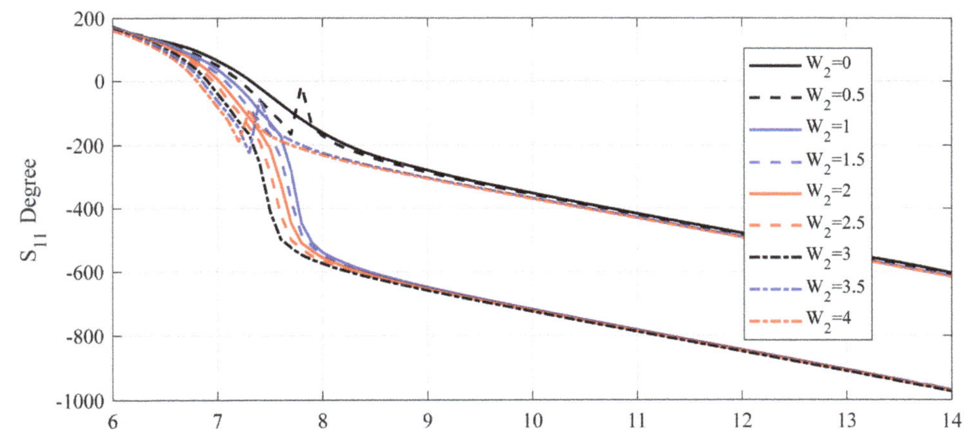

(f)

Figure 10.3. (Continued.)

10.4 Sampling strategies for gathering data points

As mentioned before, gathering the samples for the training and testing process of surrogate models is a crucial step, and for projects with a high number of dimensions, this problem becomes even more challenging. In this section, different sampling methods are studied along with their computational costs, and their success in creating surrogate models with a series of algorithms will be studied in the next section. In table 10.1, the design variations and their range of variations are presented. Here it should be mentioned that, during the data gathering simulations, for each geometrical design sample frequency, a step size of 0.1 GHz is taken for the

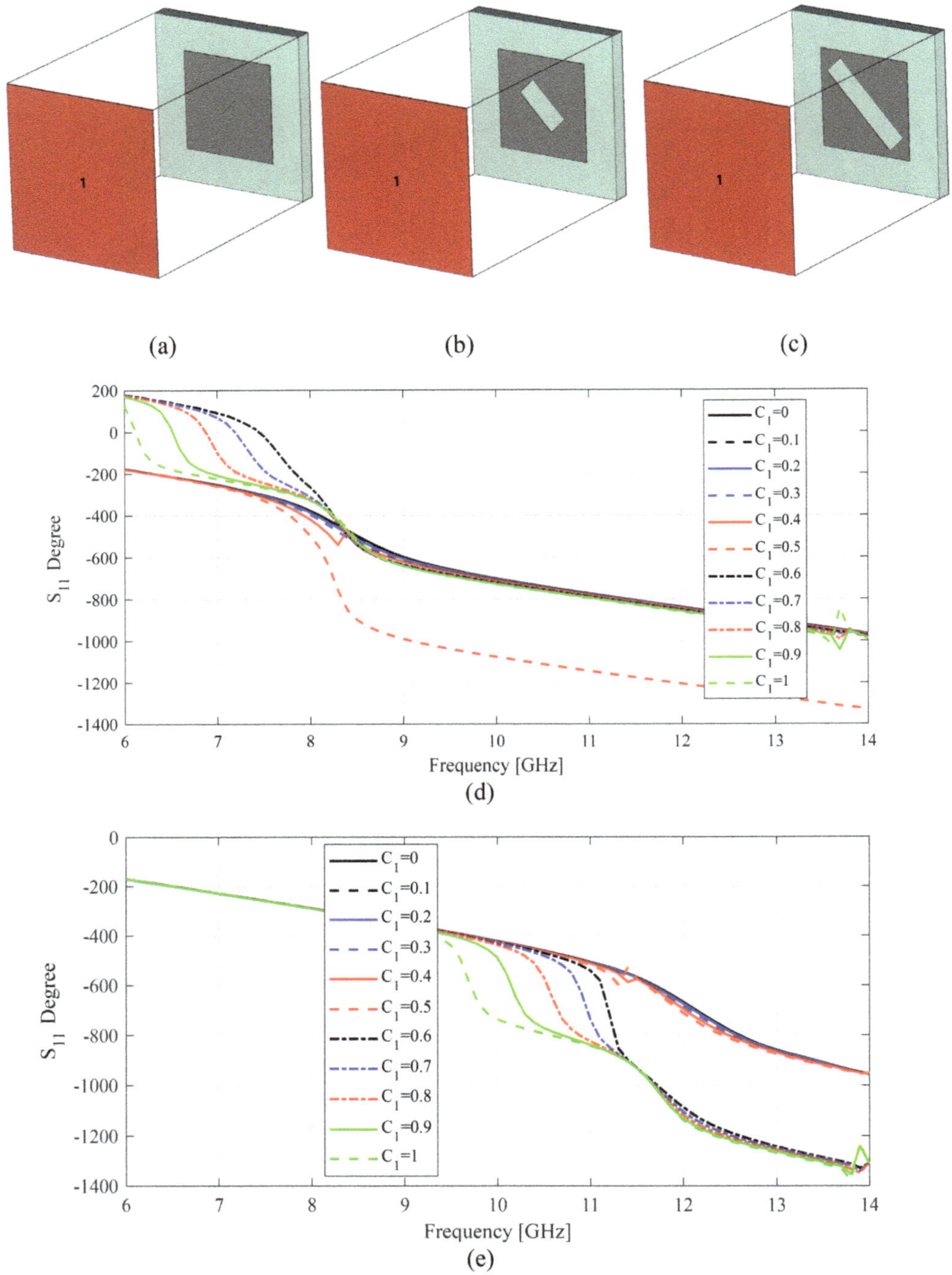

Figure 10.4. Different design geometries with respect to the variation of C, where (a) C = 0, (b) C = 0.5, (c) C = 1; (d) parametric analyses of the reflection phase characteristics with respect to C, other design parameters are taken as constant for this analyses as follows: W1 = 9 [mm], Rot = 45°, W2 = 2; (e) the reflection phase characteristics with respect to C while W1 = 6 [mm], (f) W1 = 1 [mm] Rot = 45°, C = 0.5.

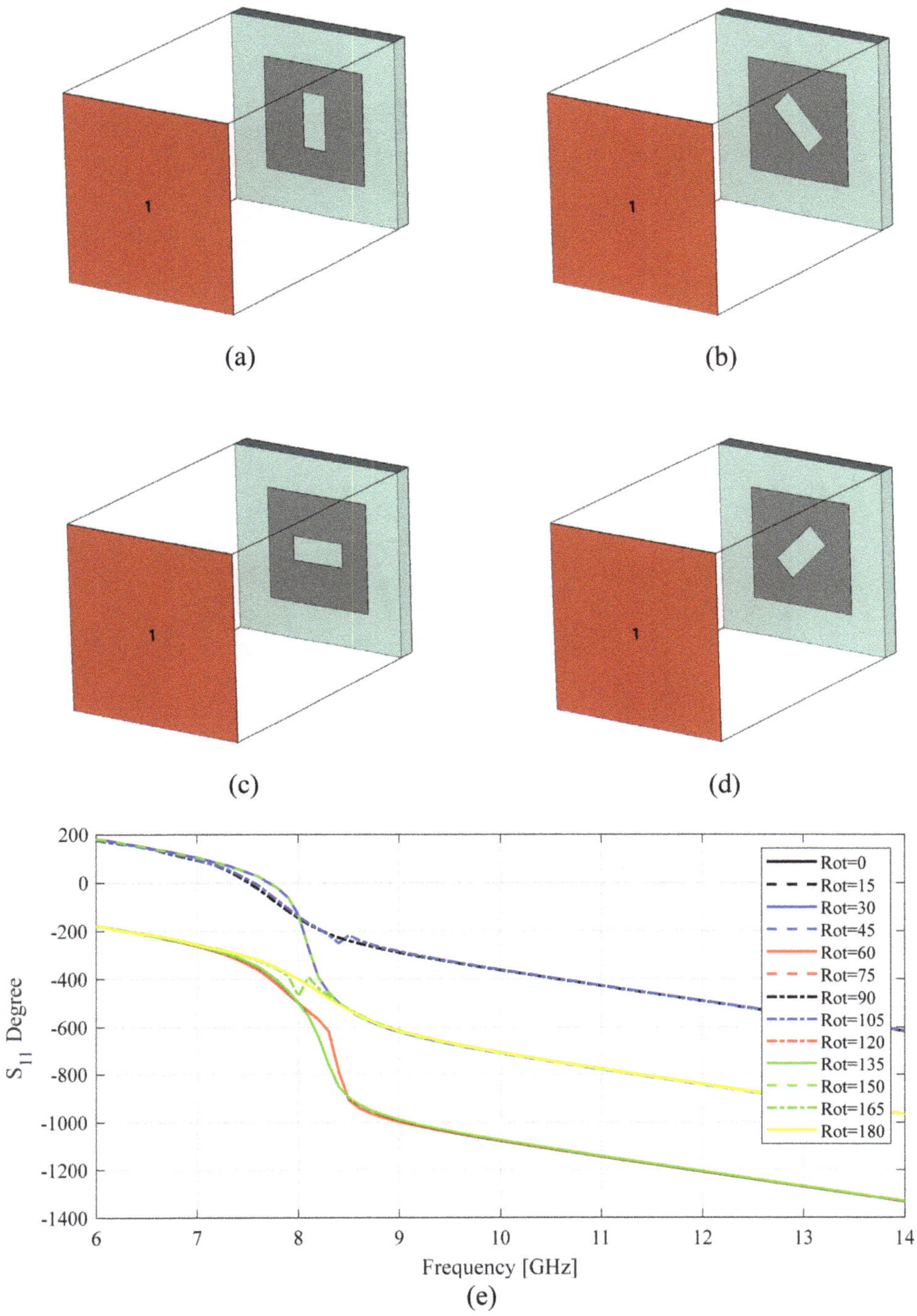

Figure 10.5. Different design geometries with respect to the variation of Rot, where (a) Rot = 0, (b) Rot = 45, (c) Rot = 90; (d) Rot = 135, (e) parametric analyses of the reflection phase characteristics with respect to W1. Other design parameters are taken as constant for this analyses as follows: W1 = 9 [mm], W2 = 2 [mm], C = 0.5.

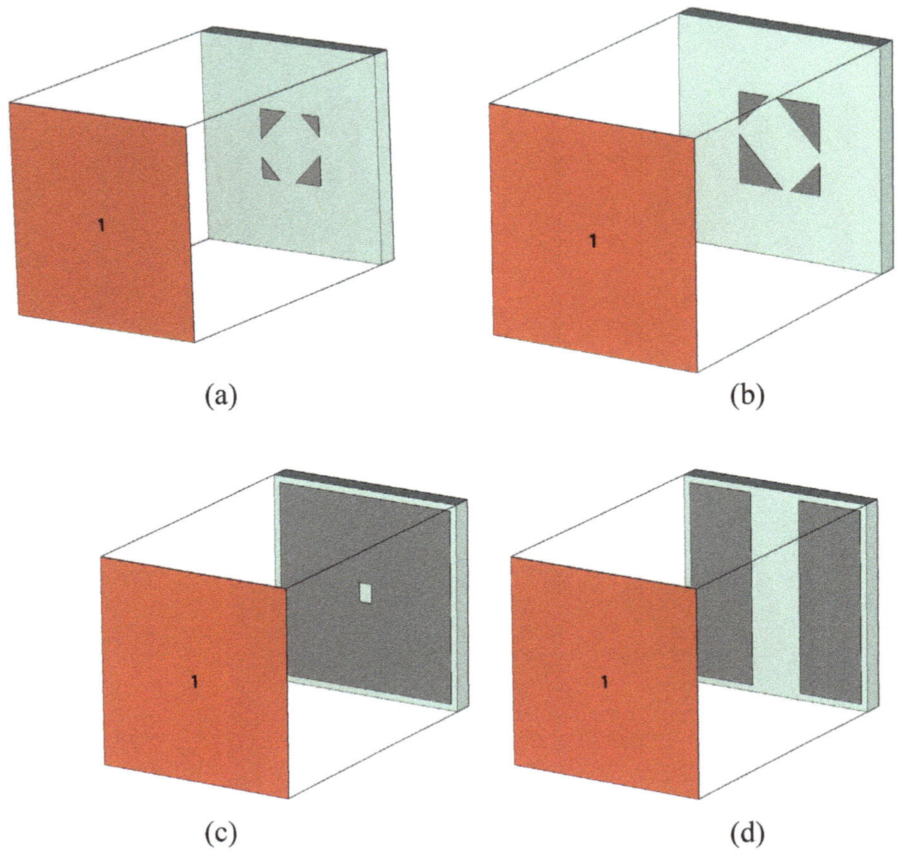

Figure 10.6. Different design geometries for (a) W1 = 5 [mm], W2 = 4 [mm], Rot = 135, C = 1, (b) W1 = 6 [mm], W2 = 4 [mm], Rot = 45, C = 0.9, (c) W1 = 14 [mm], W2 = 1 [mm], Rot = 0, C = 0.1, (d) W1 = 14 [mm], W2 = 4 [mm], Rot = 0, C = 1.

operation band of 6–14 GHz, thus each geometrical design sample would consist of 81 frequency sample points. These values are determined based on the geometrical limitations and the effect of variables on the reflection phase characteristics.

For gathering sample points, one of the most commonly used sampling methods is the linear sampling technique, which creates samples based on the requested sample intervals or step sizes. For example, a linear sampling with three sample points for variable W1 will select the following points: [4 9 14], for W2 = [0 2 4], Rot = [0 90 180], and C = [0 0.5 1]. In this case, a total of 81 samples ($3 \times 3 \times 3 \times 3$) are generated for this domain. As can be expected, in this sampling approach, two problems can come into play.

I. With an increased number of sample points for each variable, the total number of samples is exponentially increased, where for a sampling grid with 5 sample points, the total number of samples is increased to 625, for a sampling size of 7, the total number of samples is increased to 2401, and for a sampling size of 10, the total number of samples is 10 000. As it can be seen with this exponential growth, the computational cost of generating samples is also increased in such a way that after a certain value, the overall cost of the process becomes infeasible.

II. Another problem is that in this approach, it is desired that all the variables have the same effect of dominance on the studied characteristics and can be sampled equally. However, as it can be seen from figures 10.2–10.5, not only does W1 have a far higher dominance on both operation frequency and the reflection phase value compared to other variables, but its range of variation is also larger than the others. Thus, equal sampling is not a wise option for this type of problem. It can be said that one can use a higher number of samples for W1 or variables with similar characteristics; however, this is not always true. As it is presented in [24], the response of the characteristic impedance of a microstrip line is highly dominated by the width of the transition line, especially in its lower dome, where W1, and starts to lose its dominance after W1>3. Thus, instead of linear sampling, more complex approaches are a must.

Other than linear sampling, there are techniques such as random sampling, which completely determines the sample points randomly and might allow certain domains of variable space to be missed or unsampled, especially in problems with large dimension sizes. In this work, in order to check the performance of these sampling strategies, additional data sets are generated via this sampling approach equal to the size of linear sampling with sizes of 3, 4, and 5, which correspond to total sample sizes of 81, 256, and 625 using Latin hypercube sampling. Also, for the holdout data set, 100 randomly selected scenarios will be used for determining the overfit or underfit performance of the generated surrogate models.

Here it is worth mentioning that the simulation time required for a single unit element is around 63 s. Thus, as can be seen from table 10.2, with increased sampling, the total computational cost of the surrogate models increases, which is also the most dominant cost for the whole design process.

Table 10.2. The computational cost of each data set is calculated as the simulation time for a single unit element, which is 63 seconds.

Data set sample size	Required time [min]
81	85
256	269
625	656
2401	2521
10 000	10 500

10.5 Artificial intelligence based surrogate modeling

In this section, a brief explanation of commonly used and state-of-the-art regression algorithms is presented. Here algorithms such as Artificial Neural Networks (ANNs), Support Vector Regression Machines (SVRMs), Ensemble Learning (EL), Gaussian Process Regression (GPR), and Deep Neural Networks (DNNs) will be studied as potential algorithms for creating a surrogate model for mapping between the reflection phase of the RA element and its geometrical design variables over the given operation band.

10.5.1 Artificial neural networks

Computer models called Artificial Neural Networks (ANNs) are modeled after the biological neural networks that exist in the human brain [25]. They are made up of layered networks of interconnected processing units known as neurons or nodes. The artificial neuron is the basic unit of an ANN. It takes input from other neurons, processes it, and then sends its output to the layer of neurons below it. Regression, classification, and pattern recognition are just a few of the many tasks that ANNs are well suited for since they can learn complicated patterns. ANNs have been a successful and well-liked method for approximating complex relationships between input and output variables in the field of surrogate modeling. The creation of substitute models for computationally intensive simulations, such as electromagnetic simulations in microwave antenna design, has been done extensively using these techniques. The main benefit of employing ANNs as surrogate models is their capacity to generalize from sparse data and learn non-linear correlations, which makes them very good at modeling complex systems. ANNs can be taught using a variety of learning techniques, like backpropagation, which modifies the network's weights and biases to reduce the difference between expected and actual outputs. The computational cost of running full-scale simulations is considerably reduced once the ANN has been trained to quickly predict new input data.

10.5.2 Support vector regression machine

Support Vector Regression (SVR) is a type of machine learning algorithm that falls under the umbrella of Support Vector Machines (SVMs) [26]. They are known for their efficiency in handling regression problems. The SVR technique operates under the principle of structural risk minimization. Its objective is to identify the optimal function that can generalize effectively to new data while simultaneously minimizing the prediction error on the training data. The fundamental concept underlying SVR is to establish a boundary around the regression function wherein a specific proportion of data points are permitted to exist while simultaneously reducing the intricacy of the function to guarantee optimal generalization abilities. SVR has proven to be effective in surrogate modeling for approximating intricate relationships between input and output variables. This is especially true in scenarios where the underlying function is non-linear and the number of samples available is limited. SVR presents various benefits in comparison to alternative regression methods. For instance, it can effectively manage high-dimensional data and exhibit superior generalization performance by prioritizing the minimization of structural risk.

10.5.3 Ensemble learning

Ensemble learning is a widely used machine learning approach that involves the integration of predictions from multiple base models to enhance the overall accuracy and performance of predictions [27]. Ensemble learning is based on the fundamental principle that the combined knowledge of a variety of models, each with its own distinct advantages and disadvantages, can result in more precise and resilient forecasts than a solitary model. Ensemble learning is a widely used approach in machine learning that involves training and combining multiple base models. Some of the commonly used ensemble learning techniques include bagging, boosting, and stacking, which employ different strategies to achieve this goal. Ensemble learning has become a prominent technique in surrogate modeling as it has demonstrated significant potential for improving the predictive precision and generalization abilities of individual models. Ensemble learning is a technique that involves combining the predictions of multiple surrogate models, each of which has been trained on the same or distinct subsets of data, or with different algorithms. This approach enables the identification of a wider spectrum of relationships between input and output variables. The aforementioned ability holds significant value in scenarios where the underlying function exhibits complexity or non-linearity and the sample size is restricted.

10.5.4 Gaussian process regression

Gaussian Process Regression (GPR), which is also referred to as Kriging or Gaussian Process Modeling, is a non-parametric and probabilistic method for regression [28]. This approach offers predictions as well as an estimation of the associated uncertainty. GPR is founded on the fundamental principle of Gaussian Processes. These are ensembles of stochastic variables that exhibit a Gaussian distribution when considered together. The Gaussian Process is employed in regression to establish a prior distribution over functions, which is subsequently revised with observed data to yield *a posterior* distribution that can be utilized for predictive purposes. Surrogate modeling employs GPR as a versatile and potent technique to estimate intricate associations between input and output variables. GPR possesses a notable advantage over alternative regression methodologies in that it has the capacity to provide a quantifiable level of uncertainty for each individual prediction. This attribute can prove advantageous in scenarios that require decision-making and optimization tasks. Moreover, GPR has the capability to adjust itself automatically to the intricacy of the fundamental function and is notably efficient in cases where the quantity of samples is restricted.

10.5.5 Deep neural network

Deep Neural Networks (DNNs) are a type of artificial neural network that possess numerous hidden layers, enabling them to acquire hierarchical data representations through learning [29]. DNNs are capable of acquiring intricate and abstract characteristics from the input data by utilizing multiple layers of neurons. This property makes them highly effective in resolving a diverse set of problems, including regression, classification, and pattern recognition. DNNs have been successfully employed in surrogate modeling to estimate intricate associations between input and output variables of highly complex problems. This is especially useful in scenarios where the underlying function is extremely non-linear and the available data is of high dimensionality. DNNs offer various benefits compared to alternative regression methods. These include the capability to automatically acquire intricate data representations, superior generalization performance due to their deep architecture, and the ability to manage data sets of considerable size. In this work, Modified Multilayer Perceptron (M2LP) is taken as the DNN model for this problem [30].

10.5.6 Hyperparameter optimization

Hyperparameters refer to adjustable parameters that control the training procedure and overall structure of machine learning and artificial intelligence algorithms [31]. In contrast to model parameters that are acquired through the training process, hyperparameters are predetermined prior to training and have a significant impact on the models' performance and generalization abilities. Hyperparameters are a set of parameters that are not learned during the training process of a machine learning model. They include various factors such as the learning rate, number of hidden layers and neurons, regularization strength, and kernel functions in support vector machines. The significance of optimizing hyperparameters can be denoted as follows:

- Enhancing the predictive accuracy and generalization capabilities of machine learning models can be achieved by selecting appropriate hyperparameters, leading to improved model performance. An optimally calibrated model has a higher probability of accurately identifying the fundamental patterns inherent in the data set while avoiding issues of overfitting or underfitting.
- The optimization of hyperparameters can result in expedited convergence in the training process, thereby diminishing the total computational expenses and time needed for model creation.
- The optimization of hyperparameters can lead to the enhancement of model interpretability, which holds significant importance in domains where comprehending the fundamental decision-making process is critical, such as financial risk assessment or medical diagnosis.
- The robustness of a model can be enhanced by tuning it appropriately with suitable hyperparameters. This can enable the model to effectively handle variations in input data and effectively manage noise or outliers.

The optimization of hyperparameters has a direct impact on the complexity and capacity of models [32]. Augmenting the quantity of hidden layers or neurons in a neural network can enhance the model's capabilities. Nonetheless, this phenomenon may lead to a higher likelihood of overfitting if regularization techniques are not adequately implemented. Additional parameters that can be optimized include regularization and generalization. The selection of regularization hyperparameters is a pivotal factor in managing the balance between model intricacy and generalization. The selection of an appropriate learning rate in gradient-based optimization algorithms can guarantee stable and efficient convergence. In addition to the aforementioned parameters, it is imperative to take into account the algorithm-specific effects. The selection of hyperparameters can have distinct impacts on various algorithms. The type determination of decision boundaries in support vector machines can be attributed to the kernel function, while the expressiveness and generalization capabilities of decision trees can be influenced by the depth of the trees. Undoubtedly, hyperparameter optimization constitutes a crucial stage in the development of efficient machine learning and artificial intelligence models. Through the appropriate tuning of hyperparameters, professionals can substantially improve the efficiency, generalizability, and resilience of their models, resulting in heightened precision of predictions and enhanced decision-making abilities.

10.5.7 Benchmarking

The performance comparisons of the mentioned surrogate models are compared for the generated data sets in section 1.3. The hyperparameters of all given models in table 10.2 are determined via a Bayesian optimization-guided search protocol. Thus, by this means, the performance results of each surrogate model with respect to their optimal hyperparameters and size of training sets are studied in detail. Here, a k-fold validation with $K = 3$ is taken for the optimization process of hyperparameters, while a randomly generated data set with 100 samples is used to evaluate the overall performance of each model. The given results in Tables 10.2 and 10.3 are averaged after 10 independent runs to clearly present the accuracy and precision performance of each model based on the average value and standard deviation (std) of these 10 runs based on the mean absolute error (MAE) for two different sampling methods. As can be seen from the presented results, the effect of the training data sample size is extremely effective, with the overall MAE decreasing from 5.6 to 11.9 just for ANN models (table 10.3). As can be seen from tables 10.3 and 10.4, the effect of the sampling method is also a crucial parameter for the performance evaluation of models. As mentioned before, linear sampling is not a sufficient method for problems with high dimension and in cases of a dominant effect of a parameter compared to others. This effect can be clearly seen from the given MAE results in tables 10.3 and 10.4. As for the complexity of the model, as can be seen, the deep learning algorithm M2LP has the highest accuracy and precision among all studied algorithms. However, due to its complex training procedures, its average training time for a single run is higher than all the counterpart algorithms. This should be understood, but this training time is certainly a drawback in terms of the computational efficacy of the whole process. This cost of training time becomes a neglected value for repeated design procedures and optimization of unit element designs.

Table 10.3. Performance comparison of data-driven surrogate models using the linear sampling method.

Model	Hyperparameters	Data set size	Holdout average + std	Average training time minutes
ANN	Hidden layer = 2; neurons size = 15–20; activation-function tansig	81	66.3 ± 8.3	4.1
		256	34.1 ± 6.6	9.3
		625	21.9 ± 5.3	14.4
SVRM	Kernel-function: Gaussian, type = epsilon = 0.015.	81	55.1 ± 6.2	8.1
		256	28.0 ± 4.6	16.4
		625	16.1 ± 4.0	29.4
Ensemble learning	Learning rate = 0.15; depth = 6; estimators #= 3150	81	58.4 ± 10.6	3.1
		256	33.1 ± 4.4	7.9
		625	19.1 ± 3.2	12.4
Gaussian process regression	Kernel-function = matern5/2; prediction-method = block-coordinate-descent; block-size = 1000	81	59.4 ± 7.6	8.1
		256	25.7 ± 4.7	16.4
		625	16.2 ± 5.1	29.4
M2LP	Depth = 2, initial neuron #= 32	81	43.0 ± 7.1	10.1
		256	22.4 ± 3.9	19.6
		625	12.6 ± 2.3	30.2

Table 10.4. Performance comparison of data-driven surrogate models using the Latin hyper-sampling method.

Model	Hyperparameters	Data set size	Holdout average + std	Average training time (min)
ANN	Hidden layer = 3; neurons size = 10–20–30; activation-function logsig	81	56.3 ± 6.3	4.1
		256	24.1 ± 4.3	9.3
		625	11.9 ± 3.1	14.4
SVRM	Kernel-function: polynomial, polynomial order = 3.	81	35.1 ± 5.0	8.1
		256	19.0 ± 3.9	16.4
		625	8.1 ± 2.1	29.4
Ensemble learning	Learning rate = 0.10; depth = 3; estimators #= 1150	81	48.4 ± 9.4	3.1
		256	22.1 ± 3.5	7.9
		625	14.1 ± 2.5	12.4
Gaussian process regression	Kernel-function = ard-matern3/2; prediction-method = block-coordinate-descent; block-size = 2000	81	39.4 ± 4.6	8.1
		256	16.7 ± 2.7	16.4
		625	9.2 ± 3.9	29.4
M2LP	Depth = 3, initial neuron #= 128	81	33.0 ± 4.8	10.1
		256	10.4 ± 2.2	19.6
		625	7.1 ± 1.1	30.2

It should be noted that while choosing an optimization technique is crucial in general, in our case it is only of secondary importance. Any global search technique can be employed and set up with a virtually infinite computational budget if it has a quick surrogate model. Even millions of objective function evaluations would not incur significant expenses in comparison to a single EM simulation RA unit cell given a model evaluation period of roughly 1 ms. Direct EM-driven optimization, however, is typically too expensive, where the unit cell's EM simulation takes around a minute. The CPU time of the search procedure reaches 16.6 h with a computational budget of 1000 objective function evaluations.

10.6 Results and discussion

Reflectarray antennas (RAs) are a class of antenna systems that combine the advantages of parabolic reflector antennas and phased array antennas. A frequency-selective surface (FSS) is a type of metamaterial that is designed to selectively reflect or transmit certain frequencies of electromagnetic waves while blocking or absorbing others. However, design and optimization of these stages are challenging problems that require a considerable amount of time and computational power due to the high-end hardware requirements and high mesh configuration of large-scale array designs in RAs. One of the most efficient solutions for this problem is the utilization of data-driven surrogate models. In this work, for fast and accurate modeling of RA design, an artificial intelligence based data-driven surrogate modeling approach is taken into consideration for achieving the challenges of the high computational cost of design optimization of a RA unit element.

Here, parameters such as training data set size, method of sampling, and surrogate model algorithm are studied for creating a RA unit element. Based on the observed results, the effect of the training data set size is the most dominant parameter, which is a predictable fact. However, determination of the size of training data sets is another crucial matter since excessive numbers of training samples would also reduce the computational efficiency of the surrogate modeling process significantly, as can be seen from table 1.2. Another important parameter is the sampling method. As can be seen from the modeling results, linear sampling is not a sufficient method for the selected problem, where the best algorithm (M2LP) achieves a MAE value of 12.6 while the same model achieves a performance value of 7.1 for Latin hypercube sampling, which is a more suitable method for problems with a more dominant effect in the modeling space. The last studied parameter for surrogate modeling is the complexity of the modeling algorithm. With the studied AI methods, it can be clearly observed that the state-of-the-art deep learning algorithm M2LP has the highest accuracy and precision among all the studied algorithms, at the cost of having a slightly higher training time compared to the counterpart algorithms. However, it should be understood that, although this training time is certainly a drawback in terms of the computational effectiveness of the whole process. This cost of training time becomes a negligible value for repeated design procedures and optimization of unit element designs.

Thus, based on the examined results, a detailed design procedure strategy for modeling reflectarray unit elements using AI-based surrogate models is studied. As a conclusion, for modeling high non-linear problems with a high number of dimensions, the use of Latin hypercube sampling alongside an appropriate amount of training samples would not harm the overall computational efficiency of surrogate modeling. As a final observation, the usage of state-of-the-art models alongside automated training procedures for surrogate models such as M2LP+ Bayesian optimization is a significant method to achieve a surrogate model with high accuracy and precision.

10.7 Challenges and future works

In future works, the authors aim to study more complex unit element structures suitable for different applications, such as: (I) RA designs with a wider range of operation bands and multi-beam radiation patterns; (II) transmitarray models where instead of a single output (reflection phase), two outputs (transmission magnitude and phase) are required for modeling unit elements. In either of the cases, the unit element must have higher capabilities for manipulating the incoming EM waves to achieve the desired transmitting or reflection characteristics. For this to happen, the unit element needs to be more involved, either with more complex geometrical designs and/or an increased number of layers that directly increase the total number of design variables.

The main challenge of any surrogate modeling procedure is the curse of dimensionality, which usually comes into play in problems with higher numbers of design variables. This leads to two main problems: (I) low-quality sampling strategy: the training and test data sets could not accurately cover the whole design space due to the large dimensionality of the problem; (II) the surrogate model fails to create an accurate mapping between the given input and outputs due to the course of dimensionality and fails into either underfitting or overfitting. This problem can only be solved via the use of novel and involved sampling strategies alongside state-of-the-art surrogate modeling techniques such as automated deep learning methods. This is also another aim of future work of the authors, to study and propose novel automated regression algorithms based on deep learning algorithms for the modeling of highly accurate surrogate models.

References

[1] Nayeri P, Yang F and Elsherbeni A Z 2018 *Reflectarray Antennas: Theory, Designs, and Applications* (New York: Wiley)

[2] Berry D, Malech R and Kennedy W 1963 The reflectarray antenna *IEEE Trans. Antennas Propag.* **11** 645–51

[3] Yoon J H, Lee W S and So J H 2015 Broadband microstrip reflectarray with five parallel dipole elements *IEEE Antennas Wirel. Propag. Lett.* **14** 1109–12

[4] Luo Q, Gao S, Zhang C, Zhou D, Chaloun T, Menzel W, Ziegler V and Sobhy M 2015 Design and analysis of a reflectarray using slot antenna elements for Ka-band SatCom *IEEE Trans. Antennas Propag.* **63** 1365–74

[5] Ozturk E and Saka B 2021 Multilayer Minkowski reflectarray antenna with improved phase performance *IEEE Trans. Antennas Propag.* **69** 8961–6
[6] Nayeri P, Yang F and Elsherbeni A Z 2015 Beam-scanning reflectarray antennas: A technical overview and state of the art *IEEE Antennas Propag. Mag.* **57** 32–47
[7] Al-Nuaimi M K T, Mahmoud A, Hong W and He Y 2000 Design of single-layer circularly polarized reflectarray with efficient beam scanning *IEEE Antennas Wirel. Propag. Lett.* **19** 1002–6
[8] Nayeri P, Yang F and Elsherbeni A Z 2013 Design of single-feed reflectarray antennas with asymmetric multiple beams using the particle swarm optimization method *IEEE Trans. Antennas Propag.* **61** 4598–605
[9] Deng R, Xu S, Yang F and Li M 2016 A single-layer high-efficiency wideband reflectarray using hybrid design approach *IEEE Antennas Wirel. Propag. Lett.* **16** 884–7
[10] Lee S G, Nam Y H, Kim Y, Kim J and Lee J H 2022 A wide-angle and high-efficiency reconfigurable reflectarray antenna based on a miniaturized radiating element *IEEE Access* **10** 103223–9
[11] Yang W, Chen K, Zhao J, Jiang T and Feng Y 2023 A wideband high-efficiency transmit-reflect-array antenna for bidirectional radiations with distinct circular polarizations based on a metasurface *IEEE Trans. Antennas Propag.* **71** 3695–700
[12] Yang J, Xin H and Tang M C 2023 Broadband, low-profile, planar reflectarray antenna based on an achromatic metasurface *IEEE Trans. Antennas Propag.* **71** 5440–5
[13] Serup D E, Pedersen G F and Zhang S 2023 Electromagnetically controlled beam-steerable reflectarray antenna *IEEE Trans. Antennas Propag.* **71** 4570–5
[14] Ali A, Khalily M, Brown T and Tafazolli R 2023 Beam-steering capability for OAM-based reflectarray at 5G-mmWave frequencies *IET Microw. Antennas Propag.* **17** 162–8
[15] Oliveri G, Salucci M and Massa A 2022 Towards efficient reflectarray digital twins - an EM-driven machine learning perspective *IEEE Trans. Antennas Propag.* **70** 5078–93
[16] Shi L P, Zhang Q H, Zhang S H, Liu G X and Yi C 2019 Application of machine learning method to the prediction of EM response of reflectarray antenna elements *2019 Photonics and Electromagnetics Research Symposium-Fall* pp 2671–6
[17] Prado D R, Lopez-Fernandez J A, Arrebola M and Goussetis G 2018 Support vector regression to accelerate design and crosspolar optimization of shaped-beam reflectarray antennas for space applications *IEEE Trans. Antennas Propag.* **67** 1659–68
[18] Mahouti P 2020 Application of artificial intelligence algorithms on modeling of reflection phase characteristics of a nonuniform reflectarray element *Int. J. Numer. Model.: Electron. Netw., Devices Fields* **33** e2689
[19] Mahouti P, Belen M A, Çalık N and Koziel S 2022 Computationally efficient surrogate-assisted design of pyramidal-shaped 3-D reflectarray antennas *IEEE Trans. Antennas Propag.* **70** 10777–86
[20] Robustillo P, Zapata J, Encinar J A and Rubio J 2012 ANN characterization of multi-layer reflectarray elements for contoured-beam space antennas in the Ku-band *IEEE Trans. Antennas Propag.* **60** 3205–14
[21] Prado D R, López-Fernández J A and Arrebola M 2023 Support vector regression models of reflectarray unit cell in a geometrical 4-D parallelotope domain around a rectangle of stability *IEEE Trans. Antennas Propag.* **71** 5089–99

[22] Salucci M, Tenuti L, Oliveri G and Massa A 2018 Efficient prediction of the EM response of reflectarray antenna elements by an advanced statistical learning method *IEEE Trans. Antennas Propag.* **66** 3995–4007

[23] Koziel S, Belen M A, Çalişkan A and Mahouti P 2023 Rapid design of 3D reflectarray antennas by inverse surrogate modeling and regularization *IEEE Access* **11** 24175–84

[24] Mahouti P, Güneş F, Belen M A and Demirel S 2017 Symbolic regression for derivation of an accurate analytical formulation using 'big data': an application example *Appl. Comput. Electromagn. Soc. J. (ACES)* **32** 372–80

[25] Agatonovic-Kustrin S and Beresford R 2000 Basic concepts of artificial neural network (ANN) modeling and its application in pharmaceutical research *J. Pharm. Biomed. Anal.* **22** 717–27

[26] Smola A J and Schölkopf B 2004 A tutorial on support vector regression *Stat. Comput.* **14** 199–222

[27] Dong X, Yu Z, Cao W, Shi Y and Ma Q 2020 A survey on ensemble learning *Front. Comput. Sci.* **14** 241–58

[28] Schulz E, Speekenbrink M and Krause A 2018 A tutorial on Gaussian process regression: Modelling, exploring, and exploiting functions *J. Math. Psychol.* **85** 1–16

[29] Miikkulainen R *et al* 2018 Evolving deep neural networks *Artificial Intelligence in the Age of Neural Networks and Brain Computing* (New York: Academic) pp 293–312

[30] Mahouti T, Yıldırım T and Kuşkonmaz N 2021 Artificial intelligence-based design optimization of nonuniform microstrip line band pass filter *Int. J. Numer. Model.: Electron. Netw., Devices Fields* **34** e2888

[31] Claesen M and De Moor. B 2015 Hyperparameter search in machine learning *MIC 2015: the XI Metaheuristics International Conference* **14** 1–5

[32] Hutter F, Kotthoff L and Vanschoren J 2019 *Automated Machine Learning: Methods, Systems, Challenges* (Springer Nature)

Chapter 11

Metamaterials for sensing and biomedical applications

Muhamad Abuzar Baqir and Muhammad Saqlain

Symbols

ε	Permittivity
S	Sensitivity
λ	Wavelength (m)
n	Refractive Index
f	Volume fraction of Metal
k	Wave number
ω	Angular frequency

11.1 Introduction

Metamaterials are considered artificially manipulated materials that lead to several excellent properties including the use of the negative refractive index for reflection and refraction of electromagnetic waves, backward waves transmission, use of Doppler's shift in the reverse direction, improved-resolution-based imaging, accurate sensing, and negative Cherenkov radiations that are commonly not found in other natural existing mediums [1–9]. They interact with electromagnetic waves strongly. Therefore, they produce a localized field that makes them favorable for developing new devices.

Within this context, we discuss the use of metamaterials in several practical medical applications, such as imaging and bio- and chemical sensing. For sensing, several reported works show the use of metamaterials to enhance a sensor's performance by improving its quality factor and sensitivity [10–14]. Metamaterial-based sensors have been investigated from microwave to optical regimes and can sense temperature, pressure, humidity, and chemical- and bio-substances [15–23]. Their operation covers microwave, terahertz, infrared, and visible spectra of

doi:10.1088/978-0-7503-5754-8ch11

electromagnetic waves. We compare metamaterial-based sensors with conventional sensors—those are the electrochemical devices that work to detect the change in the chemical properties and concentration of the observed substance upon the chemical reaction and comprise reference electrodes and electrolyte solutions. This substance reacts with the electrode to alter the chemical properties of the electrode, and a change in the chemical properties detects the substance [24–26]. On the other hand, metasurface-based sensors are made of engineered media and usually rely on the optical properties of the wave propagation medium [27–29]. Properties such as reflection, absorption, transmission, and phase are used for substance detection. Metasurface-based sensors feature compact sizes and are cost-effective compared to their counterparts. The metasurface is periodically arranged subwavelength-sized unit cells that manipulate the incoming waves in fantastic ways, therefore becoming attractive to researchers [30–36]. It has been reported that a metamaterial is a low-cost and ubiquitous paper substrate and is used for the detection of liquid [21, 37–39]. It works on the phenomenon of change in resonance due to varying the effective capacitance of each resonator in the metamaterial-based structure [40].

Reference [41] demonstrates a ring cylindrical metasurface composed of all-dielectric materials and shows ultra-high sensitivity due to breaking the symmetry of the cylinder that produces a high Q-factor of the transmitted wave, and therefore it can be useful for the detection of biomolecular solution. Moreover, sensors composed of a metasurface made of 2D materials show high sensitivity and figure of merit (FOM). Ma *et al* recently reported that a graphene-based metamaterial sensor operating in the terahertz regime showed a high-quality factor, sensitivity, and FOM [42]. Another work shows a metasurface-based sensor made of ultra-sparse nanowires grids placed on a metallic surface operating in the near-infrared regime that shows high sensitivity of 1152 nm RIU^{-1}, with a FOM of 2768. This result indicates the high performance of this proposed refractive index-based sensor [43]. Similarly, the authors of [44] present a metamaterial-based sensor that can detect an ultra-low change in the refractive of 11^{-11} RIU using a phase change medium [44]. Besides that, metasurface-based biosensors are also useful for the detection of infectious diseases; for example, metasurface biosensors show high sensitivity and have high-throughput DNA detection for nucleic acid targets [12].

Recently, hyperbolic-shaped metamaterial for sensing applications has gained the attention of the research and development communities. Hyperbolic metamaterial/metasurface (HMM) are artificial mediums—termed hyperbolic due to their hyperbolic dispersion characteristics. They are anisotropic in nature and have shown strong interactions with propagating electromagnetic waves. They have shown metallic features in the one principal axis, and positive permittivity on the other principal axis, which make these materials interesting for sensing applications. Within this context, HMM-based sensors have attained higher sensitivity compared to sensors made of other types of metamaterials. Researchers have demonstrated that [14, 45, 46] HMM-based sensors have attained ultra-high sensitivity compared to the previously reported sensors based on metamaterials. This chapter comprises the following sections: section 11.2 presents the analytical treatment and results of several state-of-the-art biosensors using a Kretschmann configuration. In section 11.3, a nanoscale optical sensor for cancer detection is analyzed. In section 11.4, a metasurface-based nanoscale sensor for temperature sensing is analyzed. The conclusion and future work can be found in section 11.5.

11.2 Theory and analytical treatment of a prism-coupled waveguide sensor

Figure 11.1(a) shows the schematic of a sensor. The structural arrangement of the prism-coupled waveguide sensor (PCWS) comprises a measurand (solution), HMM layer, substrate of SiO_2 glass, and a prism. Figure 11.1(b) shows the HMM layer composed of periodically arranged silver (Ag) nanostrips. The different parameters such as refractive indices (RIs) of the considered solution, HMM, Si-made substrate, and prism are represented by n_s, n_{HMM}, n_g and n_p, respectively. The thickness of the

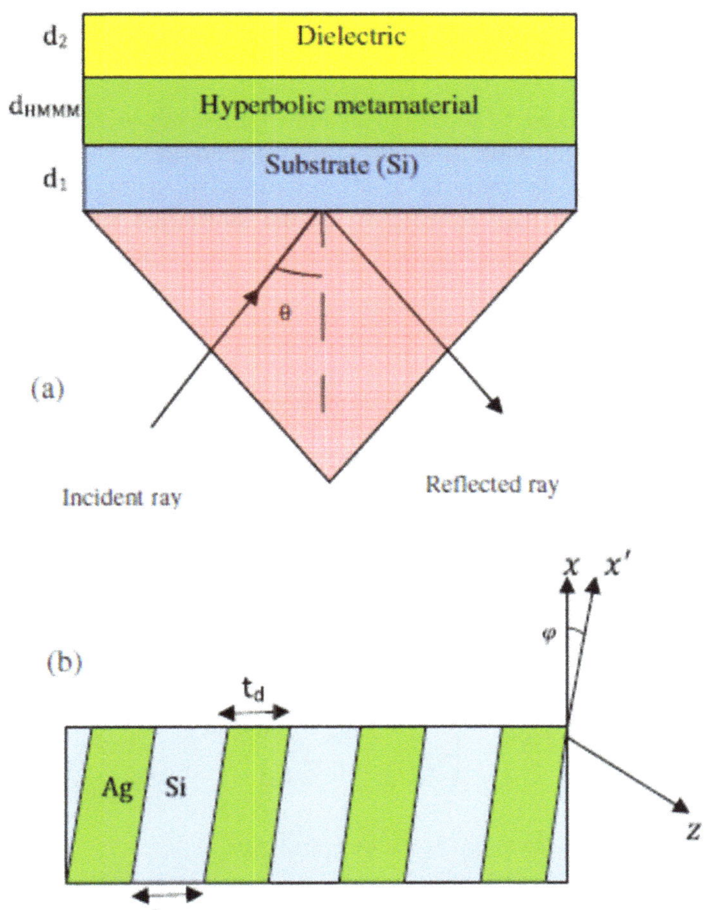

Figure 11.1. (a) Conceptual schematic representation of the sensor comprising a hyperbolic metamaterial-based structure; (b) hyperbolic metamaterial layer. Reprinted with permission from [22] ©The Optical Society.

measurand, HMM, and silicon (Si) substrate are represented by d_2, d_{HMM}, and d_1, respectively. The thickness of the measurand (solution under investigation) and glass substrates are taken at 5 μm.

Here, we present the interaction of light waves with a hyperbolic-shaped sensor structure. The light wave falls on the prism surface at 45° as shown in figure 11.1(a). The following refractive indices values are taken $n_p = 2.72$ and $n_g = 3.7$, respectively. The sensor structure is composed of an HMM nanolayer and therefore exhibits high dispersion characteristics. Moreover, effective medium theory is used for the calculation of the hyperbolic structure permittivity [47]; similarly, the Lorentz–Drude model is widely adopted to find the bulk layer permittivity [47]. It is important to note that the plasmon effect is created at the interface of the dielectric (measurand) and metal layers [48].

The wave vector of the incident wave changes after passing through different mediums, and can be expressed as

$$k_{\text{iz}} = \frac{2\pi}{\lambda_o}\sqrt{n_i^2 - n_p^2 \sin\theta} \tag{11.1}$$

Wherein subscript i is used to represent layers' numbering, θ and λ_{o} represent the incident angle of the light wave and operating wavelength.

The hyperbolic layer is made of a silver nano-strip, using an adopted model related to finding the effective permittivity of the multilayer medium. We can find their corresponding components in the longitudinal and transverse directions [47].

$$\varepsilon_{\|}(\lambda) = \varepsilon_m(\lambda)f + \varepsilon_d(1 - f) \tag{11.2a}$$

$$\varepsilon_{\perp}(\lambda) = \frac{1}{(f/\varepsilon_m(\lambda) + (1 - f)/\varepsilon_d)} \tag{11.2b}$$

Here, the above equation contains several variables that represent different parameters with the following details; $\varepsilon_m(\lambda)$, ε_d, f are permittivities of the metal-made layer and dielectric media, and the filling factor of the silver metal, further defined as $f = \frac{t_m}{t_m + t_d}$, and t_m, and t_d are nanometer sized metallic and dielectric strip widths, respectively. The mathematical representation of the permittivity of the hyperbolic structure exists in all the coordinates with non-diagonal features [49].

$$\varepsilon_2(\lambda) = \begin{pmatrix} \varepsilon_{\text{xx}}(\lambda) & 0 & \varepsilon_{\text{xz}}(\lambda) \\ 0 & \varepsilon_{\|}(\lambda) & 0 \\ \varepsilon_{\text{zx}}(\lambda) & 0 & \varepsilon_{\text{zz}}(\lambda) \end{pmatrix} \tag{11.3}$$

$$\varepsilon_{\text{xx}}(\lambda) = \varepsilon_{\|}(\lambda)\cos^2\varphi + \varepsilon_{\perp}(\lambda)\sin^2\varphi \tag{11.4a}$$

$$\varepsilon_{\text{zz}}(\lambda) = \varepsilon_{\|}(\lambda)\sin^2\varphi + \varepsilon_{\perp}(\lambda)\cos^2\varphi \tag{11.4b}$$

$$\varepsilon_{\text{xz}}(\lambda) = \varepsilon_{\text{zx}}(\lambda) = \left(\varepsilon_{\perp}(\lambda) - \varepsilon_{\|}(\lambda)\right)\sin\varphi\cos\varphi \tag{11.4c}$$

When considering polarization of the transverse magnetic (TM) mode of excitation, the existence of field due to the TM mode is represented as $H = \hat{y}H_y(z)e^{j(k_x x-\omega t)}$. Therefore, source-free Maxwell equations can be modified after inserting equation (11.3) and then the wave equation can be written as

$$\frac{d^2H_y}{dz^2} + 2j\frac{\varepsilon_{xz}(\lambda)}{\varepsilon_{zz}(\lambda)}\frac{dH_y}{dz} + \left(\frac{\omega^2}{c^2}\frac{\varepsilon_{\parallel}(\lambda)\varepsilon_{\perp}(\lambda)}{\varepsilon_{zz}(\lambda)} - \frac{\varepsilon_{xx}(\lambda)}{\varepsilon_{zz}(\lambda)}\right)H_y = 0 \tag{11.5}$$

The transfer matrix of the hyperbolic structure can be represented as [49]

$$M_2(\lambda) = e^{j(\alpha_1 d_2)}\begin{bmatrix} \cos(\alpha_2 d_{\text{HMM}}) & \frac{-j}{q_2}\sin(\alpha_2 d_{\text{HMM}}) \\ -jq_2\sin(\alpha_2 d_{\text{HMM}}) & \cos(\alpha_2 d_{\text{HMM}}) \end{bmatrix} \tag{11.6}$$

Some variables used in equation (11.6) can be defined as

$$\alpha_1 = \frac{\varepsilon_{xz}(\lambda)}{\varepsilon_{zz}(\lambda)}k_x,\ \alpha_2 = \left(\frac{\omega}{c}\right)\sqrt{\varepsilon_{\parallel}(\lambda)\varepsilon_{\perp}(\lambda)(\varepsilon_{zz}(\lambda) - \sin^2\theta)/\varepsilon_{zz}^2(\lambda)}$$
$$\text{and } q_2 = \varepsilon_{zz}(\omega)\alpha_2/\varepsilon_0\omega\varepsilon_{\parallel}\varepsilon_{\perp}.$$

When more specific to layer 1 and layer 3 of the sensor structure, we can express the transfer matrix as

$$M_k(\lambda) = \begin{bmatrix} \cos(\gamma_k d_k) & \frac{-j}{q_k}\sin(\gamma_k d_k) \\ -jq_k\sin(\gamma_k d_k) & \cos(\gamma_k d_k) \end{bmatrix} \tag{11.7}$$

wherein k=1 or 3.

Here, $\gamma_k = \left(\frac{\omega}{c}\right)\sqrt{\varepsilon_k - \sin^2\theta_i}$ and $q_k = (\gamma_k/\omega\varepsilon_o\varepsilon_k)$, where free space permittivity is represented by ε_o.

The resultant transfer matrix is the product of the transfer matrices of all three layers of the HMM sensor and therefore, is defined as

$$\begin{bmatrix} M_{11} & M_{12} \\ M_{21} & M_{22} \end{bmatrix} = M_1 M_2(\lambda) M_3 \tag{11.8}$$

We can conclude that $R(\lambda)$ can be found by using the mathematical procedure given in reference [50].

11.2.1 Results and discussion of PCWS

We discuss effective permittivity profiles of the hyperbolic-shaped metasurface layer when selecting f = 0.2 and 0.4 over the 800–2000 nm range. Figures 11.2(a) and (b), respectively, show the effective permittivity profiles when f is 0.2 and 0.4. The effective permittivity plots are derived by using the multilayer medium theory described by equation (11.2). From these corresponding figures, it is noticed that $(\Im(\varepsilon_{\parallel})$ remains slightly higher than zero, and shows an incremental trend as the wavelength increases (represented by dotted black lines). However, $(\Im(\varepsilon_{\perp}))$ remains

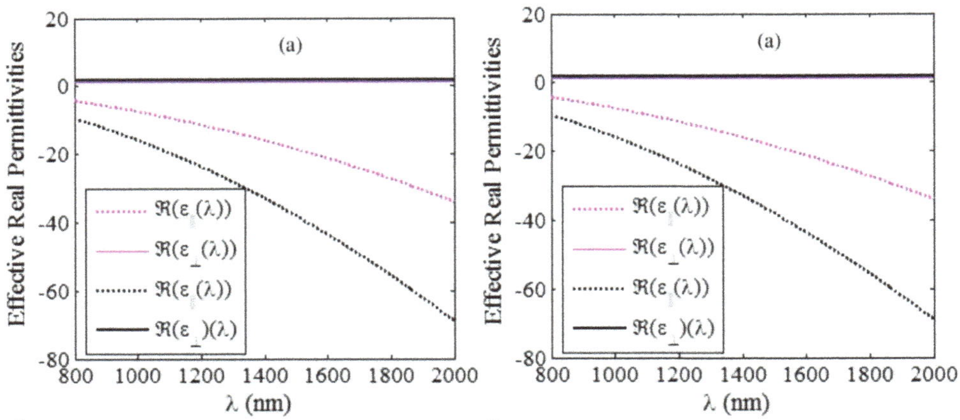

Figure 11.2. Effective permittivity profiles of the hyperbolic metasurface layer within the 800–2000 nm operating range when (a) $f = 0.2$ and (b) $f = 0.4$. Reprinted with permission from [22] © The Optical Society.

very close to zero, and it remains fixed for the entire wavelength range. Further, the corresponding plots show that $\Im(\varepsilon_{\|})$ also increases with the increase of parameter f (part (b) of figure 11.2 when f is taken 0.4. However, $(\Re(\varepsilon_{\|}))$ increases along the negative value axis when taking both values of f and increasing the operating wavelength, and this trend shows one of the prime features of the hyperbolic-shaped metasurface structure [22]. For both values of f (0.2 and 0.4), $\Re(\varepsilon_{\|})$ shows a negative increasing trend as the operating wavelength increases. We found that $f = 0.2$ $\Re(\varepsilon_{\|})$ is attained at a value of −4.5 at the operating wavelength of 800 nm, and when the operating wavelength increases and reaches up to 2000 nm, $\Re(\varepsilon_{\|})$ becomes −69. Similarly, when we take $f = 0.4$, $\Re(\varepsilon_{\|})$ remains the same as noticed in the previous case at the same operating wavelength of 800 nm. Whereas, for $\Re(\varepsilon_{\|})$ the value drops to −33 at 2000 nm. From this discussion, we can conclude that the increase in the anisotropic behavior of the hyperbolic-shaped structure is directly linked to when both operating wavelength and f increase.

Here, reflection spectra from the prism are analyzed using the PCWS. For this, it is important to investigate its sensitivity by calculating the attenuated total reflection (ATR) spectra. Notably, ATR is created by the LSPR effect at the HMM-dielectric interface. Figures 11.3(a) and (b) show the relationship of ATR when different thicknesses (200 and 300 nm) of the HMM layer are considered and the f value is 0.2. Moreover, the incident light wave is fallen on the top surface with an angle θ of 30°, whereas the slant angle is considered $\varphi = 20°$ for the silver stripes. The aforementioned parametric values of the sensor provide the best sensitivity. Considering figure 11.3(a), it is noticed that RI is 1.33 for water, and the occurrence of reflection minimum at the operating wavelength of 1695 nm is depicted by a solid black line. It can be seen from the corresponding plot (represented by a red dashed line) that a reflection minimum is achieved at the operating wavelength of 1735 nm when adding 11% of the aqueous solution of glucose and taking the refractive index value of 1.347. The ATR is also analyzed when increasing the concentration of the

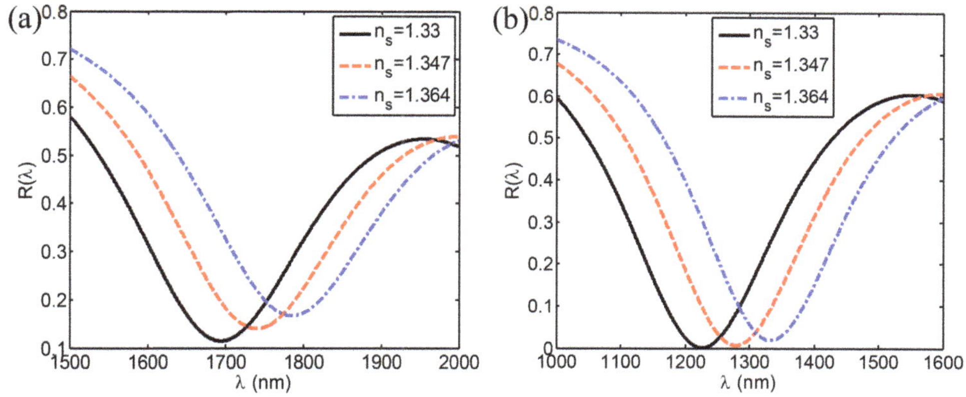

Figure 11.3. ATR for different concentration levels of glucose solution and considering different thicknesses of the hyperbolic-shaped metamaterial layer (a) 200 nm and (b) 300 nm. Reprinted with permission from [22] ©The Optical Society.

glucose solution up to 20% and considering the same refractive index value. With this parametric specification, the reflection minimum shows a shift in terms of its operating wavelength of 1792 nm, and this curve is represented in the dotted dashed blue line style. Here, the appearance of the red shift in the curve of reflection minimum is due to an increase in the concentration of the solution.

Now, we analyze ATR when varying the thickness of the hyperbolic-shaped metasurface layer as shown in figure 11.3(b), while the remaining parametric values are kept fixed. For this parametric combination, we can see a shift in the reflection minimum and a similar trend can be noticed as the concentration of the glucose solution increases when compared to the previous situation. Moreover, the reflection minimum for the water, and different concentration levels (11% and 20%) of glucose solution, are obtained at different operating wavelengths; 1227 nm, 1280 nm, and 1340 nm, respectively. In this case, an increase in shift can be observed compared to the situation discussed in figure 11.3(a).

Similarly, ATR is investigated when considering an increase in the volume fraction of the silver that is used in the composition of the HMM layer structure. First, the ATR of the PCWS is analyzed when taking f as 0.4, thicknesses of the HMM layer are taken as 200 nm and 300 nm, and their respective images are shown in figures 11.4(a) and (b). The remaining parametric values φ and θ are considered the same as discussed in the previous situation. In the present case, a change in the concentration of the glucose solution causes a reduction in the shift in the reflection minimum. Considering figure 11.4(a), it is observed for 0%, 11%, and 20% concentration glucose in aqueous solution, their respective reflection minima curves are noticed at the following operating wavelengths; 1935, 1960, and 1985 nm, respectively. Figure 11.4(b) depicts the situation wherein the HMM layer thickness is 300 nm and a blue shift occurs. Besides that, different concentration levels (0%, 11%, and 20%) of glucose cause a shift in the reflection minima at different operating wavelengths that cover a wide range of the spectra such as 1845, 1865, and 1885 nm.

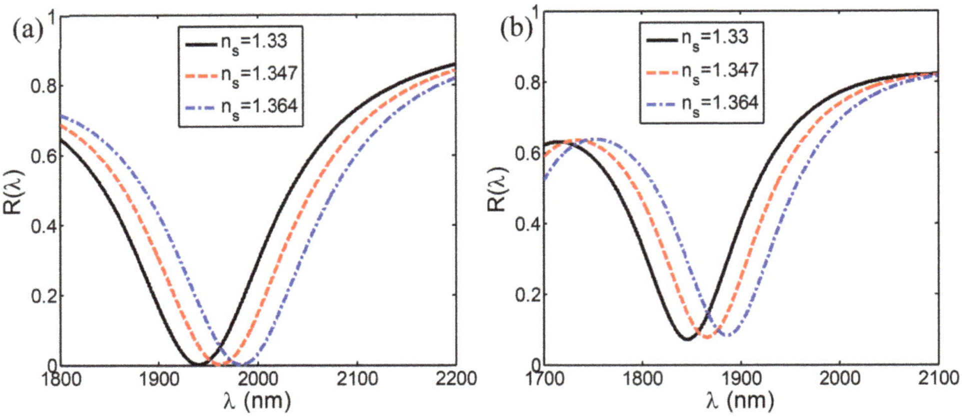

Figure 11.4. ATR for different concentration levels of glucose solution and thicknesses of the HMM layer as (a) 200 nm and (b) 300 nm. Reprinted with permission from [22] ©The Optical Society.

It can be concluded that an increase in the HMM layer thickness causes blue shift to the reflection minimum.

The sensitivity can be found due to the occurrence of a shift in the reflection minimum when the refraction index of the measurand changes and is defined as $S = \frac{\lambda_S - \lambda_W}{n_S - n_W}$,

where, λ_s and λ_w represent the wavelength of reflection minimum of the considered solution and water, respectively.

In addition to this, the sensitivity of the PCWS is analyzed when varying the thicknesses of the HMM layer over the wavelength range of 800 to 2000 nm. The thickness of the HMM layer varies in the range of 150–300 nm and figure 11.5 illustrates the sensitivity curve; however, the parameters φ and θ_{inc} are fixed at 11°. It is clear from the corresponding figure that for $d_{\text{HMM}} = 150$ nm, the sensitivity reaches up to 2000 (nm/RIU) and shows an increasing trend when the thickness increases from 150 to 250 nm. As a result, the sensitivity reaches a high value of ~ 3450 (nm/RIU); however, further incremental change in the thickness level reduces the sensitivity up to ~ 3450 (nm/RIU) when d_{HMM} reaches 300 nm.

Now, sensitivity is analyzed when considering different thickness levels of the HMM layer, light wave incidence angle, and strip slant angle (figure 11.6 corresponds to this situation). Figure 11.6(a) shows the sensitivity for different slant angles φ and different thickness levels represented by d_{HMM}. Here the incidence angle of the falling light is taken $\theta = 30°$. It is observed that sensitivity remains high for the range $5° < \varphi < 20°$. It is obvious from the corresponding figure that the slant angle varies in the range of $5° < \varphi < 20$ and the sensitivity value remains above 2000 nm/RIU. However, when the slant angles increase in the range of $\sim\varphi = 30° - 90°$, sensitivity becomes low. For instance, the sensitivity becomes low when $\varphi > 60°$ and attains a lesser value than 500 (nm/RIU). We can notice that the highest sensitivity is achieved when slant angles exist in the range of $\sim\varphi = 10° - 15°$ and also an increasing trend can be seen when the thickness level of the layer increases. In the context of sensitivity dependence on the thickness levels of the layer, it can be noted

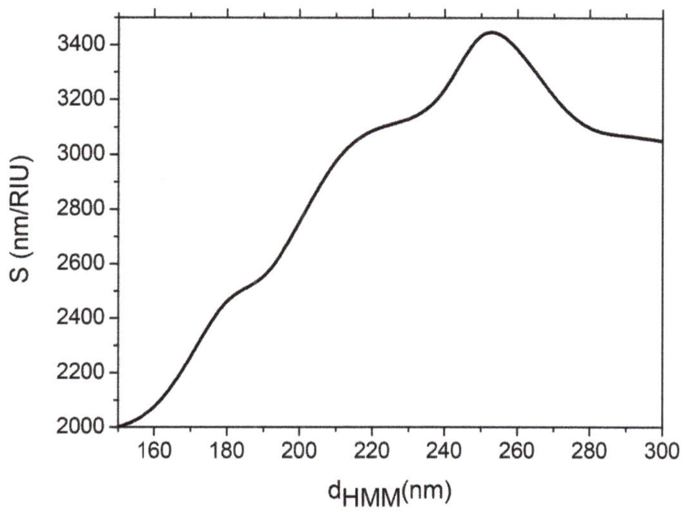

Figure 11.5. Sensitivity calculation at different thickness levels of the HMM layer and taking θ and φ of 10°. Reprinted with permission from [22] ©The Optical Society.

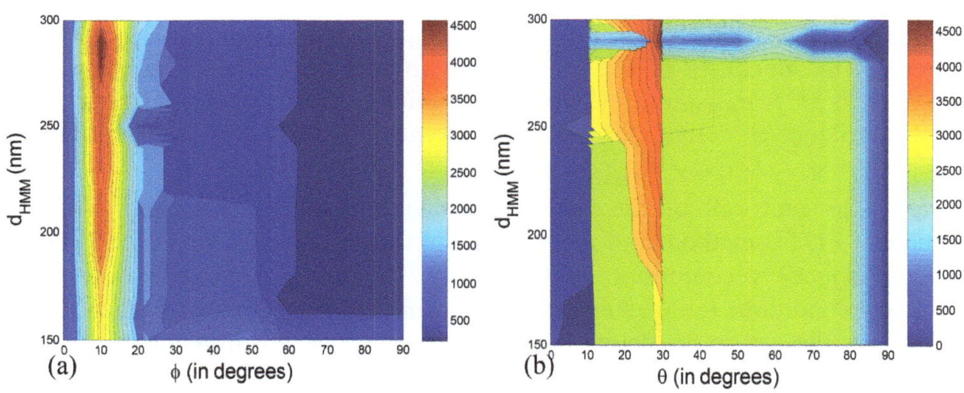

Figure 11.6. The sensitivity profile of the PCWS for different parametric changes (a) when varying layer thickness (d_{HMM}) levels and the slant angle (φ) (b) when varying layer thickness (d_{HMM}) and incidence angle (θ)n. Reprinted with permission from [22] ©The Optical Society.

from the corresponding plot that a high sensitivity value is achieved for the condition when the layer thickness remains within the ~280–290 nm and the slant angle is ~10°. From the aforesaid discussion, it can be inferred that the highest sensitivity of 4500 (nm/RIU) is at the lowest φ. Similarly, sensitivity is calculated when varying the angle of incidence of light as shown in figure 11.6(b) for the different thicknesses of the HMM layer and slant angle $\varphi = 10°$. It can be noticed from the corresponding plot that maximum sensitivity ~ 4500 nm/RIU is achieved for the incidence angle ranging from ~ 25°–30°, whereas the thickness of the HMM layer ranges from 270 to 300 nm. It is noticed that for $\theta < 10°$, and all considered thicknesses of the HMM layer, sensitivity remains lower and shows a similar trend throughout the range. By varying θ from 35° to 80°, sensitivity remains around 3000 nm/RIU for the thickness range of ~150–280 nm, whereas sensitivity is reduced for the increase in thickness of the HMM layer. It is noticed that sensitivity remains low for $\theta > 80°$ and $\theta < 15°$.

11.3 Hyperbolic metamaterial-based sensor for detection of cancer cells

A hyperbolic metamaterial/metasurface can be useful to detect various life-risk diseases, for example, to detect viral diseases and bacterial infections. A schematic of the sensor for cancer cell detection is shown in figure 11.7. The arrangement comprises a five layer structure. At the top layer, a sample of cancer cells is located on the HMM layer to detect cancer viruses. The HMM layer is made of a silver thin layer with a fraction of the wavelength sizes of the nanoholes that are arranged in a periodic fashion. After this, a thin layer of transparent conducting glass of ZrN is placed above the SiO_2 substrate. Here, variable thicknesses of the sensing medium are represented by d_s and the HMM layer thickness is represented by d_{HMM}, respectively. However, the thickness of the thin layer of ZrN is 200 nm, and the thicknesses of the silver-made layer, and SiO_2 layer are taken at 110 and 2000 nm, respectively. The dimension of the unit cell is $0.4 \times 0.4 \times l$ μm^3 for which l represents the thickness of the sensor. The volume of the nanoholes in the Ag nanolayer is $\pi r^2 \times d_{\mathrm{HMM}}$ μm^3, where r is the radius of the nanoholes. For hyperbolic metamaterial layer design, the volume fraction f of the silver is determined as $f = 1 - \pi r^2 \times \frac{h}{0.4} \times 0.4 \times d_{\mathrm{HMM}}$.

Here, the unit cell of the cancer-detecting sensor was simulated using the Lumerical FDTD method, and the xy-plane of the unit cell is bounded by applying periodic boundary conditions, and similarly, the z-axis is bounded with open boundary conditions (above and below the sample). The monitor is placed above and below the sensor to determine the reflection and transmission parameters. Furthermore, the TM mode is excited at the top of the structure.

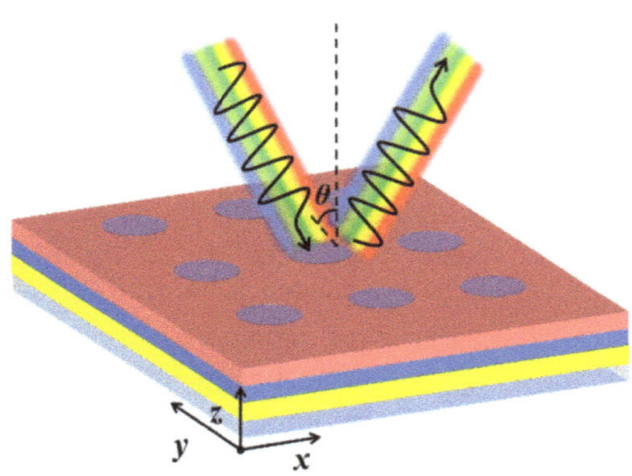

Figure 11.7. Conceptual configuration representation of the biosensor, © (2022) IEEE. Reprinted, with permission, from [23].

Generally, for the multilayer structure, effective medium theory (EMT) is widely used to find the effective permittivities. Therefore, we can find both parallel (||) and perpendicular (⊥) components based on the EMT reported in reference [47]

$$\varepsilon_{\|}(\lambda) = \varepsilon_m(\lambda)(f) + \varepsilon_d(1 - f) \tag{11.9a}$$

$$\varepsilon_{\perp}(\lambda) = \left(\frac{f}{\varepsilon_m(\lambda)} + \frac{1 - f}{\varepsilon_d}\right)^{-1} \tag{11.9b}$$

Here, wavelength-dependent parameter $\varepsilon_m(\lambda)$ represents the permittivity of the bulk metal and dielectric medium permittivity is represented by ε_d. The $\varepsilon_m(\lambda)$ can be deduced from the Lorentz–Drude model [48],

$$\varepsilon_m(\omega) = 1 - \frac{f_1 \omega_p^2}{\omega(\omega - j\gamma_1)} + \sum_n \frac{f_n \omega_p^2}{\omega_n'^2 - \omega^2 + j\omega\gamma_n'} \tag{11.11}$$

Here, the parameters used in equation (11.9) are: f_1, f_2, ω_p, γ_1, ω_n', and γ_n', representing oscillator strengths, plasma frequency, damping constant, frequency of oscillation, and operation bandwidth, respectively. It is important to note that both mediums (metallic and dielectric) have nonmagnetic properties when operating in the optical domain, i.e, using permeability $\mu = 1$. The refractive index of ZrN is deduced from reference [24].

11.3.1 Results and discussion

The hyperbolic property of this multilayer structure is determined by analyzing the constitutive parameters. As an illustrative case, the constitutive parameters are analyzed by considering equations (11.2a) and (11.2b) for $f = 0.5$. The effective permittivity curves (real and imaginary parts) of the hyperbolic-shaped metamaterial operating in the 500–800 nm range are shown in figure 11.8. From the plot, we can see that the transverse component of the real part of the permittivity $\Re(\varepsilon_{\perp})$ shows no change with a constant value of 2.45 over the entire operating region and features non-dispersive behavior represented by the red solid line. Whereas its tangential component shows a negative values trend when the wavelength increases and the parallel component features dispersive property. Notably, this situation with the condition when $\Re(\varepsilon_{\perp}) > 0$ and $\Re(\varepsilon_{\|}) < 0$ attribute to the hyperbolic nature of the metamaterial. Whereas the imaginary part of $\varepsilon_{\|}$ shows ~0.45 and a non-dispersive nature. When the effective permittivity shows a behavior of the following condition; $\Re(\varepsilon_{\perp}) > 0$ and $\Re(\varepsilon_{\|}) < 0$, that makes the hyperbolic-shaped metamaterial structure highly reflective.

Here, we present the reflection spectra of the cancer-detecting sensor which are considered for different cell samples. Commonly, the refractive index of a normal human cell is taken as 1.35. From figure 11.9(a), it can be observed that the minima reflection dip is achieved at the resonance wavelength of 631 nm when a light wave falls on a multilayer structure with a normal incidence angle. The parametric values of the sensor are kept as $r = 55$ nm, $d_{\mathrm{HMM}} = 150$ nm, and $d_s = 1100$ nm.

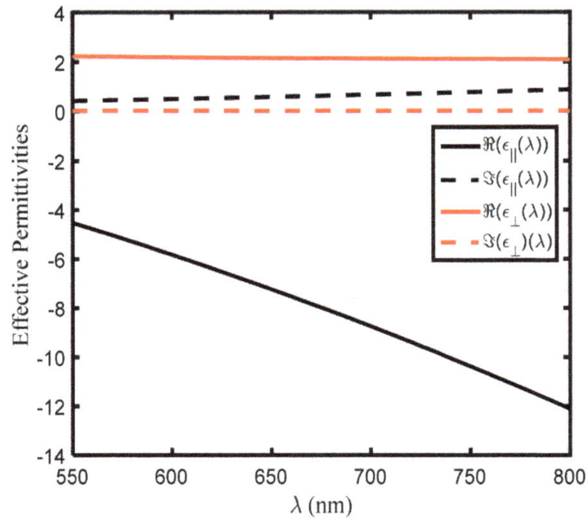

Figure 11.8. Constitutive parameter of the effective permittivity of the HMM, © (2022) IEEE. Reprinted, with permission, from [23].

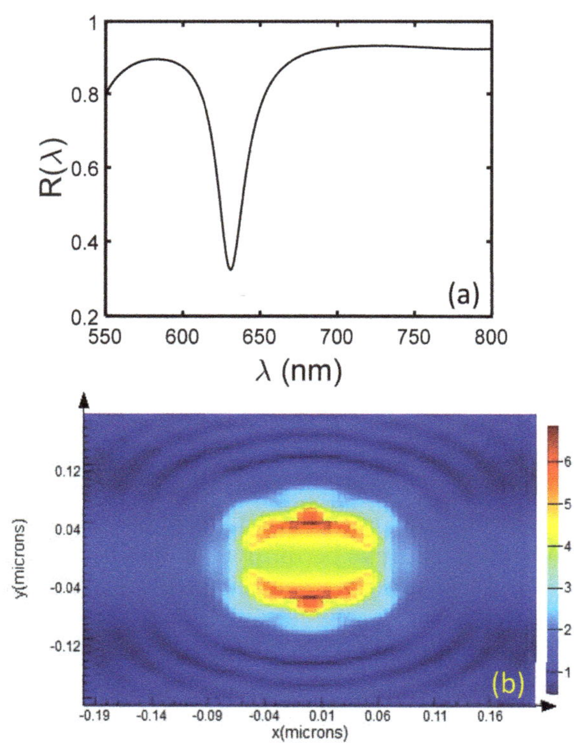

Figure 11.9. (a) Reflectance spectra of the proposed structure under normal incidence, and (b) electric field distribution corresponding to $\lambda = 631$ nm, © (2022) IEEE. Reprinted, with permission, from [23].

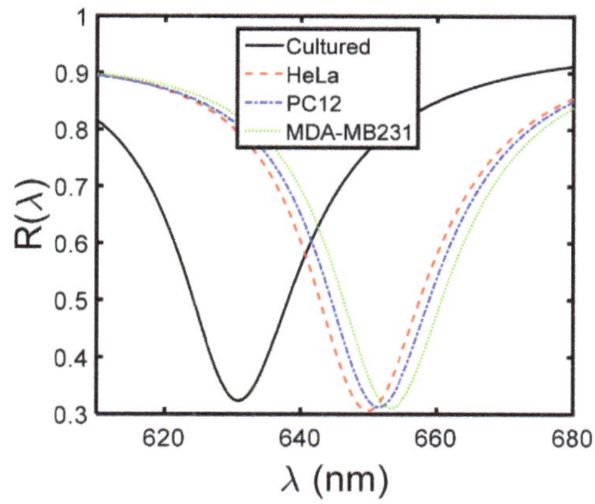

Figure 11.10. Reflection for the normal incidence of light for different cancer cells, © (2022) IEEE. Reprinted, with permission, from [23].

Notably, the existence of a reflection dip is a result of LSPR at the taken sample and HMM layer interface.

Similarly, figure 11.9(b) shows the surface electric field at resonance wavelength (λ_0 = 631 nm). It is noticed that the electric field is lesser and remains confined in the nanoholes, and it is maximum at the edges of the nanohole. It is observed that the electric field is greatly reduced when moving away from the nanohole. Here, the electric field has a dipole-like pattern, henceforth, it can be concluded that resonance is due to the electric resonance. This dipole-like field pattern causes the localized surface plasmon resonance (LSPR), and this LSPR produces strong light–matter interaction. The full-width-half-maximum (FWHM) of the ATR in figure 11.9(a) is 15 nm. The quality factor is $Q = \lambda_0/\text{FWHM}$, the figure of merit that depends on the quality factor. The quality factor of the sensor is achieved as 42.07. The Q-factor of the sensor is fairly high, and it is suitable for sensing applications. This high Q-factor is due to strong LSPR caused at the interface of the sensing and HMM layers.

The reflection spectra are analyzed by considering three different measurands: HeLa, PC 12, and MDA-MB231 of the cancer cells. The refractive indices (RIs) of HeLa, PC 12, and MDA-MB231 are 1.392, 1.395, and 1.399, respectively. Figure 11.10 demonstrates ATR versus wavelength for the normal incidence of light. The black solid line illustrates reflection spectra for a normal human cell. The results depict a significant shift in the reflection dip by replacing the normal cell with a cancer cell. The results show the red shift of the position of their reflection minima and can be seen at 650, 652, and 654.5 nm, respectively. This reflection minima are due to LSPR, which can make the sensor useful for the detection of cancerous cells.

Figure 11.11 shows the relationship of λ_0 while varying the radius r of the nanohole of the proposed sensor. For this purpose, a cultured human cell is also used to determine the unit cell. A red shift is observed in the resonance with the increase of r as can obviously be seen in the corresponding plot.

The spectral features of the sensor for the obliquity of the incidence angle θ is shown in figure 11.12. FWHM is analyzed over the range of θ. These results

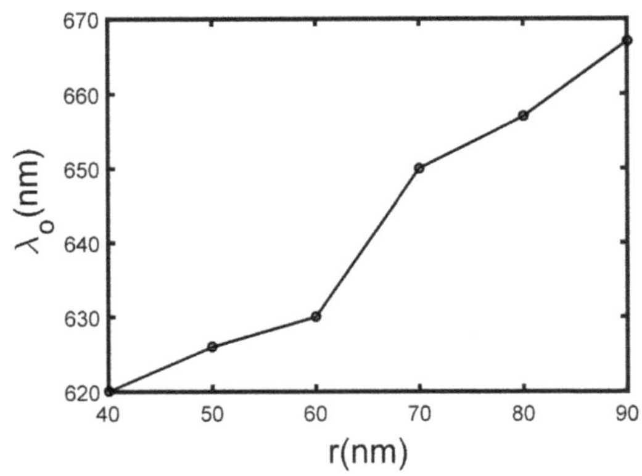

Figure 11.11. Resonance wavelength λ_0 of the sensor vs radius r of the nanohole, © (2022) IEEE. Reprinted, with permission, from [23].

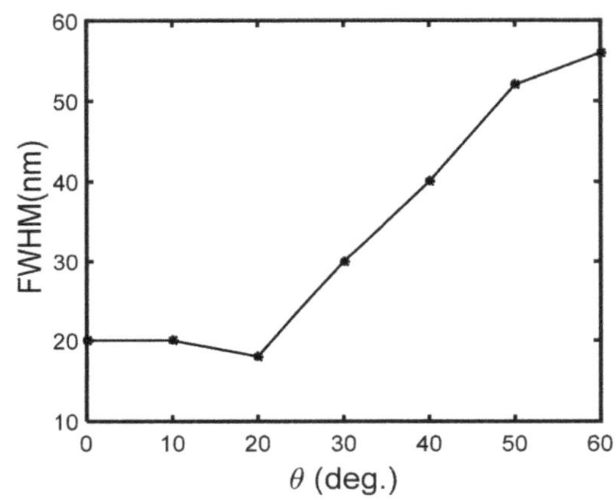

Figure 11.12. FWHM vs obliquity of the incidence angle θ of the light, © (2022) IEEE. Reprinted, with permission, from [23].

correspond to the situation wherein a normal human cell is used as a measurand. It is noticed that as the incidence angle θ approaches above 20°, an improved spectral performance of the sensor is noticed, and an increase of FWHM is observed. Whereas for $\theta \leqslant 20°$, the proposed sensor shows stable operation.

Sensitivity S and figure of merit (FoM) are analyzed for different operating conditions. To analyze the performance of the biosensor, FoM is an important metric. The sensitivity is defined in equation (11.11). Further, the FoM of the proposed sensor can be given as $\mathrm{FoM} = \frac{S}{\mathrm{FWHM}}$.

The parameters S and FoM are analyzed versus θ and d_s, respectively, and these results can be seen in figures 11.13(a) and (b). In figure 11.13(a), the parametric values of the sensor are kept as $r = 55$ nm, $d_s = 1000$ nm, and $d_{\mathrm{HMM}} = 150$ nm; however, in figure 11.13(b), the incidence angle is taken as $\theta = 0°$ and the thickness of the sensing medium d_s is varied. It is obvious in the corresponding plot when $\theta > 20°$, the S and FOM values show a decreasing trend. It is clearly seen that S and

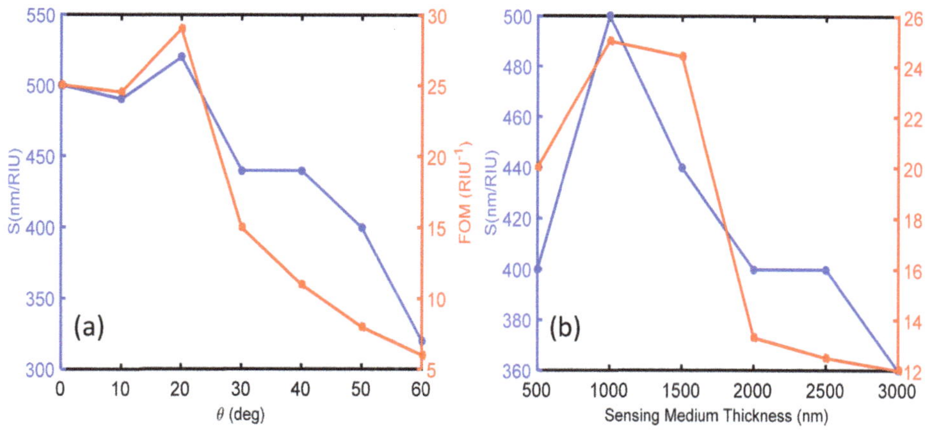

Figure 11.13. Sensitivity S and FoM for different parametric values (a) angle of incidence θ, and (b) thickness of measurand (or sample) d_s, © (2022) IEEE. Reprinted, with permission, from[23].

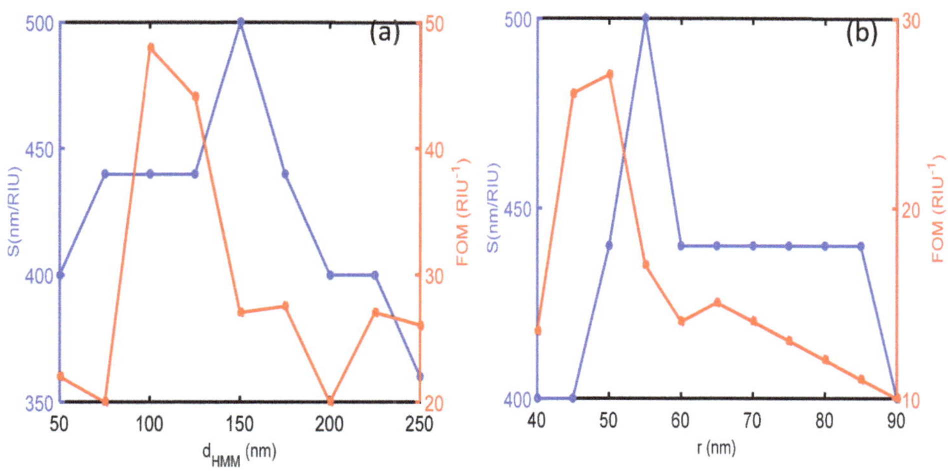

Figure 11.14. Plots of S and FoM for different values of (a) HMM thickness, and (b) radius of nanoholes (in the HMM), © (2022) IEEE. Reprinted, with permission, from [23].

FOM are high at $\theta = 20°$ with their respective values of 520 nm/RIU and 29 RIU^{-1} (as shown in figure 11.13(a)). Furthermore, figure 11.13(b) shows the highest S and FoM values when $d_s = 1000$ nm.

Figures 11.14(a) and (b), respectively, show sensitivity and FOM while considering different values of the HMM thickness d_{HMM} and the radius of nanoholes r (of HMM) under the normal incidence (i.e., $\theta = 0°$) of light waves. Keeping $d_s = 500$ nm and varying the d_{HMM} in a range of 50–250 nm, the highest sensitivity of 500 nm/RIU is observed for $d_{\mathrm{HMM}} = 150$ nm (as shown in figure 11.14(a)). For this situation, FOM is FoM = 26 RIU^{-1}. The maximum FoM of 48 RIU^{-1} is attained for $d_{\mathrm{HMM}} = 100$ nm. Furthermore, we can observe the effect of the nanohole radius of the HMM structure while keeping $d_s = 1000$ nm. The results reveal that when $r = 55$ nm, the largest sensitivity of 500 nm/RIU (figure 11.14(b)) is attained, whereas r = 50 nm provides the maximum FoM of ~85 RIU^{-1}.

Generally, the performance of the sensor can be determined in terms of quality factor, sensitivity, and the FoM. The higher these parameters, the greater the performance of the designed sensor. Given this, considering the different parametric and operating conditions, the sensor performance is optimized. As stated before, the optimized results show the highest values of S and FoM corresponding to $\theta = 20°$, and using $d_s = 1000$ nm, d_{HMM} and r in the ranges of 110–150 and 50–55 nm, respectively.

11.4 Nanoscale sensor for temperature sensing

11.4.1 Theory and design of a temperature sensor

In this section, a metasurface-based temperature sensor is considered. Figure 11.15 illustrates the schematic of the temperature sensor composed of VO_2 coated over a planar multilayered stack of Al_2O_3-SiO_2 backed with a silver nanolayer. The silver thin layer of thickness 50 nm acts as a perfect reflector to the incoming light wave. The parametric values of the unit cell of the proposed sensor are as follows: In the SiO_2-Al_2O_3 stack, the thicknesses of the SiO_2 and Al_2O_3 are taken as 0.3 and 0.2 μm, respectively. The total periods in the stack are kept as 3. The thickness of the top VO_2-based metasurface is h, which is analyzed for 30, 40, 50, and 60 nm. The unit cell size of 0.2 × 0.2 μm^2 was simulated using a commercially available simulator of Lumerical FDTD. For this, the periodic boundary conditions are applied along the

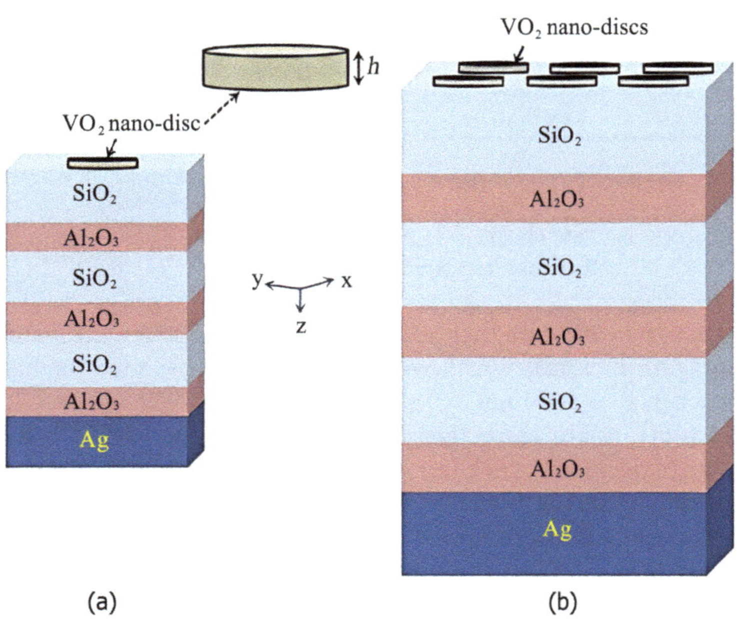

Figure 11.15. 3D view of the VO_2-based temperature sensor (a) unit cell of the sensor, (b) sensor comprised of 2 × 3 unit cells. Reprinted with permission from [17] ©The Optical Society.

xy-plane, and perfectly matched boundaries are employed along the z-axis, above and below the device to calculate the reflection and transmission spectra. Monitors are set above and below the unit cell to determine the transmission parameters over the operating wavelength range. The source is placed below the reflection monitor.

To study the optical features of the top metasurface, the effective medium theory is employed to study the anisotropic nature of the top VO_2-based metasurface of the proposed temperature sensor. The effective permittivity theory of the proposed metasurface is defined as in [51]. The tangential component $\varepsilon_{\|}(\lambda)$ and normal component $\varepsilon_{\perp}(\lambda)$ of the effective permittivity can be expressed as

$$\varepsilon_{\|}(\lambda) = \frac{\varepsilon_m(\lambda)(1+f) + \varepsilon_d(1-f)}{\varepsilon_m(\lambda)(\lambda)(1-f) + \varepsilon_d(1+f)} \tag{11.11a}$$

$$\varepsilon_{\perp}(\lambda) = \varepsilon_m(\lambda)(f) + \varepsilon_d(1-f) \tag{11.11b}$$

where f is the volume fraction of VO_2 in the top metasurface, ε_d is the permittivity of the dielectric in the top metasurface which is air, hence permittivity is unity in this situation, $\varepsilon_m(\lambda)$ is the wavelength-dependent permittivity of the VO_2. The temperature-dependent permittivity of the VO_2 is used as reported in [52, 53].

The components of the effective permittivity of the top VO_2-based metasurface are deduced by following equation (8.11). The wavelength-dependent permittivity of the Ag nanolayer can be deduced by the Lorentz–Drude (LD) model, defined as [48].

$$\varepsilon_m(\omega) = 1 - \frac{f_1\omega_p^2}{\omega(\omega - j\Gamma_1)} + \sum_n \frac{f_n\omega_p^2}{\omega_n'^2 - \omega^2 + j\omega\Gamma_n'} \tag{11.12}$$

in equation (8.12) ω_p is the plasma frequency, and Γ_1 is the damping constant, f_1 and f_n are the oscillator strength, and, ω_n' and Γ_n' are the oscillator frequency and bandwidth, respectively. It is essential to note that wavelength-dependent permittivity is deduced by the LD model that provides the best approximation with the experimentally deduced results.

For this structure, the permittivity of the SiO_2 is 2.25. The wavelength-dependent refractive index of the Al_2O_3 has been deduced by following ref. [54], and defined as

$$n^2 - 1 = \frac{1.4313493\lambda^2}{\lambda^2 - 0.0726631^2} + \frac{0.65054713\lambda^2}{\lambda^2 - 0.1193242^2} + \frac{5.3414021\lambda^2}{\lambda^2 - 18.028251^2} \tag{11.13}$$

The temperature-dependent sensitivity can be defined as

$$S = \frac{\Delta\lambda}{\Delta T} = \frac{\lambda_m - \lambda_r}{T_m - T_r} \tag{11.14}$$

in equation (8.13), S defines the sensitivity, λ_m and λ_r accordingly represent the reflection minima for measurand temperature (T_m) and reference temperature (T_r), respectively.

The effective permittivity of the top metasurface is analyzed for different temperature values for wavelengths from 700 to 1600 nm and using the aforesaid effective medium approximation theory based on equation (11.11). In this discussion, the volume fraction of VO_2 is kept as $f = 0.635$. Furthermore, the components of the effective permittivity are studied for different ambient temperatures, namely 30 °C, 70 °C, and 85 °C. Figure 11.16(a) shows the real part of the effective permittivity of the tangential component $Re(\epsilon_{||}(\lambda))$ and the normal component $Re(\epsilon_{\perp}(\lambda))$ of the top metasurface. It is noticed that at temperature 30 °C, $Re(\epsilon_{||}(\lambda))$ remains fixed at ~3 for wavelength from 700 to 1600 nm as represented by a steric line. The normal component of the real permittivity $Re(\epsilon_{\perp}(\lambda))$ has a value of ~6 at $\lambda = 700$ nm, and gradually increases with the increase of the operating wavelength, and it reaches ~7 at $\lambda = 1100$ nm. Moreover, a dip is observed at $\lambda = 1150$ nm with its permittivity value of ~7; however, the further increase in the operating wavelength increases the permittivity and reaches ~7.3 at 1600 nm. Hence, it can be concluded that $Re(\epsilon_{\perp}(\lambda))$ has a dispersive nature. With the increase of temperature to 70 °C, the $Re(\epsilon_{||}(\lambda))$ value reaches ~2.8 at $\lambda = 700$ nm; however, the increase of λ shows little effect on the permittivity, and it reaches ~3.1 at $\lambda = 1600$ nm. Considering the normal component of permittivity $Re(\epsilon_{\perp}(\lambda))$, it is noticed that permittivity has a value of ~4 at $\lambda = 700$ nm, with the increase of the wavelength, it shows less dispersion and reaches 3.8 at $\lambda = 1600$ nm. With the increase of temperature 85 °C, it has been noticed that the $Re(\epsilon_{||}(\lambda))$ value is ~ 2.7 at $\lambda = 700$ nm, and it increases with the increase of λ to reach its value of 5 at $\lambda = 1600$ nm (results are represented by the line of blue asterisks). Whereas the normal component of permittivity $Re(\epsilon_{\perp}(\lambda))$ significantly lowered with the increase of λ as shown by the blue circles; it has a maximum $Re(\epsilon_{\perp}(\lambda))$ value of ~3 at $\lambda = 700$ nm and a minimum value of ~ −4.3 at $\lambda = 1600$ nm.

Now, considering the imaginary permittivity of the metasurface, It is observed that for 30°C longitudinal permittivity $Im(\epsilon_{||}(\lambda)$ is small as ~0.2 (at $\lambda = 700$ nm), and it approaches zero at $\lambda = 1600$ nm as shown in figure 11.16(b). Whereas the

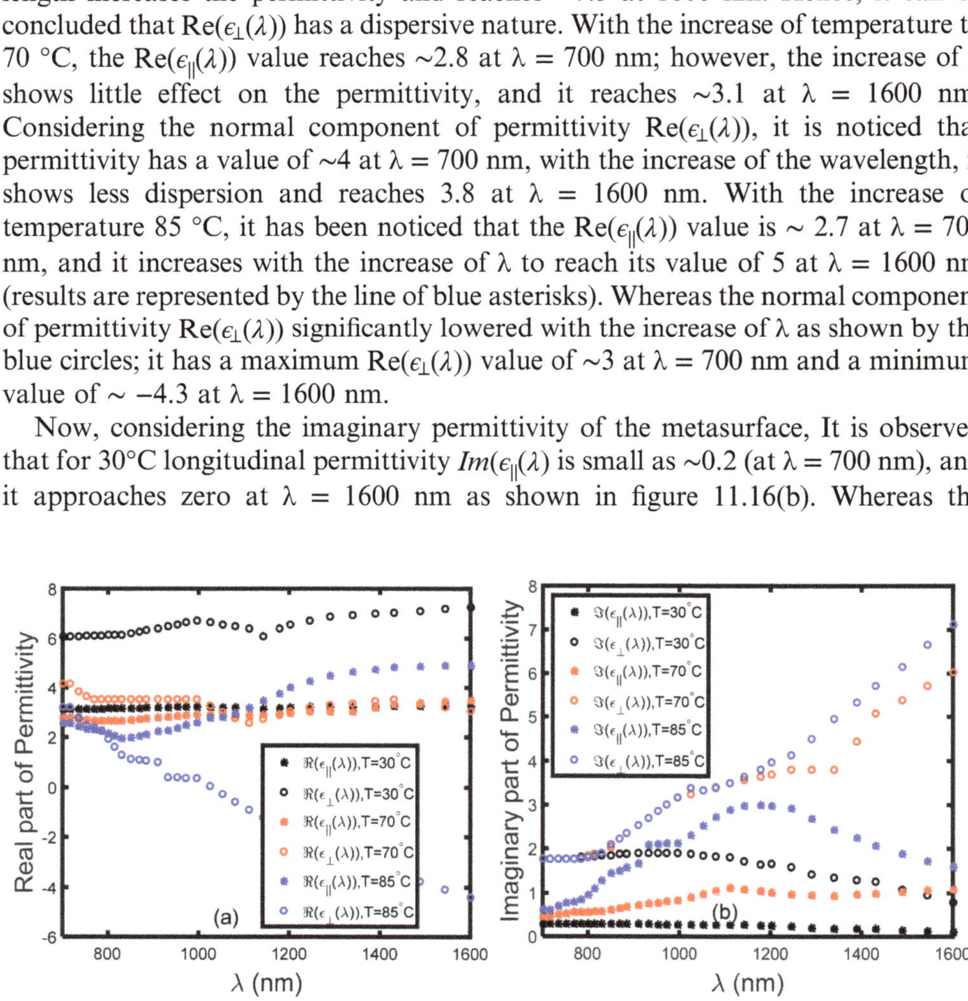

Figure 11.16. Demonstration of the effective permittivity of a VO_2-based metasurface in the near-infrared regime for different operating temperatures; (a) real permittivity, (b) imaginary permittivity. Reprinted with permission from [17] ©The Optical Society.

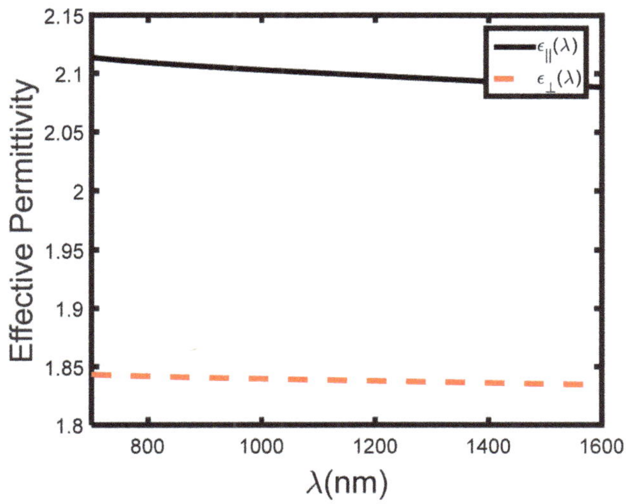

Figure 11.17. Parallel and perpendicular components of effective permittivity of the layered SiO_2-Al_2O_3 medium.

transverse imaginary part $Im(\epsilon_{\perp}(\lambda))$ has its value of ~1.8 at λ = 700 nm, and a little increment with the increase of λ is noted, and it approaches ~ 1 at λ = 1600 nm. For 70 °C, it has been observed that $Im(\epsilon_{||}(\lambda)$ has a value of ~0.5, and it shows an increment with the increase of λ, and it reaches ~1 at λ = 1600 nm. Similarly, $Im(\epsilon_{\perp}(\lambda))$ increases from ~1.6 to ~6 with the increasing wavelength from 700 nm to 1600 nm. For a high-temperature value of 85 °C, it has been noticed that $Im(\epsilon_{||}(\lambda)$ increases from 0.5 (at λ = 700 nm) to attain its value of 3 (at λ =1200 nm), then again, it gradually decreases to attain its value of 1.6 (at λ = 1600 nm). Considering $Im(\epsilon_{\perp}(\lambda))$, the minimum value of ~1.8 has been observed at λ = 700 nm, and it considerably increases to reach ~7.1 at λ = 1600 nm.

Here, the effective permittivity of the multilayered Al_2O_3-SiO_2 stacked structure is considered as shown in figure 11.17. The permittivity is analyzed by using equation (11.12). The volume fraction of the Al_2O_3-SiO_2 in the stacked multilayered structure is kept as $f = 0.4$. Figure 11.17 shows that $\epsilon_{||}(\lambda)$ is not very much dependent on the operating wavelength (λ), it has a maximum value of ~2.12 at λ = 700 nm, and it reaches ~2.08 at λ = 1600 nm. Considering $\epsilon_{\perp}(\lambda)$, it has been noticed that $\epsilon_{\perp}(\lambda)$ shows little variation over the entire range of λ, the maximum value of ~1.845 at λ = 700 nm, and shows a small change in the value of ~1.83 at λ = 1600 nm as shown by the dashed red line.

11.4.2 Results and discussion

In this section, attenuated total reflection (ATR) is analyzed in terms of change in ambient temperature. The sensitivity of the temperature sensor is studied by changing the radius of the nano-disk and radii of the VO_2-based nano-disks and operating in the near-infrared regime. The shift in the reflection dip shows the change in measured temperature that is ultimately used to calculate the sensitivity of the proposed sensor.

Figure 11.18 shows the ATR of the proposed sensor by altering the radius of the VO_2 disk in the unit cell; while keeping the thickness of the top metasurface fixed at 30 nm and the incidence angle at $\theta = 20°$. First, the nano-disk radius is kept at $r = 60$ nm (with $f = 0.29$), and the reflection minimum is observed at 970 nm for $T = 30$ °C. However, for 70 °C and 85 °C, the reflection minima are attained at ~ 950 nm, which shows that the given operating conditions do not work for the temperature sensor. Figure 11.18(b) illustrates the situation wherein the nano-disk radius is increased to 70 nm, in the current situation the volume fraction of VO_2 is $f = 0.38$, whereas the other operating conditions are kept fixed. The given results for this case are not useful to measure the sensitivity. Then, the VO_2 nano-disk radius is increased to 80 nm and $f = 0.50$, figure 11.18(c) corresponds to this situation. It is noticed that the shift in the reflection minima is increased with the increase in temperature. Notably, the reflection minima at ~980, ~960, and ~950 nm have respective temperature values of 30 °C, 70 °C, and 85 °C. Thereby, the proposed structure is useful for sensing in the given operating conditions. Figure 11.18(d) corresponds to the situation wherein $r = 90$ nm and $f = 0.68$; it is depicted that the shift in the

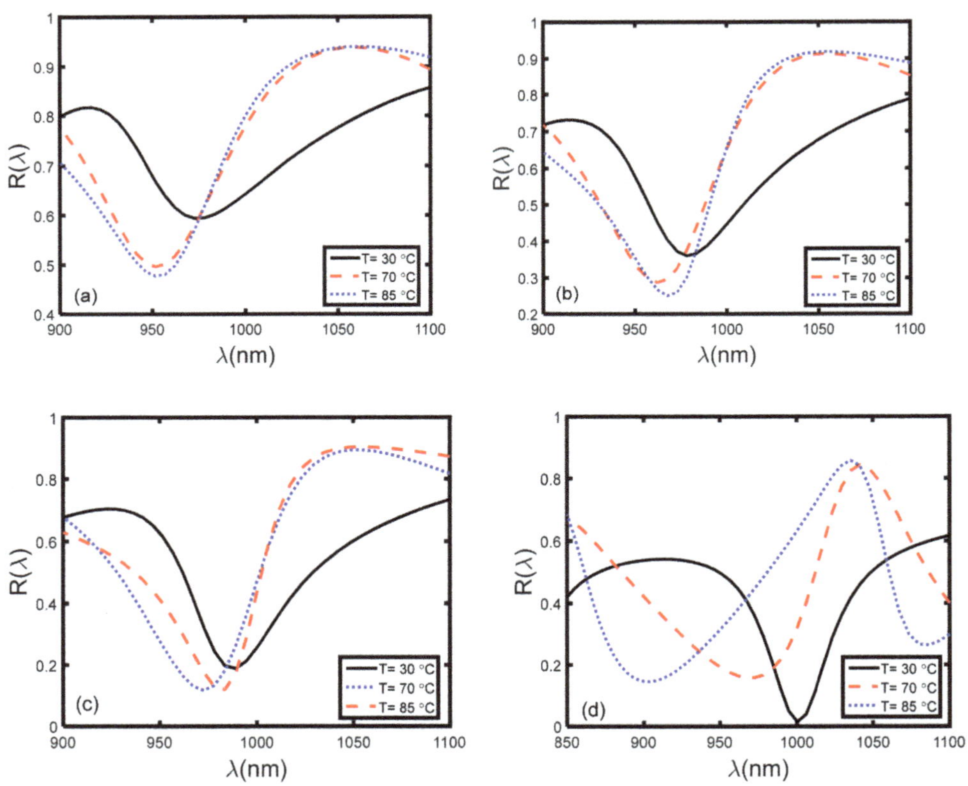

Figure 11.18. Attenuated total reflection spectra of light with the radius of the nano-disk as (a) $r = 60$ nm, (b) $r = 70$ nm, (c) $r = 80$ nm, and (d) $r = 90$ nm. Reprinted with permission from [17] ©The Optical Society.

reflection minima is considerably increased. The reflection minima for 30 °C, 70 °C, and 85 °C are, respectively, at ~1100, ~960, and ~900 nm. It can be inferred that an increase in volume fraction f of the VO_2 increases the sensitivity of the proposed sensor.

Similarly, ATR spectra are calculated by altering the thickness of the top metasurface of VO_2; in this scenario, the radius of the nano-disk is $r = 90$ nm and the angle of incidence is 20°. Consider the thicknesses $t = 30$, 40, 50, and 60 nm with the respective figures 11.19(a)–(d). Figure 11.19(a) corresponds to the situation wherein the thickness of the top metasurface is 30 nm; in this case, the reflection dip is attained at ~ 1270 nm. A blue shift in the reflection dip is observed. For the further increase in temperature, for example, 70 °C, the reflection minimum can be seen at ~1170 nm. For a temperature of 85 °C, the reflection dip is shifted at 1110 nm. As the thickness of the top metasurface is increased to 40 nm and keeping all other operating conditions fixed, the reflection minima are attained at ~995, ~960, and ~ 911 nm with their respective temperature values of 30 °C, 70 °C, and 85 °C (as can be clearly seen in figure 11.19(b)). Figure 11.19(c) corresponds to the situation wherein the thickness is kept at 50 nm; for this case, the reflection dip is

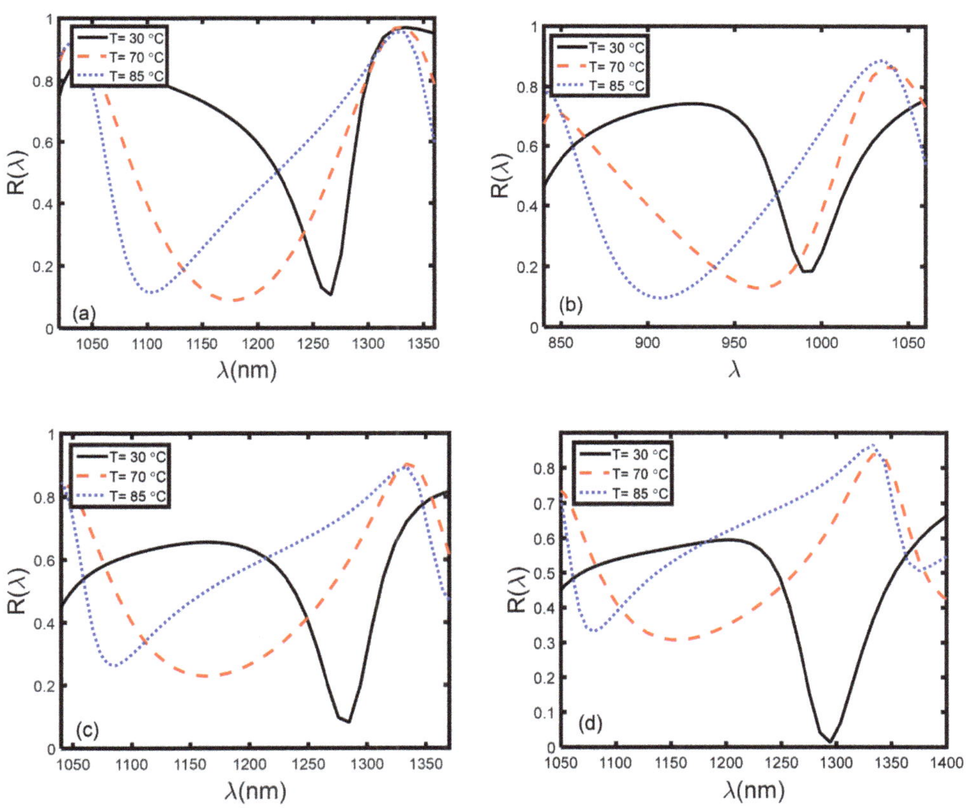

Figure 11.19. Attenuated total reflection spectra of light with thickness of the metasurface as (a) 30 nm (b) 40 nm, (c) 50 nm, and (d) 60 nm. Reprinted with permission from [17] ©The Optical Society.

achieved at ~1280 nm at 30 °C, whereas for 70 °C, the reflection dip is at ~1145, and 85 °C at ~1175 nm. Further, the thickness is increased to 60 nm. It is observed that the reflection dips are noticed at 1295, 1150, and 1175 nm for the operating temperatures of 30 °C, 70 °C, and 85 °C, respectively.

Comparing figures 11.19(a), 11.19(b), 11.19(c) and 11.19(d), it is quite obvious that the shift in the reflection dips become large when the thicknesses of the top metasurface are 50 and 60 nm. Thereby, high sensitivity is achieved at 50 and 60 nm thicknesses.

Now, taking an insight into the aforesaid discussion of the ATR spectra for different situations, the maximum sensitivity is 2.5 (wavelength/°C) in the case of the radius of the nano-disk kept as $r = 90$ nm, the thickness of top the metasurface of 50 nm, and incidence angle of 40° and 60° as shown in figures 11.19(c) and 11.19(d).

11.5 Conclusion and future work

In this chapter, we have analyzed the different biomedical and temperature sensing operations using metasurface-based sensors within the optical regime. Concerning bio-sensing, an HMM-based sensor has been used for the early-stage diagnosis of cancer cells in the human body when noticing a high reflection shift within the operating spectra due to different measured cancer cell samples on the top of the metasurface. The performance metrics such as sensitivity and FoM results show an excellent performance of the HMM-based biosensor for early-stage cancer detection. An ultrasensitive HMM-based prism-coupled waveguide sensor (PCWS) has been used to measure the concentration levels of the glucose solution. A small change in the refractive index and adjustment of the structural changes in the HMM layer alter the LSPR effect. As a result, a minimum shift in the reflection spectra curve shows the variation in the concentration level of the measured solution. Similarly, a VO_2 metasurface-based temperature sensor made of nano-disks shows excellent performance when varying the temperature-dependent permittivity of the metasurface and structural changes. These changes impact the resonance conditions and show noticeable variations in the reflection characteristics under different operating temperature scenarios.

From a future perspective, biomedical sensing in the optical regime will be a favorable choice due to several features including nanostructures, control of incident light interaction with the nanostructure while using metaphotonics technology, and improved sensitive detection of samples. Another near-future aspect is the use of machine learning (ML) in the development of novel metasurface-based sensor designs that would lead this area of research to the highest level.

Author contributions

Conceptualization, M. A. B.; methodology, M. A. B, and M. S.; data generation, M. A.; investigation, M. A. B, and M. S.; benchmarking models, M. A. B; writing—original draft preparation, M. A. B, and M. S.; writing—review and editing, M. A. B, and M. S.; visualization, M. A. B, and M. S.; supervision, M. A. B.; project administration, M. A. B.; All authors reviewed the manuscript.

Data availability statement

The datasets generated during and/or analyzed during the current study are available from the corresponding author on reasonable request.

Contact person: abuzar@cuisahiwal.edu.pk.

Acknowledgments

The authors would like to thank Prof. Muharrem Karaaslan for discussion to write the draft.

Conflicts of interest

'The authors declare no conflict of interest.'

References

[1] Lemoult F, Kaina N, Fink M and Lerosey G 2013 Wave propagation control at the deep subwavelength scale in metamaterials *Nat. Phys.* **9** 55–60

[2] Kadic M, Bückmann T, Schittny R and Wegener M 2013 Metamaterials beyond electromagnetism *Rep. Prog. Phys.* **76** 126501

[3] Rupin M, Lemoult F, Lerosey G and Roux P 2014 Experimental demonstration of ordered and disordered multiresonant metamaterials for Lamb waves *Phys. Rev. Lett.* **112** 234301

[4] Kaina N, Lemoult F, Fink M and Lerosey G 2015 Negative refractive index and acoustic superlens from multiple scattering in single negative metamaterials *Nature* **525** 77–81

[5] Smith D R, Padilla W J, Vier D C, Nemat-Nasser S C and Schultz S 2000 Composite medium with simultaneously negative permeability and permittivity *Phys. Rev. Lett.* **84** 4184

[6] Farmani A, Yavarian M, Alighanbari A, Miri M and Sheikhi M H 2017 Tunable graphene plasmonic Y-branch switch in the terahertz region using hexagonal boron nitride with electric and magnetic biasing *Appl. Opt.* **56** 8931–40

[7] Pendry J B 2000 Negative refraction makes a perfect lens *Phys. Rev. Lett.* **85** 3966

[8] Zhai S L, Zhao X P, Liu S, Shen F L, Li L L and Luo C R 2016 Inverse Doppler effects in broadband acoustic metamaterials *Sci. Rep.* **6** 1–11

[9] Brun M, Guenneau S and Movchan A B 2009 Achieving control of in-plane elastic waves *Appl. Phys. Lett.* **94** 061903

[10] Chiu N F, Huang T Y, Kuo C C, Lin C W and Lee J H 2013 Organic-based plasmonic emitters for sensing applications *Appl. Opt.* **52** 1383–8

[11] Park S J, Hong J T, Choi S J, Kim H S, Park W K, Han S T and Ahn Y H 2014 Detection of microorganisms using terahertz metamaterials *Sci. Rep.* **4** 1–7

[12] Sreekanth K V, Alapan Y, ElKabbash M, Ilker E, Hinczewski M, Gurkan U A and Strangi G 2016 Extreme sensitivity biosensing platform based on hyperbolic metamaterials *Nat. Mater.* **15** 621–7

[13] Vasilantonakis N, Wurtz G A, Podolskiy V A and Zayats A V 2015 Refractive index sensing with hyperbolic metamaterials: strategies for biosensing and nonlinearity enhancement *Opt. Express* **23** 14329–43

[14] Sreekanth K V, Alapan Y, ElKabbash M, Wen A M, Ilker E, Hinczewski M and Strangi G 2016 Enhancing the angular sensitivity of plasmonic sensors using hyperbolic metamaterials *Adv. Opt. Mater.* **4** 1767–72

[15] Aylo R, Banerjee P P, Ghosh A K and Verma P 2011 Design of metamaterial based sensors for pressure measurement *Integrated Optics: Devices, Materials, and Technologies XIV* (SPIE) vol 7604 pp 266–73

[16] Kim H G, Hajra S, Lee H, Kim N and Kim H J 2023 Additively manufactured mechanical metamaterial-based pressure sensor with tunable sensing properties for stance and motion analysis *Adv. Eng. Mater.* **25** 2201499

[17] Baqir M A and Choudhury P K 2019 On the VO_2 metasurface-based temperature sensor *JOSA* B **36** F123–30

[18] Hu S, Chen J, Liang J, Luo J, Shi W, Yuan J and Luo Y 2022 Hyperbolic-metamaterials-based SPR temperature sensor enhanced by a nanodiamond-pdms hybrid for high sensitivity and fast response *ACS Appl. Mater. Interfaces* **14** 42412–9

[19] Kim H S, Cha S H, Roy B, Kim S and Ahn Y H 2018 Humidity sensing using THz metamaterial with silk protein fibroin *Opt. Express* **26** 33575–81

[20] Ni B, Wang Z Y, Zhao R S, Ma X Y, Xing Z Q, Yang L S and Zhang D B 2017 Realisation of a humidity sensor based on perfect metamaterial absorber *Opt. Quantum Electron.* **49** 1–7

[21] Abdulkarim Y I, Deng L, Luo H, Huang S, Karaaslan M, Altıntaş O and Al-badri K S L 2020 Design and study of a metamaterial based sensor for the application of liquid chemicals detection *J. Mater. Res. Technol.* **9** 10291–304

[22] Baqir M A, Farmani A, Fatima T, Raza M R, Shaukat S F and Mir A 2018 Nanoscale, tunable, and highly sensitive biosensor utilizing hyperbolic metamaterials in the near-infrared range *Appl. Opt.* **57** 9447–54

[23] Baqir M A and Choudhury P K 2022 Hyperbolic metamaterial-based optical biosensor for detecting cancer cells *IEEE Photonics Technol. Lett.* **35** 183–6

[24] Simoska O, Gaffney E M, Minteer S D, Franzetti A, Cristiani P, Grattieri M and Santoro C 2021 Recent trends and advances in microbial electrochemical sensing technologies: an overview *Curr. Opin. Electrochem.* **30** 110762

[25] Santoro C, Bollella P, Erable B, Atanassov P and Pant D 2022 Oxygen reduction reaction electrocatalysis in neutral media for bioelectrochemical systems *Nat. Catal.* **5** 473–84

[26] Funari R and Shen A Q 2022 Detection and characterization of bacterial biofilms and biofilm-based sensors *ACS Sens.* **7** 347–57

[27] Litchinitser N M, Pires D G, Gao J, Tsvetkov D, Sedeh H B, Chandra N and Kravchenko I 2023 Topology of light and darkness in engineered optical media *Complex Light and Optical Forces XVII* (SPIE) p PC124360N

[28] Barati Sedeh H and Litchinitser N M 2023 Singular optics empowered by engineered optical materials *Nanophotonics* **12** 2687–716

[29] Yan X, Wang G and Deng D 2019 Engineering electromagnetic wave propagation in periodically layered gyromagnetic metamaterials with an external magnetic field *Plasmonics* **14** 1243–51

[30] Baqir M A and Choudhury P K 2017 Hyperbolic metamaterial-based UV absorber *IEEE Photonics Technol. Lett.* **29** 1548–51

[31] Baqir M A, Choudhury P K, Farmani A, Younas T, Arshad J, Mir A and Karimi S 2019 Tunable plasmon induced transparency in graphene and hyperbolic metamaterial-based structure *IEEE Photonics J.* **11** 1–10

[32] Bilal R M H, Baqir M A, Iftikhar A, Naqvi S A, Mughal M J and Ali M M 2022 Polarization-controllable and angle-insensitive multiband Yagi-Uda-shaped metamaterial absorber in the microwave regime *Opt. Mater. Express* **12** 798–810

[33] Wei Z, Jiang Y and Wang J 2023 Frequency magnetically tunable terahertz perfect absorber based on graphene and silica layered dielectric *Crystals* **13** 553

[34] Palermo G, Sreekanth K V, Maccaferri N, Lio G E, Nicoletta G, De Angelis F and Strangi G 2020 Hyperbolic dispersion metasurfaces for molecular biosensing *Nanophotonics* **10** 295–314

[35] Vafapour Z 2019 Polarization-independent perfect optical metamaterial absorber as a glucose sensor in food industry applications *IEEE Trans. Nanobiosci.* **18** 622–7

[36] Shahzad W, Hu W, Ali Q, Raza H, Abbas S M and Ligthart L P 2000 A low-cost metamaterial sensor based on DS-CSRR for material characterization applications *Sensors* **22** 1–11

[37] Sadeqi A, Rezaei Nejad H, Owyeung R E and Sonkusale S 2019 Three dimensional printing of metamaterial embedded geometrical optics (MEGO) *Microsyst. Nanoeng.* **5** 16

[38] Jun S Y, Izquierdo B S and Parker E A 2019 Liquid sensor/detector using an EBG structure *IEEE Trans. Antennas Propag.* **67** 3366–73

[39] Sadeqi A, Nejad H R and Sonkusale S 2017 Low-cost metamaterial-on-paper chemical sensor *Opt. Express* **25** 16092–100

[40] Chai F, Fang B, Li C, Hong Z and Jing X 2022 Highly sensitive biosensors based on all-dielectric nanoresonators *Nanoscale* **9** 4972–80

[41] Ma S, Zhang P, Mi X and Zhao H 2023 Highly sensitive terahertz sensor based on graphene metamaterial absorber *Opt. Commun.* **528** 129021

[42] Liao Y L and Zhao Y 2020 Ultra-narrowband dielectric metamaterial absorber with ultra-sparse nanowire grids for sensing applications *Sci. Rep.* **10** 1–7

[43] Youssef J, Zhu S, Crunteanu A, Orlianges J C, Ho H P, Bachelot R and Zeng S 2022 Highly sensitive plasmonic biosensors with precise phase singularity coupling on the metastructures *Biosensors* **12** 866

[44] Iwanaga M 2021 High-sensitivity high-throughput detection of nucleic acid targets on metasurface fluorescence biosensors *Biosensors* **11** 33

[45] Sreekanth K V, Mahalakshmi P, Han S, Mani Rajan M S, Choudhury P K and Singh R 2019 Brewster mode-enhanced sensing with hyperbolic metamaterial *Adv. Opt. Mater.* **7** 1900680

[46] Shekhar P, Atkinson J and Jacob Z 2014 Hyperbolic metamaterials: fundamentals and applications *Nano Converg.* **1** 1–17

[47] Rakić A D, Djurišić A B, Elazar J M and Majewski M L 1998 Optical properties of metallic films for vertical-cavity optoelectronic devices *Appl. Opt.* **37** 5271–83

[48] Ouyang Q, Zeng S, Jiang L, Hong L, Xu G, Dinh X Q and Yong K T 2016 Sensitivity enhancement of transition metal dichalcogenides/silicon nanostructure-based surface plasmon resonance biosensor *Sci. Rep.* **6** 1–13

[49] Olyaee S, Seifouri M and Mohsenirad H 2016 Label-free detection of glycated haemoglobin in human blood using silicon-based photonic crystal nanocavity biosensor *J. Mod. Opt.* **63** 1274–9

[50] Rostami A, SalmanOgli A, Farhadnia F, Dolatyari M, Rostami G and Pişkin E 2015 Design of a portable nanosensor for easy breast tomography *RSC Adv.* **5** 19002–13

[51] Ferrari L, Wu C, Lepage D, Zhang X and Liu Z 2015 Hyperbolic metamaterials and their applications *Prog. Quantum Electron.* **40** 1–40

[52] Kana J K, Ndjaka J M, Vignaud G, Gibaud A and Maaza M 2011 Thermally tunable optical constants of vanadium dioxide thin films measured by spectroscopic ellipsometry *Opt. Commun.* **284** 807–12

[53] Kana J K, Ndjaka J M, Ngom B D, Fasasi A Y, Nemraoui O, Nemutudi R and Maaza M 2010 High substrate temperature induced anomalous phase transition temperature shift in sputtered VO_2 thin films *Opt. Mater.* **32** 739–42

[54] Maliston I H 1972 Interspeciman comparison of the refractive index of fused silica *J. Opt. Soc. Am.* **55** 1205–9

IOP Publishing

Advanced Metamaterials for Engineers

Lulu Wang and Muharrem Karaaslan

Chapter 12

Metamaterial signal absorbers and applications

Emin Ünal

In an electromagnetic wave absorber structure, metamaterials are generally used to increase the absorption rate, but by using metamaterials the bandwidth and number of bands can also be increased and even resonance frequencies can be adjusted. In this chapter, electromagnetic wave absorbers and their various application areas are discussed. The criteria for determining design parameters such as operating frequency, bandwidth, number of bands, reflection and transmission properties, and polarization properties are explained in metamaterial absorber design studies. Then, in order to provide these properties, the criteria for determining the resonator shape and dielectric layer properties are discussed. Since the absorption and reflection parameters of metamaterial absorbers generally depend on the resonator structure, shape and properties of the dielectric substrate, the most important point in metamaterial absorber studies is the resonator design. The metal elements that make up the resonator electrically correspond to a lumped component. Thus, each resonator cell can be represented as an RLC microwave circuit and analyzed using a circuit simulator to obtain the S-parameters. Therefore, simulation, production and measurements are discussed along with these topics in this chapter. According to the application areas, metamaterial absorbers can be single or multi-band, as well as narrow or wide bands. Metamaterial signal absorbers are also tunable. Metamaterial absorbers can be used in many areas such as energy harvesting, solar cells, sensing, communication, radar, shielding, etc, according to their design features. In the second part of this chapter, absorber applications, especially absorber designs for energy harvesting, absorbers for solar energy, absorbers for sensor applications and tunable metamaterial absorbers are discussed in detail.

12.1 Introduction

Electromagnetic analysis is principally concerned with three properties of matter, called constitutive parameters. It is well known that constitutive parameters such as dielectric permittivity, ε, magnetic permeability, μ and (σ) conductivity in a medium

doi:10.1088/978-0-7503-5754-8ch12

determine the response and reaction of such medium or material when exposed to external time-varying electromagnetic fields [1–3]. All three parameters are macroscopic parameters and are averages of macroscopic measurements of atomic or molecular oscillators affected by external fields, depending on frequency. This idea can also be applied to larger inhomogeneous volumes.

The refractive index also depends on these constitutive parameters. The refractive index determines how much of it is reflected or refracted when an electromagnetic wave is incident on a material. When an electromagnetic wave passes through a medium, some part of it is always absorbed. To take this into account, a complex refractive index needs to be defined ($n = n + j\kappa$). The real part of a refractive index (n) is the ratio of the free space speed of light to the phase speed of an electromagnetic wave in the medium, while the imaginary part κ is called the absorption coefficient.

The index of refraction is specified as a complex expression to explain the losses, even expressed as a tensor when describing anisotropic materials, but it was not questioned for many years that the index sign could be negative [4]. In 1968, Russian scientist Veselago theorized that if the dielectric (ε) and magnetic permeability (μ) parameters, which indicate the EM property of a material, are negative in a certain frequency range, the medium may also have a negative index of refraction [5]. Although Snell's inverse refraction law, inverse Doppler effect and inverse Cerenkov propagations were introduced by Veselago for this theoretical environment, his work did not attract attention for many years because there is no such environment in nature. After nearly 30 years, Pendry *et al* in 1996 and 1999, carried out an experimental study to confirm Veselago's theoretical work and showed that these macroscopic parameters can be negative simultaneously. Veselago's work was later confirmed by Smith *et al* and Shelby *et al* in the 2000s, and these materials are named metamaterials. Metamaterials, also called negative refractive index materials, are produced using two separate structures, one with negative (ε) and the other negative (μ) properties. Due to the negative feature of this couple, the phase and group velocities of the EM waves entering this medium move in the opposite direction [6–12].

The metamaterials are periodic structures, their unit cells are smaller than the excitation wavelength, one cell size must be smaller than a quarter wavelength to ensure effective homogeneity condition [13]. Since the unit cells forming the periodic structure are smaller than the wavelength, they appear homogeneous for the incident wave and metamaterials can be defined as their effective parameters that can be controlled by the geometric structure of the unit cells [14]. In addition, if the structures that disrupt the homogeneity in a medium are much smaller than the EM wavelength to be applied, but larger than atoms and molecules, these structures can be defined as metamaterials [4].

12.2 Absorption mechanism

Metamaterials have a wide range of applications such as optical filters, medical devices, aerospace applications, sensor sensing, absorbers, energy harvesting, super lenses, cloaking, radomes, antennas, etc. In this chapter, signal absorbers—which is one of the important application areas of metamaterials—and studies in this field are examined. The concept of a metamaterial signal absorber was first experimentally presented by Bilotti *et al* in 2006 [15]. Another metamaterial absorber proposed by N. I. Landy in 2008 is a kind of thin and near-perfect electromagnetic absorber. The signal absorber concept is still a subject of interest and remains so to date [16]. In recent years, many studies on signal absorbers have been carried out, using different metamaterial resonators in various frequency spectra ranging from acoustic to microwave, terahertz and optical bands [17–22]. However, little attention has been paid to discussions of absorption theories or mechanisms that are crucial to the potential investigation of metamaterial signal absorbers [23]. An important widely accepted theoretical work in this regard is the impedance matching theory by Tao *et al* [24]. This theory proposes that the periodic metal resonator structure provides an electrical response to control the permeability $\varepsilon(\omega)$ of the metamaterial and, furthermore, the combined magnetic response of the resonator and ground plane controls the magnetic permeability $\mu(\omega)$ of the metamaterial.

In order to better understand and verify the working principle of metamaterial signal absorber structures, theoretical studies should also be verified with numerical and experimental studies.

Absorption can be calculated with the following formula,

$$A(\omega) = 1 - |\Sigma S_{11}|^2 - T(\omega), \tag{12.1}$$

where $A(\omega)$, $R(\omega) = |S_{11}|^2$ and $T(\omega) = |S_{21}|^2$ denote the absorption, reflection and transmission coefficients, respectively. In the case of impedance matching at the resonant frequency, it is generally desired that $A(\omega)$ be maximum and $R(\omega)$ and $T(\omega)$ values should be minimum. In metamaterial signal absorber studies, there is no transmission because the back plate of the metamaterial is covered with a metal plate. Therefore, the absorption value is calculated as $A(\omega) = 1 - R(\omega)$, ignoring transmission. As it is understood, absorption depends on reflection, so the main purpose of absorption studies is to achieve perfect absorption by reducing the reflection to almost zero. Reflection is minimized if the effective impedance of the metamaterial $\sqrt{\mu(\omega)/\varepsilon(\omega)} = Z(\omega)$ matches the free space impedance Z_0. On the other hand, if the metamaterial is high loss, that is, the transmission is low, then near-unit absorption occurs at a certain frequency. When $\varepsilon(\omega)$ and $\mu(\omega)$ values are adjusted as desired, it may be possible to absorb both incoming electric and magnetic fields [21, 25].

The absorption mechanism of the metamaterial absorber can also be analyzed using transmission line theory and equivalent circuit models, as suggested by Wen *et al* and Pang *et al* [26–28]. In these studies, the metamaterial resonator combined with the ground plane is first considered as an RLC circuit model, and then its reflection and transmission properties are analyzed using transmission line theory. Since the transmission line theory is valid only in the normal incidence case, it is not appropriate when we need to discuss the oblique incidence condition. On the other hand, recently, a different analysis method based on interference theory has been proposed, which is little affected by electric/magnetic responses, but is associated with the phase change caused by the thickness of the dielectric substrate and the S-parameters of the metamaterial resonator [23]. Interference theory, also known as multiple reflection theory, is not a new theory; it has been studied in EM wave theory [29], microwave engineering, [30] optics [31] and even a kind of absorbance body called the Salisbury screen [32], whose configuration is similar to a metamaterial absorber. For this reason, interference theory exhibits the advantage to study the metamaterial absorber through highly accurate quantitative analysis.

12.3 Multiple reflection

The extended interference theory model, applicable to both normal and oblique incidence conditions, is applied to figure 12.1, assuming layer 1 is zero-thickness. With the effect of multiple reflection in medium 2, the waves reflected from layer 1 to medium 1 are seen in figure 12.1(b) and can be calculated using multiple reflection theory [33].

Using the theory of multiple reflection, the first reflected wave and the transmitted wave from medium 2 through layer 1 to medium 1 are superimposed to obtain the total reflected wave; this can be expressed as:

$$\Gamma = \Gamma_1 + T_1 T_2 \Gamma_3 e^{-j2k_2 d} \sum_{n=0}^{\infty} \left(\Gamma_2 \Gamma_3 e^{-j2k_2 d}\right)^n, \tag{12.2}$$

using the series expansion

$$\sum_{n=0}^{\infty} x^n = \frac{1}{1-x}, \; |x| < 1 \tag{12.3}$$

Equation (12.2) can be written as

$$\Gamma = \Gamma_1 + \frac{T_1 T_2 \Gamma_3 e^{-j2k_2 d}}{1 - \Gamma_2 \Gamma_3 e^{-j2k_2 d}} \tag{12.4}$$

Since the back layer is metal, the reflection from the metal plate is −1

$$\Gamma_3 = -1 = e^{-j\pi}. \tag{12.5}$$

S-parameter equivalents of other transmission and reflections seen in figure 12.1(b) are

$$\Gamma_1 = S_{11},\ \Gamma_2 = S_{22},\ T_1 = S_{21},\ T_2 = S_{12}. \tag{12.6}$$

By substituting the S-parameters, equation (12.4) can be written as:

$$S_{11\ \text{total}} = S_{11} + \frac{S_{21} S_{12} e^{-j(2k_2 d + \pi)}}{1 - S_{22} e^{-j(2k_2 d + \pi)}}. \tag{12.7}$$

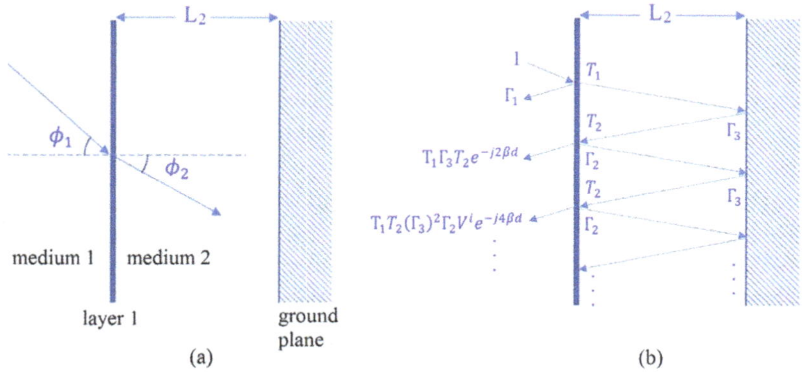

Figure 12.1. (a) Mediums in oblique incidence, (b) multiple reflection model applied to find the total reflected wave in normal and oblique incidence cases. Inspired by figure 3 in [33].

Complex S-parameters in phasor form

$S_{11} = | s_{11} | e^{j\theta 11}$ is the reflection coefficient of layer 1 from medium 1 to medium 1,

$S_{21} = | s_{21} | e^{j\theta 21}$ is the transmission coefficient of layer 1 from medium 1 to medium 2,

$S_{12} = | s_{12} | e^{j\theta 12}$ is the transmission coefficient of layer 1 from medium 2 to medium 1,

$S_{22} = | s_{22} | e^{j\theta 22}$ is the reflection coefficient of layer 1 from medium 2 to medium 2.

where θ_{11}, θ_{21}, θ_{12} and θ_{22} are the phase angles of S_{11}, S_{21}, S_{12} and S_{22}, respectively. k_2 is the wave number in region 2. d is the propagation distance of the transmitting wave from layer 1 to the ground plane and L_2 is the thickness of medium 2 as shown in figure 12.1.

The propagation distance of the transmitting wave from the 1st layer to the ground plane can be obtained using Snell's refraction rule. Assuming medium 1 and medium 2 are not magnetic material,

$$k_1 \sin \phi_1 = k_2 \sin \phi_2 \tag{12.8}$$

$$\sqrt{\varepsilon_1} \sin \phi_1 = \sqrt{\varepsilon_2} \sin \phi_2 \tag{12.9}$$

$$\phi_2 = \sin^{-1}\left(\sqrt{\frac{\varepsilon_1}{\varepsilon_2}} \sin \phi_1\right) \tag{12.10}$$

$$d = L_2/\cos \phi_2 \tag{12.11}$$

where ε_1 and ε_2 are the dielectric permittivity of medium 1 and medium 2, respectively [22].

12.4 Absorber applications

Considering the above-mentioned analysis results, we can see that the main contribution of the ground plane is only to reflect electromagnetic waves, and the absorption property of the metamaterial absorber is related to the scattering properties of the metamaterial resonator [23].

The main point in absorber studies is to propose a new and effective resonator layer design. Many metamaterial structures such as split rings, rings, strips, omega-shaped and Jerusalem cross structure can achieve very good absorption. In general, in metamaterial absorber design studies, first of all, design parameters such as operating frequency, bandwidth, number of bands, reflection and transmission properties and polarization properties are determined, then the resonator shape and dielectric layer properties that provide these properties are determined. Simulation is performed using an electromagnetic simulation software and the dimensions are optimized. Next, the structure is fabricated and measurements are carried out. Then the results are compared. If the desired features cannot be obtained, the above-mentioned processes are repeated by changing the design shape or dimensions.

Depending on the application areas, metamaterial absorbers can be single or multi-band, and these bands can also be narrow or wide band. Metamaterial signal absorbers can also be tunable; that is, switching between different frequency bands can be realized. Metamaterial absorbers can be used in many areas such as energy harvesting, solar cells, sensing, communication, radar, shielding, etc, depending on their design features.

In recent years, metamaterial absorber studies have continued with a wide range of purposes and design features. One of these studies is based on an omega (Ω) resonator with a gap and octa-star stripe configuration, as shown in figure 12.2, and proposed by Dincer *et al* [18]. In order to investigate the effect of the gap in the omega resonator, the results for the omega structures with and without the gap are obtained separately as seen in figure 12.3. A single resonance is observed at

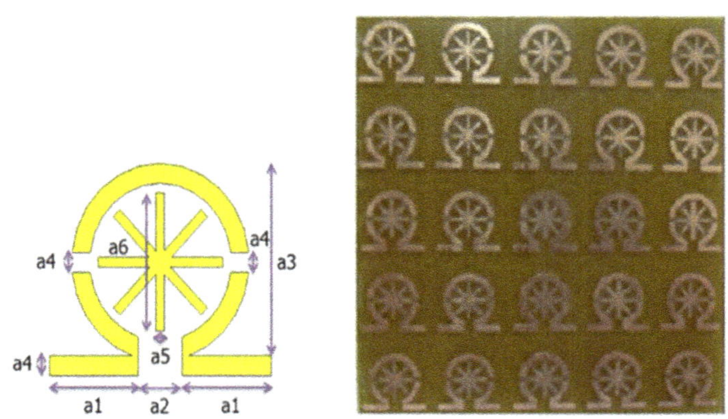

(a) (b)

Figure 12.2. (a) Unit cell, (b) sample picture. Reproduced from [18]. CC BY 4.0.

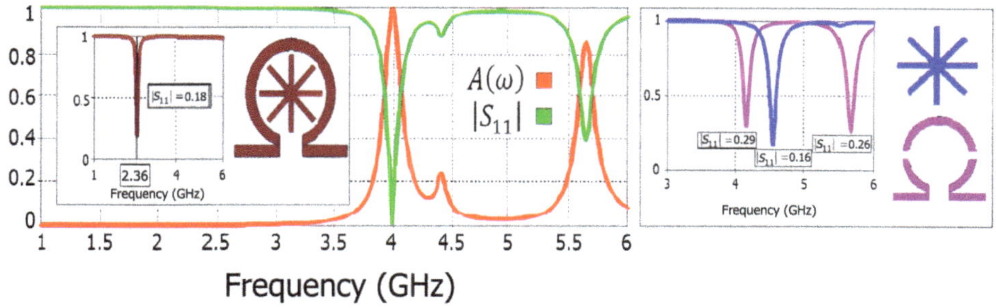

Figure 12.3. (a) Simulated reflection and absorption with and without gap (inset picture), (b) reflection of resonator and octa-star stripe structure individually. Reproduced from [18]. CC BY 4.0.

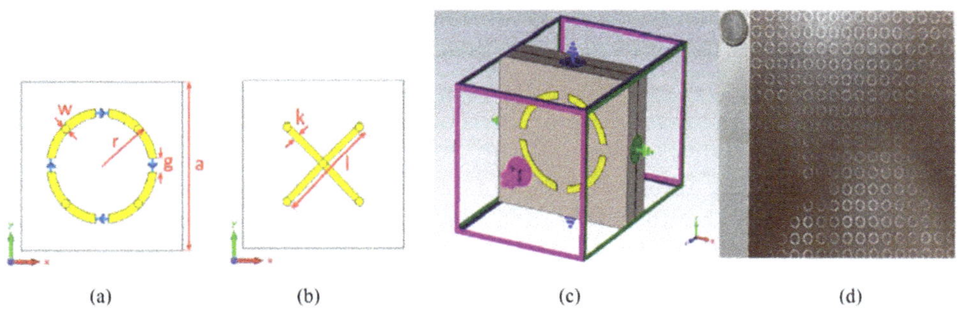

Figure 12.4. (a) Top layer, (b) middle layer, (c) simulation setup, (d) fabricated top layer [34], John Wiley & Sons. (2018).

2.36 GHz in structures without gaps. If there are gaps in the omega resonator, two resonances occur at higher frequencies, at 4.0 and 5.6 GHz, and moreover, the reflection coefficient decreases slightly.

In this study, the effect of angle of incidence on absorption is investigated and very good results are obtained in all angles of incidence. On the other hand, the effect of wave polarization on absorption is also investigated and it is observed that the proposed structure is almost polarization independent. As a result, it is confirmed by the results obtained by simulation and measurement that the proposed structure provides a strong resonance effect and perfect absorption at 4.0 GHz.

Another polarization-independent broadband signal absorber design for microwave applications was carried out by Bağmancı *et al* [34]. In this study, a polarization-independent broadband metamaterial absorber structure based on split-ring resonators loaded with lumped elements is investigated.

Figure 12.5 shows the absorption graph for transverse electric (TE), transverse magnetic (TM) and transverse electromagnetic (TEM) modes. Although the absorption value remains above 80% in the band between 6.2 and 13.5 GHz, better absorption is obtained for TEM at frequencies close to the lower part of the band, and for TE and TM modes at frequencies close to the upper part of the band. In addition, approximately the same absorption is obtained in the TE and TM modes.

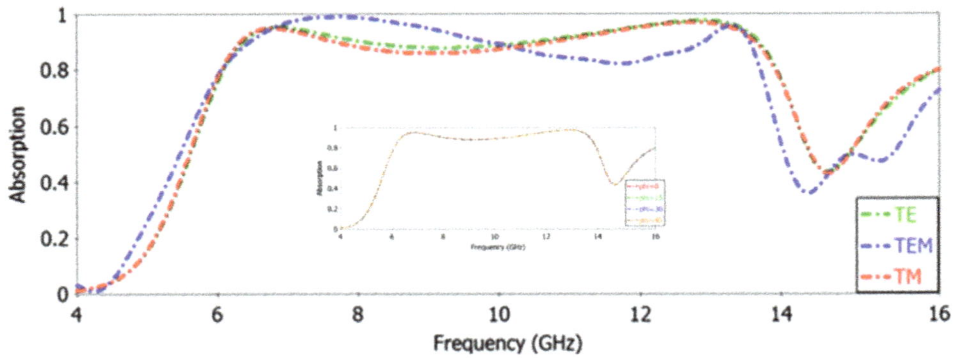

Figure 12.5. Absorption for TE, TM and TEM modes [34], John Wiley & Sons. (2018).

On the other hand, it is stated that there is no big difference when the polarization angle is changed, as shown in the inset picture of figure 12.5.

To show the effects of resistors (220 ohms) used across the gaps of the resonators in the upper layer, the magnitude of the absorption with and without resistance is compared, as shown in the figure 12.6.

As shown in figure 12.6, when resistors are not used, there are two resonances with a bandwidth of about 150 MHz at 5.2 and 14.0 GHz, and there is no absorption between 6.2 and 13.5 GHz. When using a 220 ohm resistor, absorption takes place in a very wide band between 6.3 and 13.5 GHz, with a bandwidth of about 7.2 GHz and an absorption value of over 80%. Therefore, the effect of resistors is very important in increasing the bandwidth of absorption. According to figure 12.6, when 220 Ω resistors are used as lumped elements in the proposed broadband metamaterial absorber, the absorption bandwidth is 45 times larger than in the case without a lumped element.

Another broadband microwave signal absorber design was carried out by Bilal *et al* [35]. In this study, metallic split-ring resonators surrounded with an E-shaped fractal-based excellent metamaterial absorber with very broad band absorption in the K- and Ka-bands of the microwave regime are investigated. The fabricated structure and unit cell are shown in figure 12.7.

The normalized effective impedance of the proposed metamaterial structure is calculated using the $Z = (1 + S_{11})/(1 - S_{11})$ relationship [36] and its graph is given in figure 12.8. When figure 12.8 is examined, the normalized impedance at 29.5 GHz becomes one; that is, impedance matching occurs and therefore the reflection value is zero. In addition, the impedance exhibits values close to one at some other frequency values such as 20.4, 21.0, 23.7 and 26.0 GHz.

In order to investigate the effect of fractal structure on absorption, the fractal structure is evaluated in four different stages. Figure 12.9 shows the graph of the absorptivity versus frequency at four different stages of the fractal geometry, taking into account the TE and TM incidence polarizations.

In the first three cases shown in figure 12.9(a), (b) and c, very high absorption is observed in the 20–21 GHz frequency range, insensitive to polarization. Considering

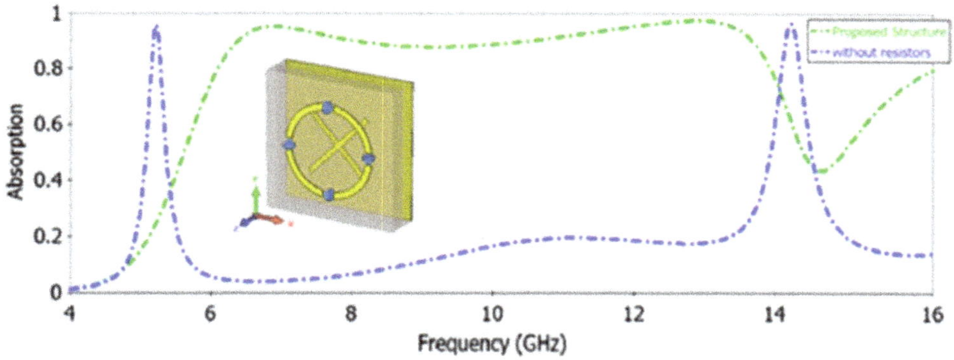

Figure 12.6. Absorption response of the proposed structure with and without a resistor [34], John Wiley & Sons (2018).

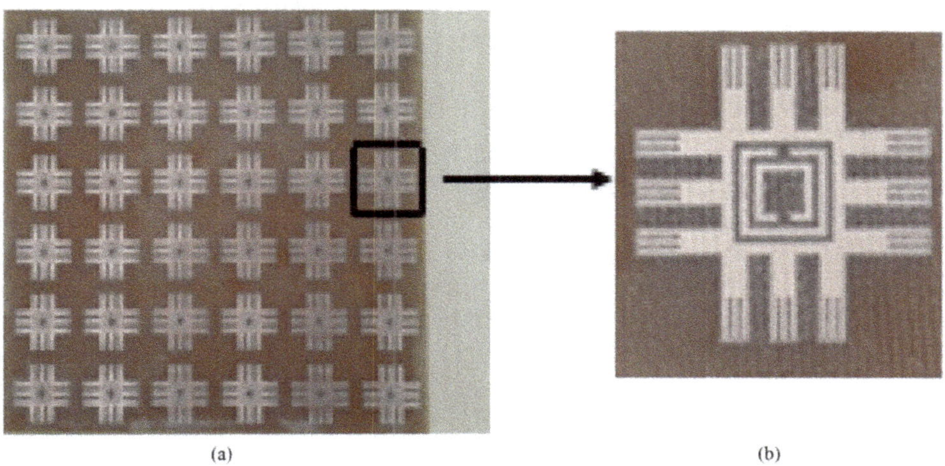

Figure 12.7. (a) Fabricated sample, (b) unit cell. Reproduced from [35]. CC BY 4.0.

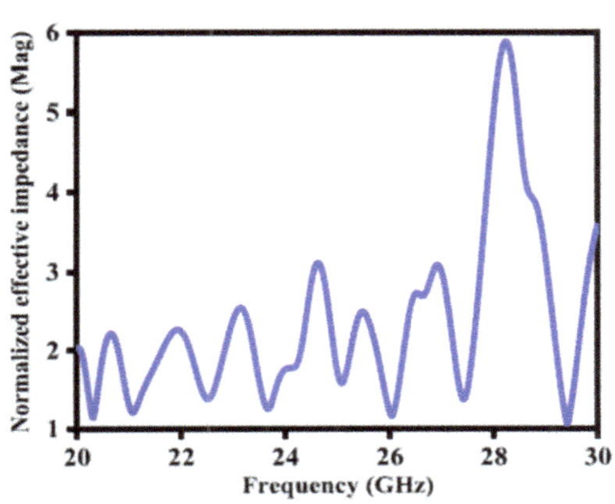

Figure 12.8. Normalized effective impedance. Reproduced from [35]. CC BY 4.0.

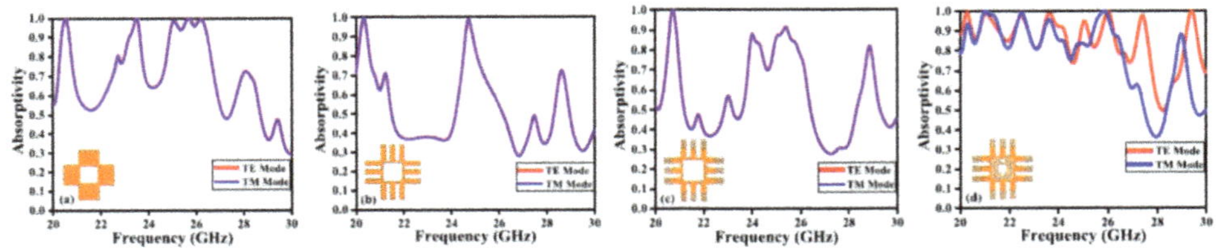

Figure 12.9. Absorptivity versus frequency at four different stages of the fractal geometry. Reproduced from [35].

the figure 12.9(a)–(c), high absorption occurs at frequencies where the impedance characteristic in figure 12.8 shows a value close to one. Then, by placing two asymmetric metallic split-ring resonators in the central gap area of the E-shaped fractal geometry seen in figure 12.9(c), the 4th geometry in figure 12.9(d) is obtained. In this case, the metamaterial absorber is no longer polarization independent, as the absorption spectra corresponding to the TE- and TM-polarized incidence radiations are now different, as seen in figure 12.9(d).

12.5 Absorber designs for energy harvesting

Energy harvesting is a technology for harvesting small amounts of energy from a variety of unconventional sources, such as EM waves, that exist in the immediate vicinity of the device. Unlike large-scale solar and wind installations that generate large amounts of energy, energy harvesters collect small amounts of energy from their immediate surroundings. When incorporated into mobile phones or other small devices, they allow these devices to generate their own power source.

For energy harvesting applications, an absorber design providing broadband absorption has been proposed by Akgöl *et al* [37]. In this study, a new type of broadband metamaterial absorber based on three types of resistive resonators is designed and analyzed.

The individual absorption responses of small, medium and large split coin resonators and their combined responses are investigated separately for the TEM mode. It can be observed that the absorption of the proposed structure is much better than that of only small, medium and large split coin resonators in almost the entire operating frequency range. Although the individual absorption response of split coin resonators is not satisfactory, the combination of all split coin resonators that make up the proposed structure has near-perfect absorption across the entire operating spectrum.

Due to the absorption-enhancing effect of the resistive loads in this structure, the power efficiency of the proposed structure is also investigated. Therefore, for the power accepted by the split coin resonators (PA), the dissipative power (PR) at resistive loads, and the stimulated power (PS) of the proposed structure for the TEM mode, PA and PR show a similar frequency response and are very close between 2.0 and 3.5 GHz. This means that the proposed structure significantly transforms PA into PR. That is, most of the power taken from the incident wave is successfully transmitted to resistive loads and they consume this power as real power.

A multilayer signal absorber design for microwave energy harvesting applications has been carried out by Karaaslan *et al* [38]. In this study, the energy harvesting capabilities of multilayered square discrete rings for wireless communication are investigated. The unit cell of the proposed structure has three resonator layers, small, medium and large. Each layer that makes up the unit cell consists of a metal resonator placed on a dielectric substrate. The back side of the third layer, that is, the layer containing the largest resonator, is covered by a metal to prevent transmission.

Since the sizes of the resonators in each layer are different, they resonate at different frequencies close to each other. The overall response of the structure can then be obtained by summing all the overlapping frequency responses corresponding to each layer.

The third layer with the largest resonator is responsible for the small resonance frequencies. The absorption values of the third layer with the largest resonator are 80%, 90% and 85% at 2.8, 5.4 and 10.5 GHz, respectively. These frequencies are within the frequency bands used in wireless communication such as GSM, WIMAX and satellite communication. In this study, the effects of important parameters such as gap width, line width and angle of incidence on the absorption level and frequency response of the proposed structure are also investigated.

12.6 Absorber for solar energy

Solar energy is directly converted into electrical energy, but the efficiency in this conversion is very small because solar cells can only absorb in a certain frequency band, although solar radiation is in a wide frequency band. This is the reason for the low efficiency. Since metamaterial-based solar cells absorb in a wider frequency band, it is thought that the absorption capacity of solar cells will improve.

For solar cell applications, an absorber design that provides absorption in the solar spectrum has been proposed by Dinçer *et al* [20]. In this study, a new type of metamaterial absorber for solar cell applications including the solar spectrum is designed and analyzed.

The proposed design has been experimentally tested with very good absorption at microwave frequencies (C-Band) as shown in figure 12.10(b). When the experimental results are compared with the simulation results given in figure 12.10(a), it is seen

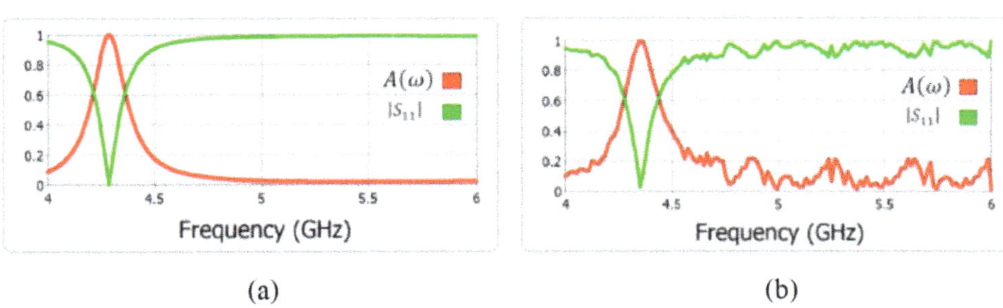

Figure 12.10. (a) Simulated reflection and absorption in C-band, (b) measured reflection and absorption in C-band. Reprinted from [38], copyright (2017), with permission from Elsevier.

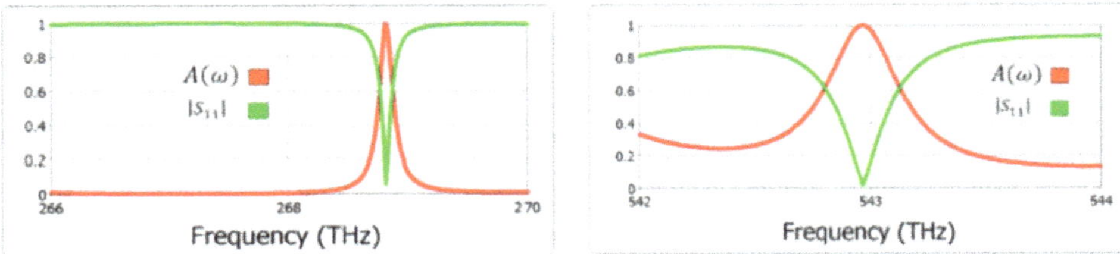

(a) (b)

Figure 12.11. (a) Simulated reflection and absorption at infrared, (b) measured reflection and absorption at infrared. Reprinted from [38], copyright (2017), with permission from Elsevier.

that they are in good agreement. It is thought that the fluctuations in the experimental results are due to production and measurement errors. Considering that the proposed model provides flexibility to adjust the metamaterial properties, it is stated that the dimensions of this structure can be easily rescaled for the infrared and visible frequency regions.

In this study, some simulations are also performed using the proposed structure for infrared and visible frequency ranges, and the results are given in figure 12.11. Maximum absorbances for the infrared and visible region are 99.66% at 268.82 THz and 99.9% at 542.97 THz, respectively, as shown in figure 12.11.

In addition, the effects of some parameters such as the gap (g) between the branches in the structure on the absorption parameters are also investigated. When the gap width changes between 2–10 mm, it is seen that the resonance frequency also changes between 3.85–6.15 GHz as shown in figure 12.12. Thus, in order to adjust the resonance frequency to the desired value, it has been stated that mechanical adjustability can be obtained by changing the gap length.

Another broadband polarization-independent metamaterial absorber design has been carried out by Bağmancı *et al* [39]. In this study, a broadband polarization-independent metamaterial absorber for solar harvesting applications is investigated. The proposed metamaterial structure has three layers as seen in figure 12.13. The top layer is the resonator consisting of nickel, the middle layer is silicon dioxide (SiO_2) as the dielectric and the back layer is the Ni layer to prevent transmission.

Numerical simulations are performed to explore the absorption mechanism of the proposed metamaterial absorber. For numerical analysis, if the electric, magnetic field components and propagation direction of the incident wave are assumed to be in the x, y, and z directions, respectively, the boundary conditions of the proposed structure are chosen as electrical boundaries for the x direction, magnetic boundaries for the y direction and open boundaries for the z direction, as shown in figure 12.13(c).

In order to understand the effect of each resonator patch, the individual absorption responses of the plus-shaped patch and cross-patches are generated and analyzed separately. These individual responses, and the response of the proposed construct, are compared and are shown in figure 12.13.

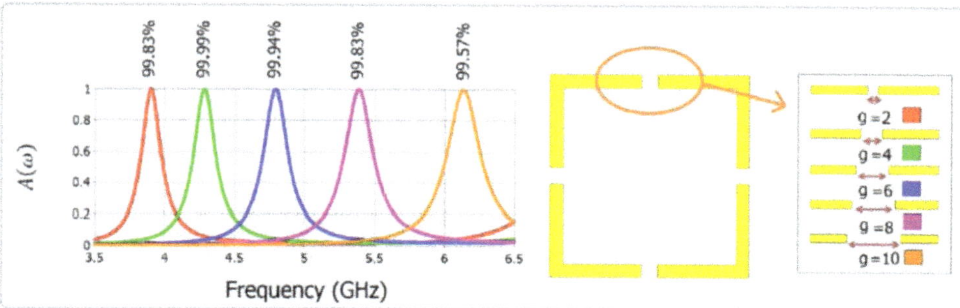

Figure 12.12. Absorption spectra for different gap values. Reproduced from [20]. CC BY 4.0.

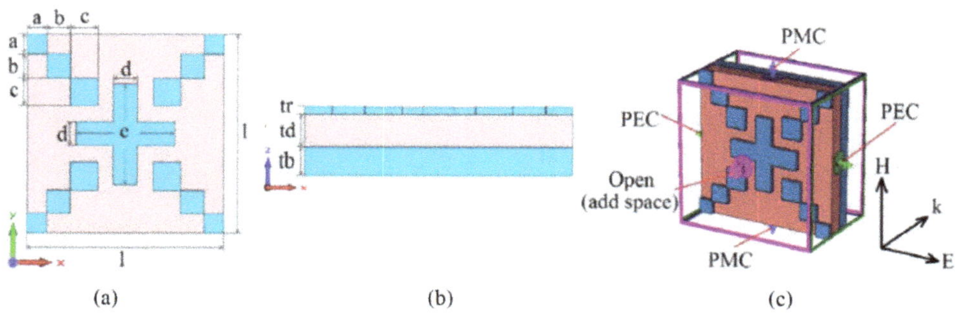

Figure 12.13. Unit cell geometry: (a) front view (b) side view and (c) boundary condition. Reproduced from [20]. CC BY 4.0.

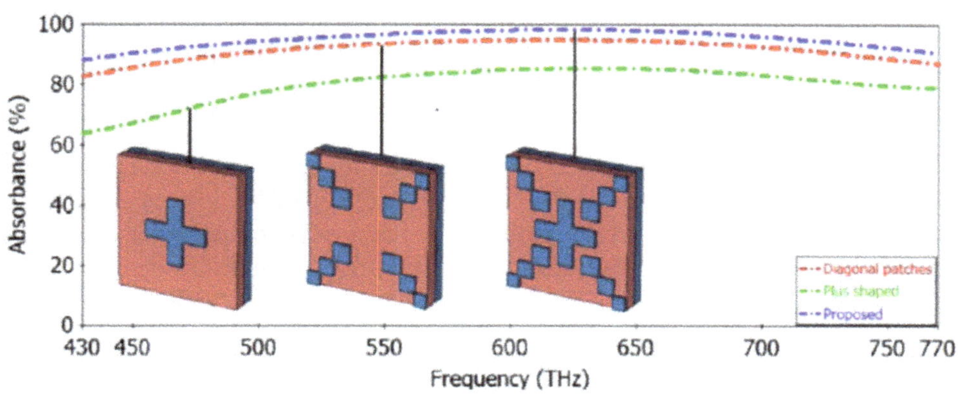

Figure 12.14. The individual responses, and the response of the proposed structure. Reproduced from [20]. CC BY 4.0.

When figure 12.14 is examined, it is seen that the absorption property of the proposed structure is better than the individual absorption properties of the other two structures, which are part of the proposed structure. Then, the absorption properties are investigated by using different metals such as nickel(Ni), aluminum (Al), copper(Cu), gold(Au) and silver(Ag) as the resonator in the proposed structure and the results are plotted on the same graph as seen in figure 12.15.

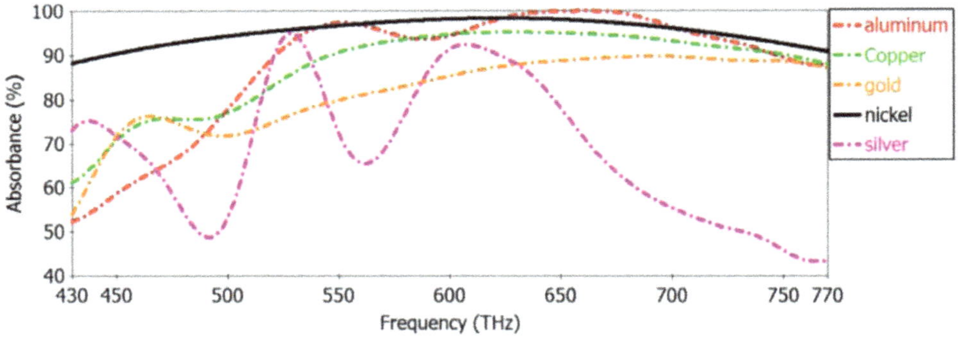

Figure 12.15. Responses of different metals as resonators. Reproduced from [20]. CC BY 4.0.

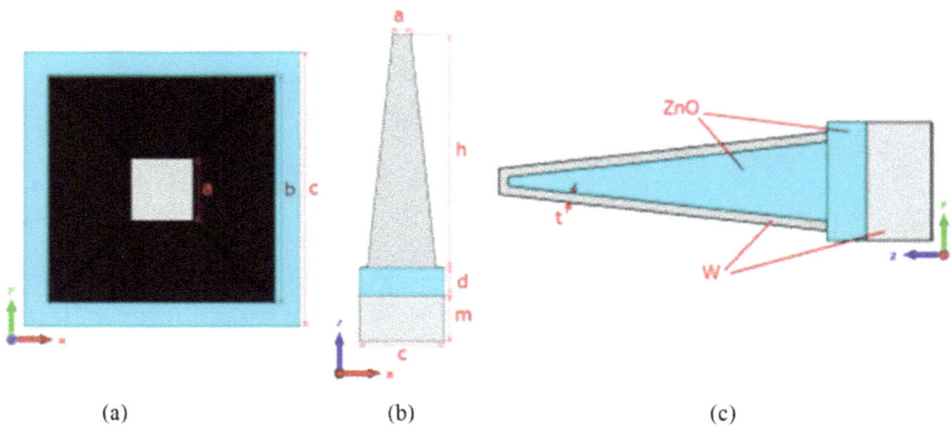

Figure 12.16. Unit cell geometry: (a) top view, (b) side view and (c) sectional view. Reprinted from [39], copyright (2017), with permission from Elsevier.

When the results are compared, it is seen that the highest absorption in the visible spectrum is achieved with resonator structures consisting of Ni. It is seen that as the frequency increases, the absorption rates of the resonators composed of Al, Cu and Au also increase. It is clear from these statements that, with an absorption rate of over 88% across the entire visible spectrum, Ni is a very good candidate for solar cell technologies, real invisibility research and light sensing sensor applications.

A pyramidal metamaterial absorber design for microwave energy harvesting applications has been carried out by Ünal *et al* [40]. In this study, ultra-broadband zinc oxide-tungsten-based pyramid structures for solar harvesting applications is investigated. Figure 12.16 shows the top, side and sectional view of the pyramid structure.

Figure 12.17 shows the absorption properties of the proposed pyramid structure for the TEM mode and the TE and TM polarizations. Figure 12.17 also shows that the proposed pyramid structure is frequency independent not only for the TE and TM polarizations, but also for the TEM mode. Thus, the proposed solar absorber is frequency and polarization independent, with an absorption of over 98% in the band

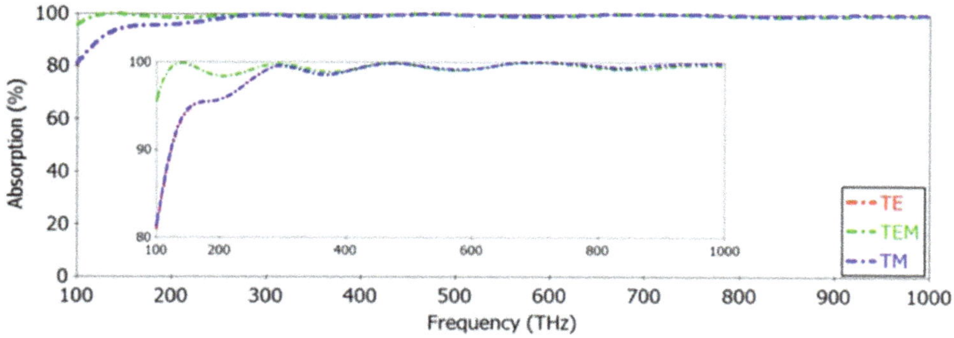

Figure 12.17. Absorption characteristic for the TEM mode and the TE and TM polarizations. Reprinted from [39], copyright (2017), with permission from Elsevier.

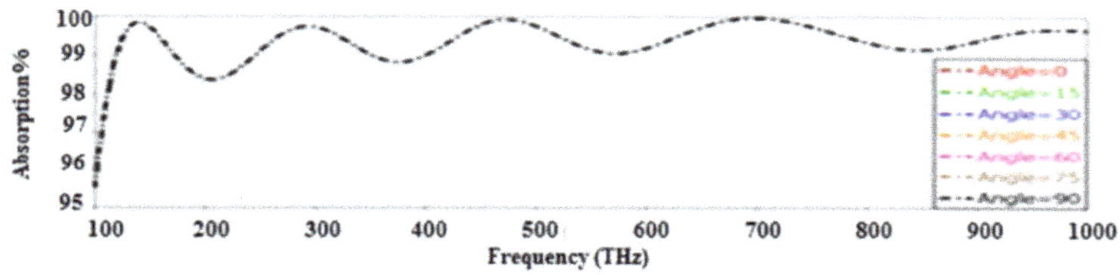

Figure 12.18. The effect of the angle of incidence on the absorption characteristic for the TEM mode. Reprinted from [39], copyright (2017), with permission from Elsevier.

between 300 and 1000 THz, where the spectral irradiance of sunlight is at a maximum of 2.5 W m^{-2}.

In this study, the effect of the angle of incidence on the absorption characteristic is also investigated for the TE and TM polarizations and the TEM mode. Figure 12.18 shows the effect of the angle of incidence on the absorption characteristic for the TEM mode.

As seen in figure 12.18, the proposed pyramidal metamaterial absorber is independent of the angle of incidence for the TEM mode with over 98% absorption. In this study, it is also stated that absorption is not affected by the angle of incidence for TE and TM polarizations.

12.7 Absorber for sensor applications

Metamaterial absorber based sensors find a wide area of use due to their high sensitivity, cost and production possibilities, as well as their robustness, compactness and reusability features. Metamaterial based sensors can be used in a wide range of applications such as medicine, engineering, biomedical, physics, chemistry, food and agriculture. Absorber-based sensors are not only used to detect physical parameters such as temperature, humidity, pressure, strain, density, rotation and thickness, but can also be used to determine and characterize the electromagnetic properties of materials [41].

A metamaterial absorber design for shielding and sensor applications has been carried out by Sabah *et al* [42]. In this study, a polarization and incidence angle

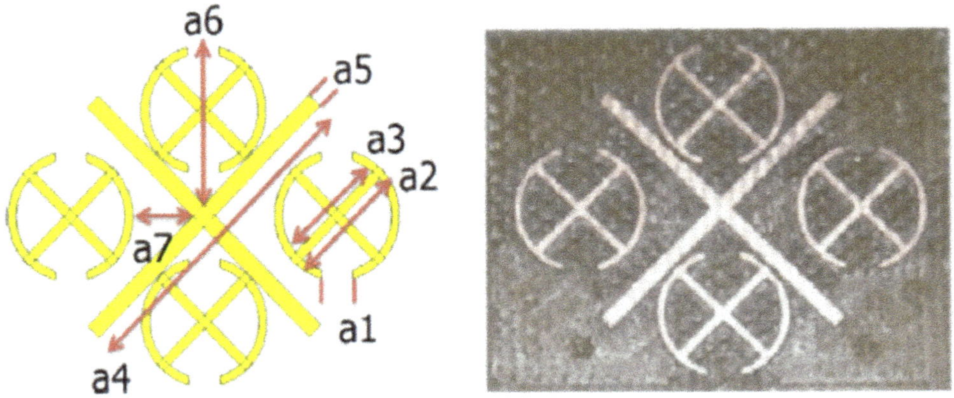

(a) (b)

Figure 12.19. (a) Dimensions of the structure, (b) fabricated sample. Reproduced with permission from [40].

independent perfect metamaterial absorber for shielding and sensor applications is investigated. The resonator structure of the proposed metamaterial absorber design consists of ring segments and cross wires connecting them, as shown in figure 12.19.

After the metamaterial absorber geometry was designed, numerical calculations were performed to obtain the absorption properties of the proposed absorber. The results show that the proposed structure can be used for excellent absorption applications. The measurements of the fabricated absorber were measured using a vector network analyzer and two horn antennas. The simulation and measurement results are shown in figure 12.20(a) and (b). When the results are compared, it is seen that the measurement results are in good agreement with the simulation results. It is assumed that the fluctuations in the experimental results are due to fabrication tolerances and dielectric dispersion of the substrate layer.

Also, the fractional bandwidth (FBW) calculations of the resonance region are examined to decide the quality of the proposed metamaterial absorber. Important in many applications, FBW is calculated as the ratio of half power bandwidth to center frequency. The simulation and measurement results of the resonance frequency and FBW for the proposed absorber are obtained as $f_o = 2.76$ GHz, FBW = 4.42%, and $f_o = 2.82$ GHz, FBW = 3.54%, respectively. In addition, it is stated that the proposed model has an acceptable FBW according to many studies in the literature [18].

In order to understand the absorption performance of the proposed structure, the effect of the distances between the cross wires and segmented rings as shown in figure 12.21 is also examined for comparison.

As a result, the structure is simulated according to some geometric parameters (a6 and a7 shown in figure 12.19) and the effect of the change of these parameters appears as a shift in the resonance frequency and an increase or decrease in the absorption level.

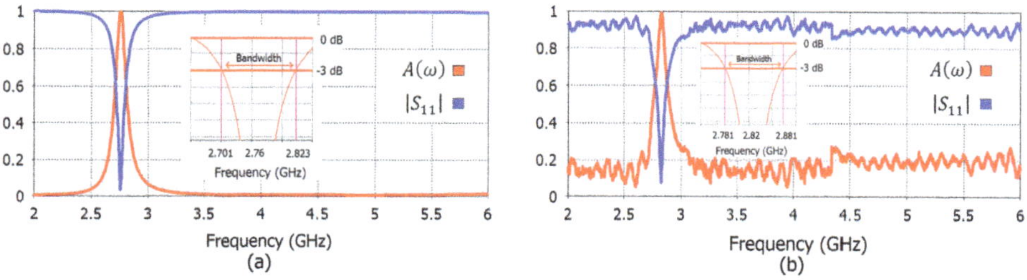

Figure 12.20. Reflection and absorption: (a) simulated, (b) measured. Reproduced with permission from [40].

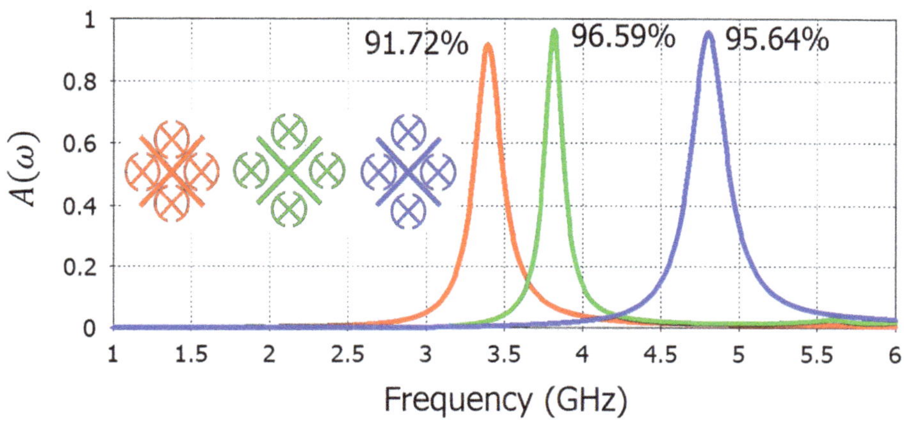

Figure 12.21. The effect of the distances between the cross wires and segmented rings. Reproduced with permission from [40].

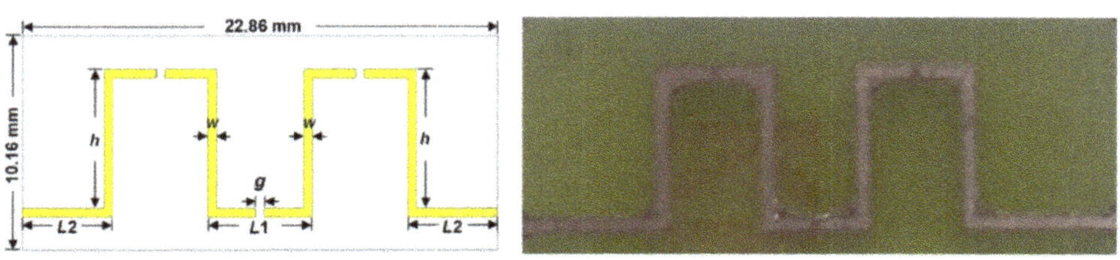

Figure 12.22. Dimensions of the structure (left), fabricated sample (right). Reprinted from [42], copyright (2014), with permission from Elsevier.

Another metamaterial absorber based on a meander-line resonator for multi-sensor application has been carried out by Akgöl *et al* [43]. In this study, the applications of a multifunctional sensor based on a metamaterial absorber structure in the microwave frequency regime are investigated. The geometry and dimensions of the meander line and the fabricated sample are shown in figure 12.22.

The gap width of the meander line is changed from 0.1 to 0.7 mm to see its effect on the absorber properties. As can be seen from figure 12.23, the effect of the gap

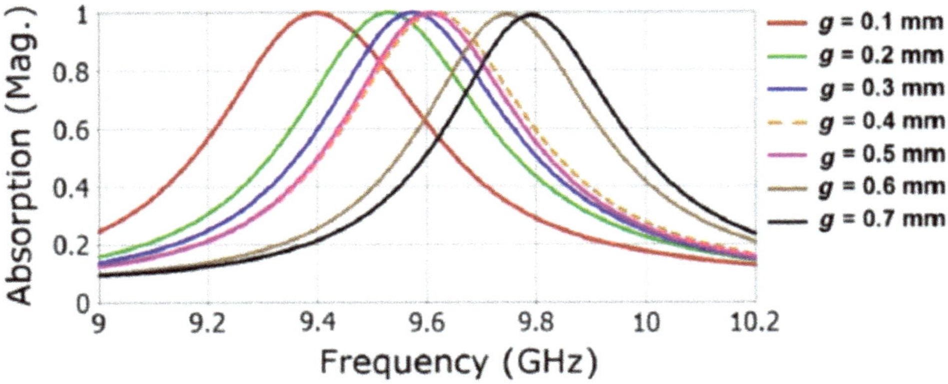

Figure 12.23. The effect of gap on resonance frequency and absorption. Reprinted from [42], copyright (2014), with permission from Elsevier.

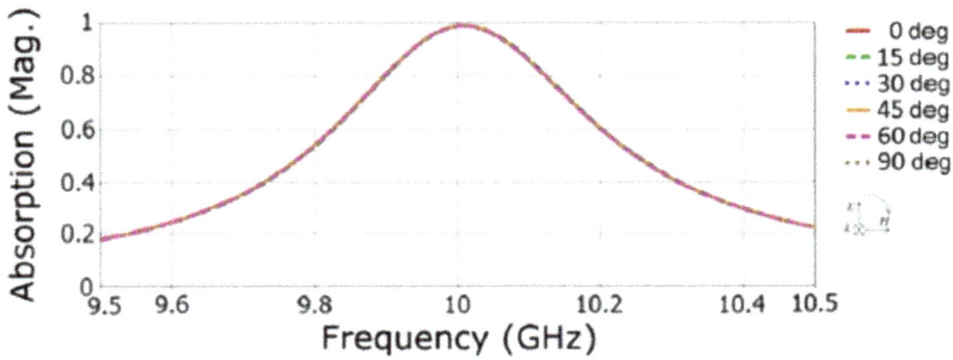

Figure 12.24. The effect of incidence angle, resonance frequency and absorption. Reprinted from [42], copyright (2014), with permission from Elsevier.

width on the absorption value is insignificant, but it is seen that the resonance frequency increases as the width increases. It has been stated that this shift in the resonant frequency can be explained by the general capacitance changes of the resonator.

In this study, the effect of the angle of incidence on the absorption characteristics of the structure is also investigated. As can be seen from figure 12.24, the proposed sensor operates as a polarization-insensitive structure at oblique incidence.

When the gap width is set to $g = 0.4$ mm, the real and imaginary parts of the normalized internal impedance and the absorption values are given in figure 12.25. When the real part of the normalized internal impedance is one, that is, when impedance matching is achieved, the absorption is maximum, and the imaginary part of the internal impedance is zero as expected. Since the imaginary part of the impedance represents the loss, it is understood that if the imaginary part is zero, there is no reflection, that is, all the incident power is absorbed by the absorber.

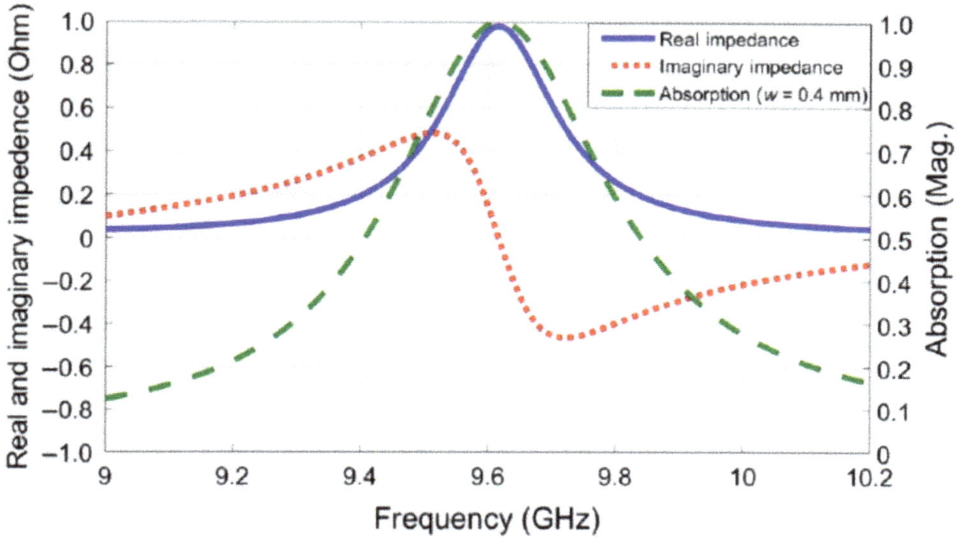

Figure 12.25. The real and imaginary parts of the normalized internal impedance and the absorption. Reprinted with permission from [43] © SPIE.

12.8 Tunable metamaterial absorber

Materials with tunable dielectric permittivity and magnetic permeability can find a wide range of potential uses, especially in wireless communication and radar technology. Various methods have been studied to develop tunable metamaterials or to change the absorption frequency. In general, these methods can be grouped under two strategies, either from a structural or material point of view [44, 45]. Structurally, microelectromechanical systems [45], microfluidics [46] or mechanically tunable dielectrics and conductors can be used to obtain tunable metamaterials [47, 48]. Therefore, structural tunability is proposed as a simple way for real-time control of EM properties such as polarization, directionality and amplitude [45]. From the material point of view, tunable metamaterials can be obtained by using phase change materials, liquid crystals and ferrites as substrate materials [49, 50].

A multi-band tunable metamaterial absorber design has been carried out by Dinçer *et al* [51]. In this study, a tunable and dual/multi-band metamaterial based perfect microwave absorber is investigated. The resonator structure of the proposed metamaterial absorber design consists of ring segments and four crossed wires in the form of a star in the center as shown in figure 12.26.

The measurements of the fabricated absorber are measured using a vector network analyzer and two horn antennas. The simulation and measurement results are shown in figure 12.27(a) and (b). When the results are compared, it is seen that the measurement results are in good agreement with the simulation results. It is assumed that the fluctuations in the experimental results are due to fabrication tolerances, dielectric dispersion of the substrate layer and measurements errors.

Figure 12.26. (a) Unit cell with dimensions, (b) fabricated sample. Reprinted with permission from [43] © SPIE.

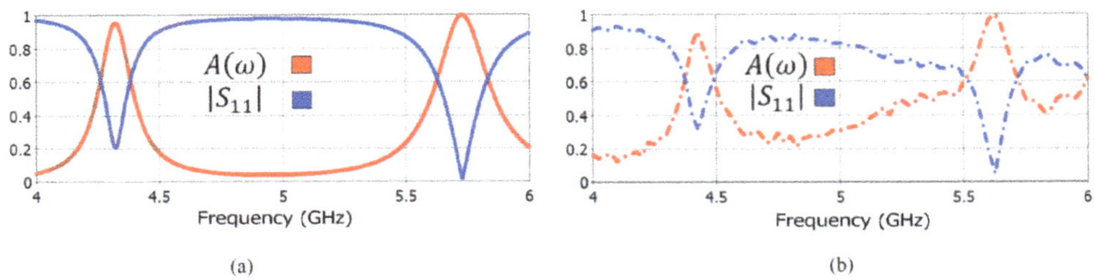

Figure 12.27. Reflection and absorption: (a) simulated, (b) measured. Reprinted with permission from [43] © SPIE.

Then, the fractional bandwidth (FBW) calculations of the resonance region, which determine the quality of the proposed metamaterial absorber, are examined. FBW is calculated as the ratio of half power bandwidth to center frequency [18]. According to the simulation results in figure 12.27(a), the bandwidths of the two resonances occurring at 4.32 and 5.73 GHz are 0.16 and 0.28 GHz, respectively. Accordingly, the FBW values of the two resonances occurring at 4.32 and 5.73 GHz are calculated as 3.70% and 4.88%, respectively. On the other hand, considering the measurement results, the bandwidths of the two resonances occurring at 4.42 and 5.63 GHz are 0.33 and 0.20 GHz, respectively. Accordingly, the FBW values of the two resonances occurring at 4.42 and 5.63 GHz are calculated as 7.46% and 3.55%, respectively. Considering the studies in the literature, it is stated that the proposed model has an acceptable FBW [18, 52, 53].

Also, in this study, the number and position of the gaps on the ring are changed in order to obtain multi-band and mechanically tunable properties for the proposed metamaterial absorber. The results are shown in figure 12.28. Each configuration has been assigned a different color for better understanding of its effects. Examining figure 12.28, all configurations have at least one resonance within the frequency band of interest. The configurations given here can be defined as symmetrical or unsymmetrical with respect to the applied EM wave. As can be seen from figure 12.28, the symmetrical configuration (light blue) has one resonance, the unsymmetrical configurations (black and pink) have three resonances and the others have two resonances in the operating region, and the spectral position of the resonances depends on the position of the gaps.

Finally, the proposed metamaterial absorber structure is rescaled and studied for high-frequency applications. For this purpose, the proposed metamaterial absorber

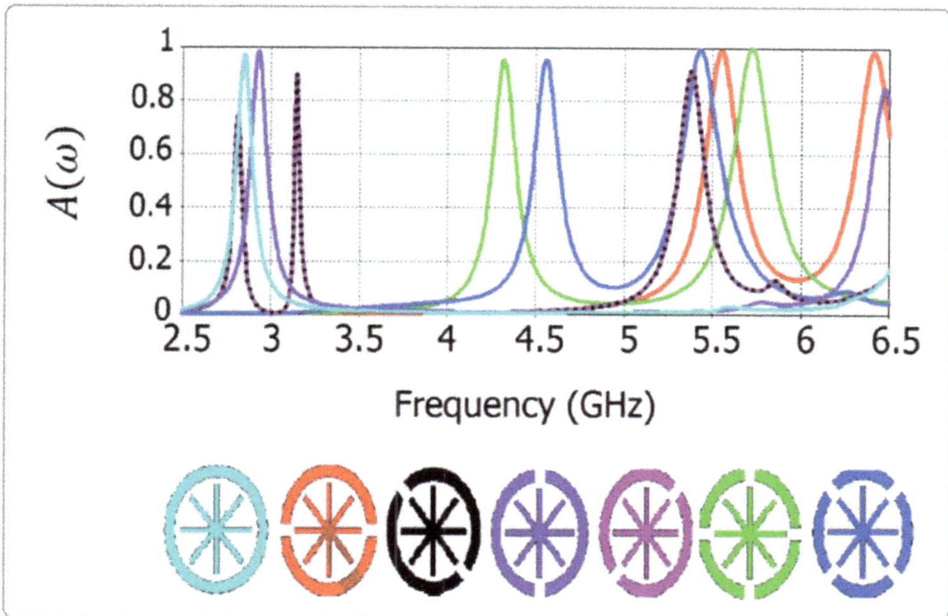

Figure 12.28. The effect of the number and position of the gaps in the ring on the absorption properties. Reprinted with permission from [43] © SPIE.

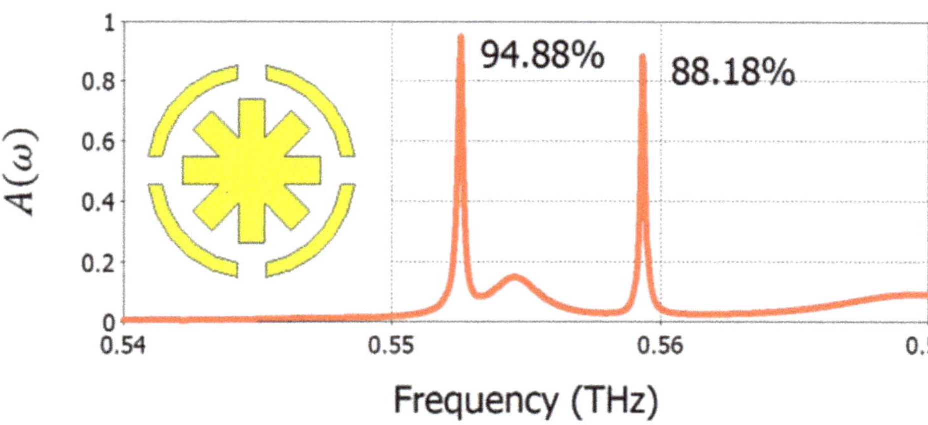

Figure 12.29. Resized, high-frequency design and its absorption graph. Reprinted from [51], copyright (2015) with permission from IOS Press.

structure is geometrically reduced to have resonance at high frequencies. Silver and quartz are used for the metallic and dielectric parts, respectively, in the simulation. The simulated absorption value for the resized, high-frequency design is shown in figure 12.29. For the rescaled model, the two maximums in absorption are obtained as 94.88% and 88.18% at 0.553 and 0.559 THz for the first and the second bands, respectively.

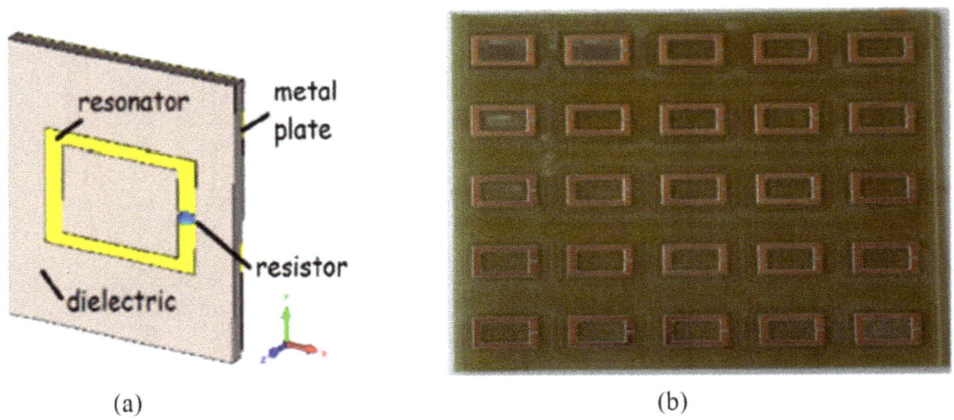

Figure 12.30. (a) Unit cell, (b) fabricated sample. Reprinted from [51], copyright (2015) with permission from IOS Press.

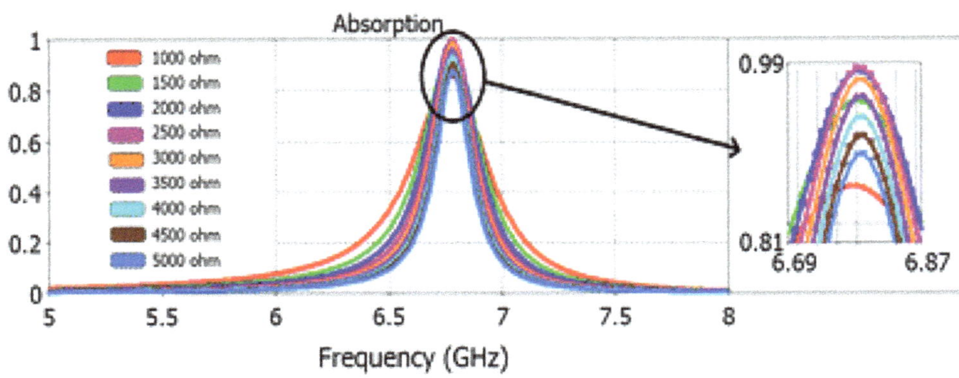

Figure 12.31. Absorption values calculated with different resistors connected across the gap. Reprinted from [51], copyright (2015) with permission from IOS Press.

Another tunable metamaterial absorber design for energy harvesting and sensor applications has been carried out by Ünal *et al* [54]. In this study, a tunable microwave absorber using the golden ratio is investigated for energy harvesting and sensor applications. The resonator of the proposed structure is a split rectangular ring as shown in figure 12.30.

Various resistance values are connected across the gap in the rectangular structure with the golden ratio, and the maximum absorption values are calculated and shown in figure 10.31. Numerical studies on the absorber structure designed for energy harvesting show that the resistor connected across the gap of the structure collects the energy stored in the gap, while at the same time maximizing the energy absorbed from the incident wave.

Figure 12.31 shows that the maximum absorption occurs at a resistance value of 2500 Ω. In this study, it is stated that the proposed structure can be used for excellent absorption applications and also, as shown in figure 12.31, the proposed model is a very good candidate for perfectly tunable metamaterial absorbers.

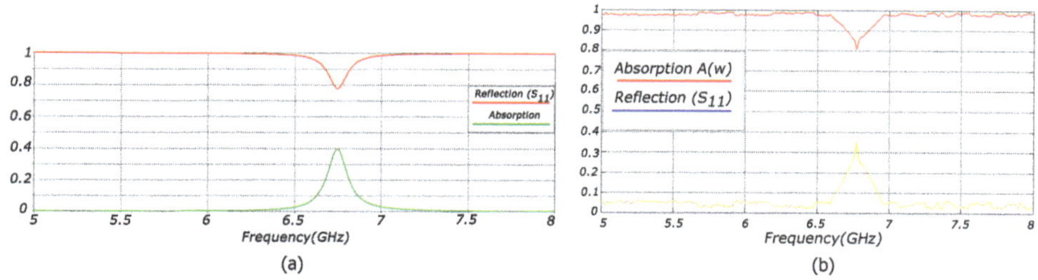

Figure 12.32. Absorption and reflection without a lumped element: (a) simulated results, (b) measured results. Reprinted from [51], copyright (2015) with permission from IOS Press.

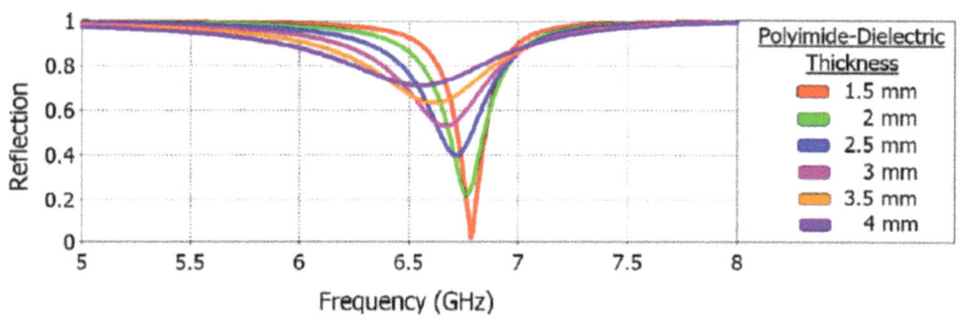

Figure 12.33. Simulated reflection with different dielectric thicknesses. Reprinted by permission from Springer Nature [54], Copyright (2015) with permission from IOS Press.

In order to support the numerical results of the energy harvesting application, the measurements of the proposed structure between 5 and 8 GHz are carried out with a vector network analyzer. Simulated and measured reflection and absorption values are presented in figure 12.32.

When the simulated and measured results are compared, it is seen that the two results are compatible with each other. As shown in figure 12.32(a), at the resonant frequency of 6.74 GHz, the reflection and absorption values are 0.77 and 0.39, respectively. Compared to the absorption value given in figure 12.31, a dramatic decrease is observed in the absorption value given in figure 12.32(a). As mentioned above, the resistor connected between the gap is the reason for this decrease because in figure 12.32(a) and 12.32(b) simulation and measurements are performed without a resistor.

In this study, it is also investigated whether the proposed structure can be used in sensor applications. For this purpose, the reflection properties of the structure are investigated numerically by changing the dielectric thickness of the proposed tunable metamaterial absorber structure. In this context, the effect of dielectric thickness variation on reflection values is shown in figure 12.33.

While keeping the other parameters constant, the dielectric thickness of the proposed structure is changed between 1.5 and 4 mm, and the effect of this change on the reflection is investigated numerically. It is seen that this change in dielectric thickness affects both the resonance frequency and the reflection value. For this reason, it is stated that the proposed structure can be used in thickness sensor applications.

12.9 Conclusion

In an electromagnetic wave absorber structure, metamaterials are used not only to increase the absorption rate, but also to adjust the bandwidth and number of bands and even the resonance frequency. Since these parameters, as well as other parameters of the metamaterial absorber, depend on the resonator structure, shape and properties of the dielectric substrate, the resonator design is the most important part of the metamaterial absorber designs. In this section, besides the metamaterial absorber design criteria, absorber applications, especially absorber designs for energy harvesting, absorbers for solar energy, absorbers for sensor applications and adjustable metamaterial absorbers are discussed in detail.

References

[1] Engheta N and Ziolkowski R W (ed) 2006 *Metamaterials, Physics and Engineering Explorations* (Piscataway, NJ: IEEE-Wiley Press)

[2] Sabah C 2012 Electric and magnetic excitations in anisotropic broadside-coupled triangular-split-ring resonators *Appl. Phys.* A **108** 457–63

[3] Burlak G and Martinez-Sanchez E 2013 Change of structure of the Cherenkov emission at modulated source in dispersive metamaterials *Prog. Electromagn. Res.* **139** 277–88

[4] Karaaslan M 2009 *Negatif kırılma indisli metamalzemelerin elde edilmesi* (Doktora Tezi: Çukurova Üniversitesi Fen Bilimleri Enstitüsü)

[5] Veselago V and G 1967 The electrodynamics of substances with simultaneously negative values of ε and μ *Sov. Phys. Usp.* **10** 509–14

[6] Pendry J B 2000 Light runs backwards in time *Phys. World* **13** 27

[7] Pendry J B 2000 Negative refraction makes perfect lens *Phys. Rev. Lett.* **85** 3966

[8] Pendry J B 2004 A chiral route to negative refraction *Science* **306** 1353–5

[9] Shelby R A, Smith D R and Schultz S 2001 Experimental verification of a negative index of refraction *Science* **292** 77–9

[10] Smith D and Kroll R 2000 N. Negative refraction index in left-handed materials *Phys. Rev. Lett.* **85** 2933–6

[11] Smith D R, Schultz S, Markos P and Soukoulis C M 2002 Determination of effective permittivity and permeability of metamaterials from reflection and transmission coefficients *Phys. Rev.* B **65** 195104

[12] Smith D R, Vier D C, Koschhy T and Soukoulis C M 2005 Electromagnetic parameter retrieval from inhomogeneous metamaterials *Phys. Rev.* E **71** 036617

[13] Caloz C and Itoh T 2005 *Electromagnetic Metamaterials: Transmission Line Theory and Microwave Applications* (New York: Wiley)

[14] Plum E 2010 Chirality and metamaterials *PhD. Thesis* University of Southampton Faculty of Engineering, Science and Mathematics Optoelectronics Research Centre

[15] Bilotti F, Alu A, Engheta N and Vegni L 2006 Features of a metamaterial based microwave absorber *Proc. of the Third Workshop on Metamaterials and Special Materials for Electromagnetic Applications and TLC (Roma, Italia)*

[16] Landy N I, Sajuyigbe S, Mock J J, Smith D R and Padilla W J 2008 Perfect metamaterial absorber *Phys. Rev. Lett.* **100** 207402

[17] Wang B, Koschny T and Soukoulis C M 2009 Wide-angle and polarization-independent chiral metamaterial absorber *Phys. Rev.* B **80** 033108 4

[18] Dinçer F, Karaaslan M, Ünal E and Sabah C 2013 Dual-band polarization independent metamaterial absorber based on omega resoanator and octa-star strip configuration *Prog. Electromagn. Res.* **141** 219–31

[19] Sabah C, Dinçer F, Karaaslan M, Ünal E and Akgöl O 2014 Polarization-insensitive fss based perfect metamaterial absorbers in GHz and THz frequencies *Radio Sci.* **49** 306–14

[20] Dinçer F, Akgöl O, Karaaslan M, Ünal E and Sabah C 2014 Polarization angle independent perfect metamaterial absorbers for solar cell applications in the microwave, infrared, and visible regime *Prog. Electromagn. Res.* **144** 93–101

[21] Dinçer F, Karaaslan M, Ünal E, Delihacioğlu K and Sabah C 2014 Design of polarization and incident angle insensitive dual-band metamaterial absorber based on isotropic resonators *Prog. Electromagn. Res.* **144** 123–32

[22] Dinçer F, Karaaslan M, Ünal E, Akgöl O and Sabah C 2014 Design of polarization- and incident angle-independent perfect metamaterial absorber with interference theory *J. Electron. Mater.* **43** 3949–53

[23] Wanghuang T, Chen W, Huang Y and Wen G 2013 Analysis of metamaterial absorber in normal and oblique incidence by using interference theory *AIP Adv.* **3** 102118

[24] Tao H, Landy N I, Bingham C M, Zhang X, Averitt R D and Padilla W J 2008 A metamaterial absorber for the terahertz regime: design, fabrication and characterization *Opt. Express* **16** 7181

[25] Dinçer F, Karaaslan M, Ünal E, Akgöl O and Sabah C 2014 Multi-band metamaterial absorber: design, experiment and physical interpretation *Appl. Comput. Electromagn. Soc.* **29** 197–202

[26] Wen Q Y, Xie Y S, Zhang H W, Yang Q H, Li Y X and Liu Y L 2009 Transmission line model and fields analysis of metamaterial absorber in the terahertz band *Opt. Express* **17** 20256

[27] Pang Y, Zhou Y J and Wang J J 2011 Equivalent circuit method analysis of the influence of frequency selective surface resistance on the frequency response of metamaterial absorbers *Appl. Phys.* **110** 023704

[28] Pang Y, Cheng H, Zhou Y and Wang J J 2013 Analysis and design of wire-based metamaterial absorbers using equivalent circuit approach *Appl. Phys.* **113** 114902

[29] Kong J A 1986 *Electromagnetic Wave Theory* (New York: Wiley)

[30] Pozar D M 2005 *Microwave Engineering* (New York: Wiley)

[31] Wolf E and Born M 1999 *Principles of Optics* (Cambridge University Press)

[32] Salisbury W W 1952 Absorber body for electromagnetic waves *US Patent* **2** 944

[33] Pan W, Shen T, Ma Y, Zhang Z, Yang H, Wang X, Zhang X, Li Y and Yang L 2021 Dual-band and polarization-independent metamaterial terahertz narrowband absorber *Appl. Opt.* **60** 10

[34] Bağmancı M, Akgöl O, Özaktürk M, Karaaslan M, Ünal E and Bakır M 2018 Polarization independent broadband metamaterial absorber for microwave applications *Int. J. RF Microw. Comput. Aided Eng.* e21630

[35] Bilal R M H, Baqır M A, Choudhury P K, Karaaslan M, Ali M M, Altıntaş O, Rahim A A, Ünal E and Sabah C 2021 Wideband microwave absorber comprising metallic split-ring resonators surrounded with e-shaped fractal metamaterial *IEEE Access* 5670–**79**
[36] Smith D R, Vier D C, Koschny T and Soukoulis C M 2005 Electromagnetic parameter retrieval from inhomogeneous metamaterials *Phys. Rev. E, Stat. Phys. Plasmas Fluids Relat. Interdiscip. Top.* **71** 036617
[37] Akgöl O, Bağmancı M, Karaaslan M and Ünal E 2017 Broad band MA-based on three-type resonator having resistor for microwave energy harvesting *J. Microwave Power Electromagn. Energy* **51** 134–49
[38] Karaaslan M, Bağmancı M, Ünal E, Akgöl O and Sabah C 2017 Microwave energy harvesting based on metamaterial absorbers with multilayered square split rings for wireless communications *Opt. Commun.* **392** 31–8
[39] Bağmancı M, Karaaslan M, Ünal E, Akgöl O, Karadağ F and Sabah C 2017 Broad-band polarization-independent metamaterial absorber for solar energy harvesting applications *Phys.* E **90** 1–6
[40] Ünal E, Bağmancı M, Karaaslan M, Akgöl O, Arat I T and Sabah C 2017 Zinc oxide–tungsten-based pyramids in construction of ultra-broadband metamaterial absorber for solar energy harvesting *IET Optoelectron* **11** 114–20
[41] Prakash D and Gupta N 2022 Applications of metamaterial sensors: a review *Int. J. Microwave Wireless Technolog.* **14** 19–33
[42] Sabah C, Dinçer F, Karaaslan M, Ünal E, Akgöl O and Demirel E 2014 Perfect metamaterial absorber with polarization and incident angle independencies based on ring and cross-wire resonators for shielding and a sensor application *Opt. Commun.* **322** 137–42
[43] Akgöl O, Altıntaş O, Dalkılınç E E, Ünal E, Karaaslan M and Sabah C 2017 Metamaterial absorber-based multi-sensor applications using a meander-line resonator. *Opt. Eng.* **56** 087104
[44] Wu T, Wei Li W, Chen S and Guan J 2021 Wideband frequency tunable metamaterial absorber by splicing multiple tuning ranges *Res. Phys.* **20** 103753
[45] Tao H, Strikwerda A C, Fan K, Padilla W J, Zhang X and Averitt R D 2011 MEMS based structurally tunable metamaterials at terahertz frequencies *J Infrared Milli Terahz Waves* **32** 580–95
[46] Ling K, Yoo M, Su W, Kim K, Cook B, Tentzeris M M and Lim S 2015 Microfluidic tunable inkjet-printed metamaterial absorber on paper *Opt. Express* **23** 110
[47] Lapine M, Powell D, Gorkunov M, Shadrivov I, Marqu´es R and Kivshar Y 2009 Structural tunability in metamaterials *Appl. Phys. Lett.* **95** 084105
[48] Liu P, Yang S, Jain A, Wang Q, Jiang H, Song J, Koschny T, Soukoulis C M and Dong L 2015 Tunable meta-atom using liquid metal embedded in stretchable polymer *J. Appl. Phys.* **118** 014504
[49] Bi K, Guo Y, Liu X, Zhao Q, Xiao J, Lei M and Zhou J 2015 Magnetically tunable Mie resonance-based dielectric metamaterials *Sci Rep.* **4** 7001
[50] Li W, Wei J, Wang W, Hu D, Li Y and Guan J 2016 Ferrite-based metamaterial microwave absorber with absorption frequency magnetically tunable in a wide range *Mater. Des.* **110** 27–34
[51] Dinçer F, Karaaslan M, Ünal E and Sabah C 2015 Design of tunable and dual/multi-band metamaterial based perfect microwave absorber *Int. J. Appl. Electromagnet Mech.* **47** 729–35

[52] Yilmaz A and Sabah C 2013 Diamond-shaped hole array in double-layer metal sheets for negative index of refraction *J. Electromagn. Waves Appl.* **27** 413–20

[53] Dinçer F, Sabah C, Karaaslan M, Ünal E, Bakır M and Erdiven U 2013 Asymmetric transmission of linearly polarized waves and dynamically wave rotation using chiral metamaterial *Prog. Electro-magnet. Res* **140** 227–39

[54] Ünal E, Dinçer F, Tetik E, Karaaslan M, Bakır M and Sabah C 2015 Tunable perfect metamaterial absorber design using the golden ratio and energy harvesting and sensor applications *J. Mater. Sci.: Mater. Electron.* **20** 9735–40

www.ingramcontent.com/pod-product-compliance
Ingram Content Group UK Ltd.
Pitfield, Milton Keynes, MK11 3LW, UK
UKHW051904100926
8556IPUK00005B/764

9 780750 35752